The Silent Epidemic of Gun Injuries

The Silent Epidemic of Gun Injuries

Challenges and Opportunities for Treating and Preventing Gun Injuries

MELVIN DELGADO

OXFORD
UNIVERSITY PRESS

Oxford University Press is a department of the University of Oxford. It furthers the University's objective of excellence in research, scholarship, and education by publishing worldwide. Oxford is a registered trade mark of Oxford University Press in the UK and certain other countries.

Published in the United States of America by Oxford University Press
198 Madison Avenue, New York, NY 10016, United States of America.

Library of Congress Cataloging-in-Publication Data
Names: Delgado, Melvin, author.
Title: The silent epidemic of gun injuries : challenges and opportunities for treating and preventing gun injuries / Melvin Delgado.
Description: New York, NY : Oxford University Press, [2022] |
Includes bibliographical references and index.
Identifiers: LCCN 2022002760 (print) | LCCN 2022002761 (ebook) |
ISBN 9780197609767 (hardback) | ISBN 9780197609781 (epub) | ISBN 9780197609798
Subjects: LCSH: Firearms accidents—United States. | Firearms and crime—United States. |
Violence—Health aspects—United States. | Public health—United States.
Classification: LCC RA772.F57 D45 2022 (print) | LCC RA772.F57 (ebook) |
DDC 363.330973—dc23/eng/20220211
LC record available at https://lccn.loc.gov/2022002760
LC ebook record available at https://lccn.loc.gov/2022002761

DOI: 10.1093/oso/9780197609767.001.0001

1 3 5 7 9 8 6 4 2

Printed by Integrated Books International, United States of America

Books are collaborative creative projects, even when sole authored. That was the case with this book, with my collaborators being the victims of gun violence and the countless number of individuals who freely gave of their time to share their stories. This book is dedicated to those who have survived gun injuries and have chosen to transform their lives to help others with similar experiences. They are not only inspirational but essential in helping to shape the field of gun injury interventions now and in the future.

CONTENTS

ACKNOWLEDGMENTS

Many individuals took time from their busy schedules to aid me in this book endeavor, too numerous to mention. However, three individuals stand out. Two external reviewers for this manuscript provided expertize and guidance in improving the original manuscript. Finally, Kathy Shorr stands out for her willingness to answer my many questions about those injured by guns.

PROLOGUE

I have made it a habit to write on subjects that resonate for me because of personal experiences, positive and negative. Gun violence touches many different aspects in my life. I have lost relatives and friends to gun violence, and had them take lives, too, and the same said for injuries. Aspects of writing this book unfolded in a manner I expected. There were aspects that were surprisingly upsetting and elicited painful memories buried deep in my psychic, which I did not realize were there. It is true that writing is a process of discovery. I sincerely hope readers find aspects in this book having meaning for them, helping shape their future research, practice, and scholarship, because gun violence is not going away.

SECTION I

Grounding Gun Injuries

Gun injuries are a social and public health phenomenon. They defy simplification because of the profound consequences they bring to victims, their social network, and neighborhood. This section grounds this subject within public health; this is not the last time it will be covered in this book. This section consists of three chapters on a social topic with profound health implications for victims, their social networks, communities, cities, and the nation.

1

Context Setting

While in my car on the way to help my mother with her janitorial work, I was struck by a stray bullet that exploded the right side of my face. I remember picking up chunks of cheek, meat. Bone, teeth, and blood off my chest and thighs as I sat in the car that we were shot in. I nearly dies that night.
—Catherin Gutierrez (Shorr, 2017, p. 4)

INTRODUCTION

There are numerous preconceived notions about this topic, as well as profound concerns on how gun violence is altering the life course of residents, family members, neighborhoods, and the nation as a whole. Some of these notions will be widely embraced, while others may enjoy limited acceptance. Regardless of stance, we can acknowledge that gun violence undermines a quest for a healthy and productive life. Further, by giving this book an urban focus, race and socioeconomic class assume prominence, thus bringing a social justice and equity lens (Zakrison, Williams, & Crandall, 2021).

Gun violence is best conceptualized as a jigsaw puzzle, with deaths representing one dimension of this puzzle and injuries (visible and invisible) representing the other part, or a backdrop to the central figure(s). Further, it is a puzzle with hundreds of pieces, which makes solving it arduous, requiring both patience and skill. In the case of gun violence, it is a public health puzzle. In our case, the task is much easier because it is an "urban puzzle," with all the necessary markers to help meet this task. Nevertheless, this clarity brings a corresponding set of challenges, including intense feelings of loss, sorrow, and anger, which can be viewed through a multifaceted lens on trauma.

US gun violence, particularly in its cities and with residents of color, is commonplace and viewed as natural as the air we breathe; and, as with breathing, we cannot stop it (Denne, Baumberger, & Mariani, 2020). When residents of a neighborhood can list the number of gun injuries and fatalities but cannot identify a high school graduate, it highlights the loss of valuable human and economic capital essential for a viable and thriving neighborhood.

The Silent Epidemic of Gun Injuries. Melvin Delgado, Oxford University Press. © Oxford University Press 2022.
DOI: 10.1093/oso/9780197609767.003.0001

A brief glimpse into economic costs highlights why gun violence cannot be ignored as a major public health issue, as in the case of Chicago in 2018 because of how gun violence extracted heavy social and economic costs estimated at $774,000 (trauma/emergency room, long-term care/disability) per nonfatal shooting in direct annual cost of injuries (Crane's Chicago Business, 2019). This estimate does not cover law enforcement and incarceration. Estimates have reached approximately $91 million in hospital costs from readmission (Spitzer et al., 2019). An economic study of emergency department and gun injuries (2002–2014) found a mean per patient of Emergency Department (ED) and inpatient charges of $5,254 and $95,887, translating into an annual financial burden of almost $2.8 billion in ED and inpatient charges (Gani et al., 2017).

The saying that you only live once is wrong. You only die once, but you live anew every day. How you live places life at center stage. When life means surviving a gun injury, it shapes every aspect of a day, with some victims visible and some invisible to the world. This view is instrumental in shaping my views on gun victims and what we can achieve by helping them negotiate and find a renewed purpose in life.

Gun injuries have ascended in prominence because dramatic health care changes have occurred. These advances have occurred throughout all stages of treating gun violence, from treatment at the scene by bystanders, to transporting them to the hospital in ambulances, development of new techniques at trauma centers, and follow-up. However, saving a life from a medical standpoint is but one dimension; treating the ensuing trauma is only now starting to be recognized as important.

We are aware of the frantic nature of emergency rooms, with life and death hanging on minutes, if not seconds. The following statement elicits a sense of dissonance for anyone who has gone to an emergency room (Busch, 2019):

> If you are ever unfortunate enough to be the victim of gun violence, it is quite possible that you may not hear the word "gunshot wound" mentioned in the halls of the hospital as you are wheeled into the trauma bay, tethered to IV's, connected to monitors, and attached to breathing equipment. That's because gunshot injuries are so common, that we often simply refer to them as "GSW," followed by the location of the injury. Perhaps you are patient "GSW back" with a spinal cord injury, unable to move your legs. Maybe you are "GSW head" with a severe brain injury, or "GSW chest" unable to breathe from a collapsed lung.

The prosaic nature of gun injuries in urban centers speaks volumes on how normalized it has become, yet this description is anything but normal. Urban stories of hope versus pain and sorrow are much in need when discussing gun injures.

Developing insights into gun injuries is desperately needed for addressing African Americans/Blacks and other males of color (Rivers, 2018), and increasingly women of color. For example, to what extent can gun victims enjoy family support and their broader social network in adjusting to life postinjury, physically, financially, and emotionally? What happens when this support is simply inadequate or unavailable? To what extent can organizations fill this void, and how can we get it done in a financially feasible manner? How can we enhance community assets? Answers to these questions shape this nation's response toward injury.

Gun violence is a topic that researchers, social scientists, and providers are increasingly devoting attention to because of how it permeates this nation's marginalized communities. What can be said about gun violence that has not been already said? Well, unbelievably there is much involving urban gun injuries, ascending in importance as this field gains greater national attention.

Is it a gun injury or gun fatality? Several factors determine outcomes (Christensen, 2016): (1) where the bullet entered the body, (2) recovery time in the case of survivors, (3) racial and ethnic background, (4) place of residence, (5) geographical distance to a level 1 trauma center, (6) health status prior to shooting, (7) fitness prior to shooting, (8) caliber of gun used, and (9) victim–shooter relationship. We need to add the category of quality of care available to victims. These factors are covered in this book. Nevertheless, each factor can easily involve a book(s) of its own to do justice to the topic.

Contextual grounding of gun violence is the first logical step in developing a comprehensive picture of this major public health issue, and more so in the case of African Americans/Blacks and others of color because of how it shapes everyday life in urban America (Joseph et al., 2018). Understanding the magnitude of grief on victims and their loved ones can only transpire when there is recognition that encourages this human exchange and quest for understanding.

This starts with recognition of the profound seriousness of gun injuries and why the topic is worthy of multiple books focused on major groups and subgroups of color. The term "second victims" captures those left to mourn those killed or to care for those injured who survived (Galiatsatos et al., 2021; Petty et al., 2019). If we recognize this second set of gun victim statistics, it would be staggering, providing a more realistic comprehension of this national problem.

Time Magazine's article "Gun Violence Is Killing More Kids in the United States Than COVID-19. When Will We Start Treating It Like a Public Health Issue?," although focused on killings, is applicable to injuries, with the odds of urban children injured by a bullet higher than contracting COVID-19 (Sathya, 2020):

We in health care are tired of pulling *bullets* out of children and breaking bad news to parents. We are exhausted from COVID-19—and the fact that there is no relief in sight with respect to gun violence is disheartening and breeds a sense of futility. Gun violence was an epidemic decades ago: now it is endemic, made worse by the *economic insecurity* and ongoing lockdowns spurred by COVID-19, and disproportionately targeting underprivileged Americans, most of them people of color.

Multiple assaults on urban children and their communities highlight the immense public health challenges. The creation of a gun violence prevention and treatment plan needs to build on the goal of health equity.

This book summarizes existing knowledge and fills in the gaps regarding gun injury, a social and public health problem that is increasingly salient and often gets overlooked within coverage and services focused on gun murders in general (Cox, 2018; de Freytas-Tamura, Hu, & Cook, 2020; McLean et al., 2019; Oppel, Jr. et al., 2020; Ritter, 2019; Sandoval, 2020). More specifically, I focus on urban communities of color because of how guns have a disproportionate consequence on their lives. Are gun injuries of critical importance in other parts of the country? Yes! However, to do justice to the subject, I have narrowed it to a particular geographical area and specific population groups.

Public health advocates argue that gun violence is the challenge of our time (Dzau & Leshner, 2018; Spinrad, 2017), and only history will be the judge, with COVID-19 ravaging our nation. Nevertheless, it is not your typical public health issue. These two public health issues are not mutually exclusive (Otto, 2020). We can expect to have gun violence decrease in the advent of the COVID-19 pandemic, but that has not occurred, as in Philadelphia, making it the foremost public health epidemic of our time (Abdallah et al., 2020; Abdallah & Kaufman, 2020). COVID-19 and gun violence have been referred to as "super infections," complicating emergency care (Przybyla, 2021).

Fatalities alone do not provide a comprehensive picture of gun consequences in the United States (Sauaia et al., 2016). In 2015, gun injuries surpassed car accidents numerically, providing a view on the extent of this public health problem. Gun injuries often are more serious, bring stigma, and raise alarm within neighborhoods when they occur, which is not the case with car accidents. Automotive crashes are a health epidemic as well, and much progress has been made to increase survival rates through car safety improvements.

A study (2003–2013) of gun injury and motor vehicle crash fatalities by age group found that gun assaults and self-inflicted firearm injuries were highly lethal when viewed with a case-fatality percentage and the percentage of out-of-hospital deaths (Tessler et al., 2019). Gun injuries cause worse long-term consequences when compared to car crashes, signaling the importance of gun injuries. A Boston study of patients 6 to 12 months after discharge from three level 1 trauma centers found that those with gun injuries, when compared with survivors of car crashes and falls, had higher rates of daily pain (68%), did not return to work (59%), had posttraumatic stress disorder (PTSD) (53%), and had difficulties with daily activities (Lindsay, 2020).

Substance overdoses, too, are a public health issue, but they but do not elicit the same response. It is important to emphasize that guns and drugs often are difficult to disentangle, serving as an example of how major public health issues are interrelated. Gun deaths and injuries, however, have not ascended to this status in the United States, and that is alarming (Dodington et al., 2017; Franke, 2019; Kaufman & Richmond, 2020; Sehgal, 2020). A national study of opioid dependence and gun injury treatment outcomes found patients with dependence experienced higher rates of 30-day readmission and resource utilization, but those with lower injury acuity had lower in-hospital care and lower 1-year mortality rates (Peluso, Cull, & Abougergi, 2020b).

Grounding gun violence within a public health context brings depth and direction in the construction of research and solutions, fostering potential collaboration between community and health/social service providers in a mutually respectful manner that builds upon community assets (Wang et al., 2020). There is a desperate call for innovative strategies that build upon community voices, participation, and direction.

A PUBLIC HEALTH STANCE

Gun violence is one of this nation's primary causes for premature deaths, with over 1 million gun injury victims, or the equivalent of the population of San Jose, California, every year (Stern & Zhang, 2017). Finding a lens that makes sense of what appears to be a chaotic problem is essential in helping to solve it (Healthline, 2021): "Given that issues of mortality, lingering health impacts of a gunshot wound, and the psychological impact a gun death or injury can have on a household or community at large, why isn't this discussed as a public health crisis on par with the current pandemic impacting our lives nationwide?"

Public health brings a wide and coherent lens to make significant strides in solving gun violence (Wen & Sadeghi, 2020): "As with motor vehicle accidents and drowning, a public health approach to injury prevention can reframe firearm violence as a societal issue, rather than an individual one." A public health model acknowledges the intersections of gun violence with specific groups and their environment (Rydberg, Stone, & McGarrell, 2016).

Public health's challenge is formidable, with parallels with COVID-19 and lessons for the field (Moselle, 2020):

> The public health approach is rooted in a simple idea: Addressing gun violence is as complex as treating a contagious virus, like the one that causes COVID-19. Both can be deadly. Both are unpredictable. And it can be difficult to get individuals to follow the rules necessary for stopping the spread, whether it's wearing masks and staying 6 feet apart or solving disputes without guns. And even when people do follow those rules, there are greater social factors at work that may defy even the best efforts.

Nevertheless, a formidable challenge is never an excuse not to fight the worthy fight, and gun violence is such a fight.

Unlike other public health challenges, we do not need a novel vaccine or a conventional cure to eradicate gun violence, yet this health issue brings its own unique needs (Wen & Goodwin, 2016). Public health has shifted focus from an emphasis on a biomedical paradigm to one embracing a social ecological stance; this shift has uplifted the importance of health equity and social determinants of health, as well as more prominently recognizing the importance of culture and

trauma (Golden, 2019). Gun violence has been a direct beneficiary of this paradigm shift.

A public health approach for addressing gun violence is a shift away from
a political and criminal stance, according to Dr. Christopher Barsotti, CEO of
AFFIRM, as quoted by Spearman (2020): "Each one of us has a role we can play
right now, to bend the arch of this problem away from gunshot wounds and
heartbreak, and begin treating our national health crisis of firearm injury with
hope trust and partnership." A public health stance consists of four phases: (1)
obtaining an accurate count; (2) identifying risk factors; (3) identifying solutions
with a high probability of success; and (4) implementing solutions taking into
account local circumstances.

These approaches will be interwoven throughout this book. We can take a
different approach by asking: (1) Where do these injuries transpire? (2) Who is
likely to be a victim? (3) How they are injured? Answers to these questions go a
long way toward helping public health officials develop necessary interventions
across an entire continuum of practice.

However, we must endeavor not to associate a public health response to urban
violence as a means of making race synonymous with criminality (Rosbrook-
Thompson, 2019); instead, we can view police actions as undermining public
health. Police violence makes it more difficult to develop a comprehensive approach
to gun violence (Calvert, Brady, & Jones-Webb, 2020; Cooper & Fullilove, 2020;
Gomez, 2016). One has only to look at the Black Lives Matter movement in 2020
to witness the connection between urban public safety and public health, with
gun violence playing a prominent role.

Public health's reach on gun violence is broad (Dodson & Hemenway, 2020), and
it is useful in addressing police violence, racial profiling (Gibbs, 2020; Laurencin
& Walker, 2020), and criminal justice (Feder & Angel, 2020). Public health's epidemiology and criminology (EpiCrim) perspective on urban youth gun violence
is starting to receive attention because of how behavioral risk factors are stressed
that cross these two fields, allowing the sharing of information with public health
to bring an early identification potential (McMillan, 2020).

There is a big difference between public health and criminal justice approaches,
with the former approach having the potential to achieve success at a much lower
cost level (Cerdá, Tracy, & Keyes, 2018). Enlisting law enforcement as a partner in
advocating for gun violence as a public health problem is important for advancing
a public health violence agenda (Eastman, Acevedo, & McDonnell, 2020).

The medicalization of violence is an argument commonly used against public
health taking a leading role on gun violence. It stresses that such an approach
ignores violence from structural factors and explains it by attributing it to individual pathology (Riemann, 2019), as will be discussed in Chapter 7. I do not
agree with this stance, particularly when interventions involve communities in
the decision-making process on the best approach to curbing gun violence, and
they assume prominent roles in these endeavors. Participatory democracy has a
prominent place in community-centered interventions and within public health.

Those with histories of addressing urban health issues will naturally embrace a public health viewpoint because it lends itself to engagement of a variety of helping professions under one conceptual roof and involves more than one health issue. We cannot separate the effects of violence on health, with outcomes determined by a range of sociodemographic factors, and health ramifications being cumulative from physical and emotional outcomes (Rivara et al., 2019).

The promise of a public health stance on gun injuries brings an ability to gather data (quantitative and qualitative) that welcomes professions coming together in search of understanding (research) and the crafting of community-centered solutions (interventions) (Masiakos & Warshaw, 2018). This quest for a coalition or task force of helping professions has gained momentum in the past 5 years, with indications that it will gain strength this coming decade (Kalyanaraman, 2020).

A unifying vision certainly goes a long way to conceptualizing the most appropriate approaches addressing (primary, secondary, and tertiary interventions) gun injuries and associated needs (Abaya, 2019; Braga & Weisburd, 2015). The American College of Surgeons statement is applicable to other professions (Talley et al., 2019):

> Science, research, technology, and innovation are proven approaches to improve safety, reliability, and efficacy. We believe encouraging this approach is beneficial to firearm owners and those who do not own firearms. Revolutionary improvements in automobile safety have come in concert with improvements in reliability. We believe a similar approach to firearms could yield the same result—improved safety with improved reliability. Addressing intentional violence requires a robust research agenda that is supported at a level commensurate with the burden of the problem. Research, innovation, and technology are critical if we are to have effective interventions.

Research goes hand in hand with innovative interventions, setting parameters, answering key questions, and setting policies, with public health values and principles setting the agenda and bringing health-focused professions, including bringing communities together, in pursuit of a common goal.

Successful gun violence prevention requires a multifaceted approach based upon scientific data, as with public health. This approach, and in this case prominently involving social work, must follow considerations and recommendations to increase success rates (Hardiman, Jones, & Cestone, 2019):

> 1) Violence was viewed as significant, widespread and a source of despair and hopelessness; 2) Residents perceived a lack of resources and opportunities within the community; 3) There was general consensus regarding the program's importance to the community; 4) There were several obstacles and challenges to program implementation as identified by participants; and 5) Community involvement and shared responsibility were viewed as central to the reduction of gun and related violence. (p. 492)

These recommendations offer valuable insights into the prominent role that residents can play in collaborating with researchers, helping the public health field connect with other professions that value community partnerships.

The importance of a relevant research base guiding public health interventions is established when applied to gun violence, as it has on other health problems in the country (Christensen et al., 2019):

> As public health scientists we believe that efforts to address gun violence should be based on scientific evidence. Just as policy changes associated with automotive safety, other preventive health recommendations, workplace safety, and air travel, have largely evolved from an empirical evidence base, so should policy changes or other measures developed to decrease the public health toll of gun violence. Unfortunately, while reports of gun violence have become part of the daily fabric of our society, development of an evidence base to help guide efforts around gun violence prevention has until recently failed to be a priority for the scientific community, demanded by the public, or supported by funders. (p. 581)

The call for a robust public health and social science research agenda is not new, providing a foundation to proceed with new and highly innovative initiatives that are driven by community participation, building the necessary foundation to launch initiatives.

Public health has been criticized in the past because it tended to favor a top-down approach. I do see a shift toward a bottom-up approach, however. The centrality of community members sharing and shaping interventions stands out in significance because of their ability to identify potential obstacles and recommend strategies with a high likelihood of success. As a social worker, solutions start with communities assuming a prominent place at the table, helping to set the meal and deciding who is invited to dinner. The centrality of community within a public health stance may come across as "over the top" for some professionals, and I am well aware of this.

This section on public health would be incomplete without addressing the role that legal viewpoints play in shaping this response to gun violence. A public health gun violence stance brings with it a legal consideration (Second Amendment) that shapes views of guns (Ulrich, 2019):

> A public health approach to gun violence must include a public health *law* approach to the evaluation of gun regulations. . . . Yet, examination of public health law cases proves that subjugating a fundamental right to reduce the risk of harm to the public is not equivalent to ignoring the strength of that right. Indeed, fundamental rights have been restricted in the name of public health and public safety with some regularity when there is an identifiable risk to the health of the public sufficient to justify state action. To restrict the right to keep and bear arms under the same premise does not relegate it to a "second-class right." Rather, it continues the tradition, long upheld by the courts in this

country, that the right of the individual does not enable them to place other citizens at risk. (p. 114)

A strong argument is advanced that violence is a public health threat to this country as a whole (Freire-Vargas, 2018). However, within this broad declarative statement, it ascends in importance within urban centers among residents of color and their quest for health equity.

Guns represent one of this nation's greatest public health threats, calling for more legislation on gun control rather than less. Public health and allied professions are positioned well to take on this struggle in all settings, including courts as well as legislation. The introduction of assault weapons meant for war is finding increased prominence in urban streets, making guns that much more deadly, and therefore a greater threat to health and wellbeing.

Other critics of a public health model have argued that this approach has placed a great deal of emphasis on research, but insufficient attention to figuring how to navigate socially and politically by putting findings into practice (Mitchell & Ryder, 2020). Smith et al. (2017) note that public health has historically placed an emphasis on defining the victims of gun violence (host) and the environment (gun policies), without paying similar attention on the agent (guns and ammunition) or the vector (gun and ammunition manufacturers, sales outlets, and the industry lobby). The field, however, has started to make this important correction in approaches.

THE NATION AND THE CHALLENGE OF GUNS

Urban communities of color are facing a multitude of challenges in recovering from gun violence and for professionals seeking to serve these communities, as academics develop greater insights furthering their gun knowledge base. The nation is in the midst of a gun violence upsurge within urban communities. Gun violence has significantly altered freedom of movement and curtailed opportunities for a better life for all age groups, but youth and young adults bear the onus (Masho et al., 2016). Firearm injuries have also been called an "injury plague" when viewed from a fuller vantage place of how they extract a heavy multifaceted toll on victims (Nelson, 2017). Violence, in similar fashion to COVID-19, is a contagious disease (Slutkin & Ransford, 2020).

How this country responds to this trauma says a great deal about our values and priorities, and what the future has in store for victims and their neighborhoods (Wright, 2019). We live in a country that is oriented around the individual, and this value influences our perceptions of trauma and gun violence. Our views of gun violence, too, are focused on individuals, ignoring neighborhood trauma, thereby missing an important context and arena for understanding and addressing this violence (Lane et al., 2017).

A gun injury lends itself to being viewed through a lens far wider than that of one victim because a gun injury symbolizes hundreds of individuals suffering this trauma, and countless more, if we take exposure seriously. Collective trauma highlights the far-reaching dimensions of gun violence from a neighborhood standpoint (Tuller, 2020), connecting individual victims in a community web of pain and suffering.

Individuality translates into gun victims being the only victims of this violence. However, the places where these acts occur are composed of groups that have a propensity to have intergenerational and collective values, translating into a group versus individual stance. There is no single victim, but rather "victims." Services, in response, must not be individually conceptualized to be effective. An individualistic stance means that there is virtually no public recognition of the pain and grief of community members (Rubinstein et al., 2018). A public health issue of this magnitude must be contextualized as broadly as possible to be effective (Kuhls, Stewart, & Bulger, 2020; Odom-Forren, 2016).

The saying that we live in turbulent times has never been truer than it is today as the nation struggles against three significant forces, two visible (gun and police violence) and the other invisible (COVID-19), bringing tremendous consequences for communities of color (Delgado, 2021; Fernandez, 2020; Hatchimonji et al., 2020a, 2020b). The primary focus of this book is on gun injuries, with attention paid to COVID-19 and the police when it influences contextual factors such as violence, help seeking, and service provision pre and post hospital discharge. Marginalized communities have always struggled with multiple major health and social needs, including high probabilities of resulting in premature deaths and long-term health compromises.

The United States is a country where if you are White and non-Latinx, you can breathe free, but if you are of color, you cannot breathe (Everytown USA, 2020b, p. 1):

> Unprecedented increases in gun sales, combined with economic distress and social isolation due to COVID-19, are intensifying the country's long-standing gun violence crisis. The pandemic highlights the deadliness of weak gun purchase and access laws that allow firearms to fall into the wrong hands, and shedding light on existing structural inequity. The coronavirus puts vulnerable populations, including women, children, and communities of color, at heightened risk.

Immense challenges can translate into immense opportunities for social justice–informed charge, and nowhere is this most pressing than with gun injuries (Sutherland, McKenney, & Elkbuli, 2020).

Urban spaces, however configured, have symbolic and practice implications. Gun violence is ascending in importance, with several major cities highlighting this climate. Chicago stands out, garnering its share of national publicity, and it is highlighted in this book (Barnes, 2020; McLone et al., 2017; Wood & Papachristos, 2019). Former President Trump has had a particular fascination with this city (Garbarino, 2017), comparing it to El Salvador and with Operation Legend and the sending in of federal troops.

Chicago is a city that has been hit hard by gun violence; its neighborhoods of color are disproportionally affected because of health inequities, living conditions, the jobs these residents occupy, and the extent of gun violence, creating a dangerous environment (Heath, 2020; Lee, 2018). Gun violence has not stopped because of COVID-19 or any other health concerns. As this book goes out to press, 15 people were injured due to gun violence outside of a Chicago funeral parlor, and 10 of the victims were women aged 21 to 65 years, illustrating how gun violence not only has no holiday but also no regard for cultural norms or the importance of the healing process (McCann, 2020).

Is there such a thing as a season of gun violence? Yes. Summer is a season that is associated with gun violence in many of this nation's cities (Colbert et al., 2019; Kotlowitz, 2019; Williams & Bassett, 2019). Weather falls under the category of "ecological factors" alongside season (Kieltyka, Kucybala, & Crandall, 2016). Holidays, too, are correlated with increased gun violence.

It does not take a social scientist to see that weather plays a role in shootings, as in Chicago daily shootings over a multiyear (2012–2016) period, with warm weather, particularly during weekends and holidays, proving particularly violent (Reeping & Hemenway, 2020; Ruderman & Cohn, 2021). June is the most dangerous month of the year for those under the age of 18 years. Although weekends bring increased accidents and incidents of violence, no "weekend effect" was found in outcomes (mortality, amputation, length of stay, and discharge disposition), in this case involving lower extremity vascular trauma, calling attention to the presence of other factors influencing outcomes (Jundoria et al., 2020). Another study, although dated, found that weekends are 1.5 times more violent, however (Grossman et al., 2005).

In 2020, Chicago witnessed its most violent Memorial Day period in 4 years, showing that gun violence does not take a holiday, with eight people of color killed and 24 injured (Vera, 2020). Those eight fatalities engendered profound grief and significantly altered the social networks of surviving relatives and friends. The 24 individuals injured, too, face tremendous challenges that affect their social networks and required help in the recovery process. Gun violence only got worse. The 24-hour period of (5 a.m.) Friday, May 29, 2020 (7 p.m.) through Monday had the worst gun violence day in 60 years, with 25 killed and 85 wounded by gunfire (Scuba, Charles, & Hendrickson, 2020).

In keeping with Chicago, the July 4th weekend witnessed 87 individuals shot, with 17 dying, marking the third straight week that a child (7 years old) was killed by a gunshot as she played outdoors (D'Onofrio & Wall, 2020). Finally, Chicago, from January 1 to July 30, 2020, had 440 homicides and 1,800 injuries, compared to 290 homicides and 1,480 injuries during the same period in 2019 (Associated Press, 2020).

New York City, in turn, witnessed its most gun violence at the start of the summer since 1996, with 125 shootings in the first 3 weeks of June, doubling the number from the same period a year earlier, with approximately 100 suffering injuries (Southall & MacFarquhar, 2020). Anyone who has experienced fear will understand its toll on their lives; communities living in fear of gun violence magnify this fear in everyday life (O'Leary et al., 2017). Fear begets fear. When guns are introduced, the consequences can be severe (Fisher, 2019). One youth stated this existence well (Opara et al., 2020, p. 2123): "And then you have to sit there and worry about sitting in the house and your kids, your siblings, your family, are outside. You have to worry about, 'ok my kid might get shot'. . . . Bullets have no one's name on them, bullets can hit anybody."

It would be misleading to think of gun violence as solely centered in the nation's major cities. The following shootings occurred on a June 2020 day in Syracuse, New York, a city not customarily in the news concerning gun violence, but indicative of how pervasive gun violence is in this nation's cities over one summer night (Urban CNY News, 2020):

23, Male, Gunshot to groin and leg, Non-life threatening;
29, Male, Gunshot wound to the neck, Non-life threatening;
20, Female, Gunshot wound to the leg, Non-life threatening;
18, Female, Gunshot wound to the leg, Non-life threatening;
17, Male, Gunshot wound to the head, Critical;
19, Female, Gunshot wound to the leg, Non-life threatening;
37, Female, Gunshot wound to the shoulder, Non-life threatening;
22, Male, Gunshot wound to the leg, Non-life threatening;
53, Female, Gunshot wound to the back and midsection.

Seven of these involved "non-life threatening" outcomes but that obscures their long-term consequences. It is rare for a news outlet to provide this level of detail, which on the surface is minimal. The stories of surviving gun victims are rarely covered and shared with the public. Gun violence statistics are not just about numbers and sociodemographics; they are about lives.

Urban gun violence affects victims and their families, institutions such as houses of worship and schools, and those perpetrating gun crimes when apprehended, convicted, and incarcerated (Miller, 2020). Akosua (2019) makes an important observation of gun violence perpetrators: "Our community still needs healing. Hurt people hurt people. Someone who pulls the trigger is extremely hurt. That person is dealing with suffering and trauma. No one can kill and just be OK. I have sympathy and empathy for that because that person needs some type of resource. They need help."

Gun violence trends and gun carrying among African American adolescents over an extended period (2002–2015) found that degree of school engagement and satisfactory school performance serve as protective factors in countering the effects of violence, emphasizing the importance of these institutions in reducing gun violence (Khubchandani & Price, 2018). The social and economic costs go far beyond being victims, reaching perpetrators (when apprehended and sentenced), their families, and neighborhoods where this violence occurs (Fransdottir & Butts, 2020):

Medical costs, however, represent only a portion—even a relatively small portion—of the total economic impact of gun violence. Comprehensive measures of gun violence costs would need to include other factors. Families expend considerable resources dealing with the mental and emotional trauma that follows a shooting and paying for physical rehabilitation for the survivors of gun injuries. Shootings may result in unemployment and loss of income for survivors as well as their caregivers, and there are potentially enormous costs for neighborhoods in general as high rates of gun violence affect community safety, economic activity, and the value of housing and other forms of property.

Surviving a shooting is traumatic and life-altering, and what follows may be even worse, bringing financial hardships and more, as covered in Chapter 6.

The American College of Surgeons Committee on Trauma recognizes the importance of institutions having "robust rehabilitation and reintegration programs" to minimize disability and facilitate rehabilitation (Stewart et al., 2017). Successful gun injury rehabilitation should strive to achieve multiple goals, with independence taking a place alongside autonomy and social engagement to achieve a balance as a way of life (McClure & Leah, 2021). Adding a social role in advocating against gun violence enhances this process.

Vast improvements in treating gun injuries at the scene, with increased speed transferring victims to hospitals and successful resuscitation while in transit, and once admitted, increased survival rates (Kingsman, 2020). Over the period of 2005 to 2016, there were 322,599 gun victims admitted to hospitals, with 262,098 undergoing surgery. This figure is an increase of 18% (19,832 to 23,480), largely attributable to improved survival rates prior to hospital admittance, reducing mortality rates from 8.6% to 7.6% during the study period (Kingsland, 2020). Advances in treating gun injuries translate into a higher survival rate and need for a wide range of services.

"Surviving day to day is almost as bad as the event, as being shot," shared by Sara Cusimano, a gunshot survivor (Amnesty International, 2019, p. 5), highlights the importance of what follows becoming a gun victim. The social fabric of urban communities faces many challenges, with gun injuries representing an almost daily occurrence, with economic aspects permeating virtually all aspects of community life for victims and the institutions dealing with the aftermath of these injuries. Cities, it needs emphasizing, bring incredible assets that must be tapped (Tynan, Bas, & Cohen, 2018).

Trauma screening measures for children and adolescents are key in assessing and planning appropriate interventions (Eklund et al., 2018). An exposure to an urban community violence inventory has existed over 20 years (Seiner-O'Hagan et al., 1998). The Brief Symptom Inventory (BSI-18) has been found to be valid for use among African American youth exposed to community violence, holding much promise for the field of gun violence involving other groups of color (Kim, Michalopoulos, & Voisin, 2017). Trauma screening, to be meaningful in addressing youth of color, must take into account historical trauma resulting from this nation's period of enslavement and policies that furthered their marginalization (Delgado, 2020c).

SOCIAL CONSTRUCTION OF A VISIBLE VERSUS INVISIBLE EPIDEMIC

Readers will pick up on a socio-cultural-ecological conceptual approach in this book, which provides a foundation for research and collaboration that is multidisciplinary, helping to capture the complex interactions that lead to gun violence and why urban communities of color are disproportionately at risk, setting the stage for interventions that are placed based and address how racism and structural forces place urban people of color at an increased risk for gun violence.

Social phenomenon unfolds within a context and is best explained within this grounding, as covered in Chapter 4 and throughout this book. The intersection of knowledge and values is where we can engage in a debate about gun violence. The domain's social construction of gun injuries competes with nondominant perceptions, resulting in a dissonance. This subject is important enough to cover in this chapter in much greater depth.

The social construction of US gun violence is intertwined with race and class. Parham-Payne (2014, p. 752) goes to the heart of this point when arguing that gun violence is how communities of color are framed by the media "as a convergence of cultural, environmental, and individual shortcomings and immorality." Strikingly, almost all people of color (99.85%) know a gun violence victim, which speaks to why it is a major problem in urban neighborhoods (Kalesan, Weinberg, & Galea, 2016).

The vast majority of gun violence encounters cause an injury rather than death, which only enhances the importance of this public health issue. This translates into major policy decisions that ignore these communities, and fear of crime is the glue that helps associate gun violence and race (Hirner, 2019). The influence of social networks and gun violence is widely recognized (Tracy, Braga, & Papachristos, 2016). Illegal guns are at an increased likelihood of recovery in the highest risk sector of a network (Ciomek, Braga, & Papachristos, 2020). Those with gun injuries, their social network, and providers construct one reality; those outside of their world construct another reality. Bringing these two worlds together increases the likelihood that gun injuries go from invisibility to visibility, making these injuries no longer silent, bringing national attention and requisite funding when understood as an epidemic.

Scholars refer to gun violence as an epidemic in similar fashion to other epidemics (Green, Horel, & Papachristos, 2017, p. 326): "Gunshot violence follows an epidemic-like process of social contagion that is transmitted through networks of people by social interactions. Violence prevention efforts that account for social contagion, in addition to demographics, have the potential to prevent more shootings than efforts that focus on only demographics." Some critics of this nation's gun policies have even gone so far as to label gun violence endemic to this country's character, bringing to bear the ingrained and intractable nature of this problem (Pirelli & Gold, 2019).

Others scholars, although fewer in number, have argued against this analogy. I have argued in another book (Delgado, 2021) that we have many options for how to label this phenomenon. We can view gun violence as state-sanctioned violence because of how society tolerated its toll in communities of color, encouraged it, and financially benefited from this tragedy. We can label gun violence an issue, challenge, crisis, problem, or epidemic, for example (Giffords Center, 2016; Hickner, 2018). The response society undertakes will vary across this continuum. I selected epidemic as a lens because of the seriousness and long-lasting consequences of gun violence, and the fact that it is preventable (Carmichael et al., 2019).

Suicides involving guns, for example, are widely considered to have reached epidemic proportions in this country, accounting for more than half of all suicides and illustrating its primacy as a weapon of choice in this act of desperation (Drane, 2020). Guns are the most lethal method used to achieve suicide, accounting for 5% of suicide attempts but responsible for more than 50% of suicide deaths (Giffords Center, 2018). This death, and corresponding injuries resulting from failed attempts, is often overlooked in discussions of this nation's major social problems (Martínez-Alés & Keyes, 2019). Victims have loved ones, both in the successful and unsuccessful attempts, and they, too, are traumatized by these acts.

If the gun violence debate is settled on whether or not it is an epidemic, it bears pausing and attempting to understand what makes an epidemic invisible or visible for a nation. Multiple books are available on what makes an epidemic. Gun injuries clearly fall into the former, as highlighted in the title to this book, raising profound implications for how and when a local social problem becomes a national social problem.

Interestingly, the epidemic of gun violence, unlike COVID-19, is a very American epidemic. Epidemics are socially constructed and dependent upon everyday experiences and implicit social agreement, with corresponding institutional practices or collective social action (Kircher et al., 2011). One consequence of how gun injuries are socially constructed is that those injured are deserving of this outcome. Mind you, no elected official will stand up in public and say you got what you deserved. The silence on this issue, more so with communities of color, is that it is a problem of their creation—"You play with guns, you are going to get hurt."

A BRIEF OVERVIEW OF GUN INJURY STATISTICS AND RESEARCH

I can appreciate possible reader hesitancy in commencing reading this section, which only provides an overview. The following chapter goes into greater depth. A book devoted to gun violence without attention to data would provide an incomplete picture. Writing on research findings is fraught with challenges. One challenge is how to present an immense amount of data on a very complex topic that does justice to the importance of gun injuries, provides sufficient depth but does not overwhelm readers, and analyzes methodologies to better capture shortcomings. That is a tall order but one that must be embraced.

Understanding the multiplicity of factors that converge to create a toxic situation is essential in addressing gun violence (Malina et al., 2016):

> If we never address the underlying beliefs feeding guns-everywhere extremism, we will not be able to diminish its power. Too many Americans will continue to get their hands on assault weapons, too many will kill or maim other Americans, and we will continue to bicker about whether the first step is more research or better mental health care—while we continue to do nothing to cure the disease.

Gun injuries are a silent epidemic because of how they occur without garnering significant national outrage. We can argue that gun violence introduces a new dimension to American exceptionalism (Lopez, 2018; Weisser, 2018). A silent epidemic becomes an invisible epidemic, with victims ignored and without their voices ever being heard and their lives made invisible (Gallagher & Hodge Sr., 2018; McLively, 2019). Youth voices are often disregarded when discussing gun violence. We need to be prepared to have them share their feelings and thoughts, and to use modalities that best capture these voices (Lee et al., 2019).

Statistics play an important role in painting a picture of gun injuries. Quantitative data represent only one perspective, but an important one. Qualitative data help capture the nuances or shades in this picture, bringing statistics and stories to life. The intent of this section is not to introduce a plethora of statistics. Instead, it is to ground the subject on its national importance and why it will increase in the future. It is my contention that our knowledge base on this subject is very limited when considering how it changes victims' lives.

New York City, for example, witnessed 447 homicides and 1,518 shootings during 2020, increasing respectively by 41% and 97.4%, but these statistics do not capture the stories behind these shootings (Mangual, 2021):

> Of course, the data doesn't even begin to capture the tragic stories of trauma and loss experienced last year in New York, by far the country's biggest city. Statistics are poor stand-ins for lives cut short, the families torn apart, and the innocence and the souls stolen from children—like the 5-year old girl whose father was gunned down right before her eyes as they crossed a Bronx street, hand-in-hand in broad daylight.

Mangual sums up quite well, and disturbingly so, what lies behind a gun violence statistic, although I do not share in blaming that city's relaxing of law enforcement procedures and changes in policies for this upsurge.

It is critical that when talking about gun violence the conversation must go far beyond numbers (Reardon, 2020), although we live in a society that is enamored with statistics. Statistics are important. However, a social and public health issue of the magnitude of gun violence requires extensive partnership networks at local, regional, and national levels involving local law enforcement agencies, trauma centers, other key community organizations, academic institutions, and communities engendering a multifaceted understanding of firearm type and resultant injury patterns (Manley et al., 2020; Rijos et al., 2020; Wang et al., 2020).

Although this book addresses potential collaborations, community-academic partnerships must prominently be in this mix (Risser, 2020). Seeking partnerships with parents to reduce gun violence casualties may seem like a logical approach, but it is surprisingly rare in the field (Ozuna, Champion, & Yorkgitis, 2020). Our understanding of gun injuries requires a multifaceted picture to understand patterns and exceptions to generally based statistics, which provide a broader view on this complex topic and are intended to guide prevention and intervention efforts at the local level.

I selected several captions of photographs illustrated in Shorr's (2017) book (*SHOT: 101 Survivors of Gun Violence in America*) to bring this issue to life for readers, representing a range of circumstances and experiences of those injured beyond a typical focus on conventional statistics:

(LIZ HJELMSETH): *"I was eight years old the Halloween my brother shot me in a fight over a cat. As I hopped off to the bathroom to die in the bathtub, thinking it would be easier for my mom to clean up, he followed behind apologizing and telling me he didn't mean to do it. That was the very last time words were passed between us about the shooting."*

(PHILLLIP GOUAUX): *"It does hurt to die, it hurts to live."*

(DARIO BAXTER): *"When I got shot, I took it as if it was something that comes with the street life."*

(SHANNESSE PITTMAN): *"Being shot changed my life both physically and mentally. I was told that time heals all wounds. Well, with that being said I hope time will come soon. I will never forget how it felt to be fighting for my life and to catch my breath. After the first shot entered my body and pierced my lung, the second entered my neck. I could smell and feel death surrounding me. But obviously God's angel was there, too, because I survived."*

(TYREK MARQUEZ): *"I had a really bad experience going through things I went through. Kids made fun of me, having to learn how to play sports again, having to learn how to ride a bike again."*

True, these stories can be quantifiable in some form. Each caption touches on a trauma that lies beneath, but it does not capture the depth of the pain and how lives were forever changed, with no statistic doing justice to these experiences. These captions do not even attempt to capture how their loved ones, too, changed in their outlook and behavior. There are thousands of such narratives every year.

Violence generates different levels of concern depending upon the weapon used. One study found that 81% of respondents reported that an episode with a gun was severely stressful, compared to 25% when it involved another weapon or 12% when no weapon was involved (Kravitz-Wirtz et al., 2019). A study of traumatic injury found pain rates, PTSD, and other physical and mental health outcomes were "alarmingly high" for gun violence survivors and even higher when compared to survivors sustaining similar injuries in car accidents (Bridgeri, 2020).

One Philadelphia study of almost 200 adult male victims found that surviving a gunshot scars victims externally and emotionally long after the incident, with almost 100 suffering from PTSD several years after being shot, and having high unemployment and drug and alcohol use (Giordano, 2019). These findings are not unique to Philadelphia. One study of violently injured youth (8 to 18 years) found male and female youth of color displaying a propensity to report a worsening of PTSD over a period of time, which is counter to an expected decline. This tendency for symptom increase raises the importance of aiding youth in early intervention programs to prevent the onset of PTSD (Garcia et al., 2020).

Gun violence victims, for example, may react negatively to fireworks, illustrating how their experience can be triggered by an innocuous and celebratory event. The presence of hyperarousal, a critical core component of PTSD, is triggered among those who are hyperalert to any sign of threat (Javanbakht, 2020). Leaving a home can be a major struggle for gun victims and a life-altering experience, introducing a new set of dynamics in daily living (Francis, 2018). Benner et al. (2018, p. 45) argue that PTSD "is predominantly viewed in terms of the usual neuro-physiological causal models with traumatic social events viewed as pathogens with dose related effects." A broadening of views on PTSD allows a more in-depth understanding of gun injuries.

Gun injuries are a public health and social issue deserving the same attention given to gun fatalities. The scope of the consequences go beyond injured parties (Hemenway & Nelson, 2020; Magruder et al., 2016). A national gun injury perspective and a US Surgeon General Annual Report on gun injuries assist in this effort (Kurek, Darzi, & Maa, 2020). Such an effort needs framing by the latest and most accurate statistics. As addressed later in this book, there is no book devoted to gun injuries in the United States or any other country in the world.

Mass shootings generate high death tolls and national publicity, with injury totals considerably higher, as in Las Vegas in 2017 when the largest modern-day mass shooting in US history caused 58 fatalities and 500 injuries. Ironically, even mass shooting statistics eschew those injured as victims. The Congressional Research Service narrowly defines "public mass shooting" victims as those killed. Yet most victims are not killed but injured. For example, if more than 10 people are shot but only two die, this shooting incident would not be classified as a mass shooting. Degree of injury is unaccounted for, even if it causes profound life-altering implications.

There is no disputing that mass gun killings generate intense national coverage, as in the cases with the 2019 mass killings in El Paso and Dayton when there were 31 people killed and 40 injured. These events, too, are not exempt from deemphasizing gun injuries. Chicago, however, experienced a weekend of killings and injuries during that time period with seven killed and 52 injured (Ali, 2019b). Interestingly, mass shootings generate immediate extensive publicity when they occur with an emphasis on deaths. However, in similar fashion to gun injuries in general, they, too, fade into the background of media coverage.

Gun death numbers are inaccurate because statistics underestimate the extent of health compromised from long-term disability caused by gun injuries (Raza, Thiruchelvam, & Redelmeier, 2020), which are sometimes referred to as a fate worse than death itself. Although gun fatality statistics have serious flaws, our knowledge of gun injuries is in its infancy when compared to fatalities, and that must change if interventions are to have a high probability of success (Bernstein, 2017):

> Fatal gun violence is often categorized in ways that make it easy to track and study. That's how researchers know that the murder rate in the United States has declined steadily over the past three decades. But what about gun violence that does not result in death? That is far trickier to measure. That's because nonfatal gun violence has mostly been ignored. As a result, policymakers, law-enforcement officials, public-health experts, urban planners, and economists are all basing their work on information that is unproven or incomplete.

Bringing together police and clinical data holds promise for capturing the extent of gun injuries within a locale because of a lack of integrated and reliable data (Magee, 2020a). That necessitates close mutually trusting working relationships.

Gun violence has much in common with COVID-19. Media coverage focuses on death counts as a measure of the devastation caused by this virus, including difficulties in burying the victims during the height of the spread. Minimal attention is paid to the long-term health consequences of surviving this virus ("long haulers"), including potential compromised health status (Ranney et al., 2019) and how lives are changed for victims and their families. In many ways, this is understandable because death brings finality. Further, we do not know the long-term consequences of surviving a COVID-19 infection, and we are only now grasping them for gun injuries.

Nevertheless, unlike the long-term consequences of COVID-19, we have greater knowledge of gun injuries, although much remains to be learned about the long-term consequences. As with COVID-19, gun violence creates anticipatory trauma. This trauma is an emerging concept that captures the concerns of children of color growing up in and navigating gun violence (Armstrong & Carlson, 2019). Violence victimization, for example, predicts body mass index one decade later for African American young adults, showing how exposure can have an effect in the life of these victims (Assari et al., 2019). This brief introduction serves as a foundation for the following chapter that goes into much detail on various aspects of gun injuries.

A BRIEF INTRODUCTION TO TRAUMA CENTERS

Readers will find references to trauma centers throughout this book because of their prominence in treating gun injuries and their unique approach to understanding this public health problem. It is impossible to understand gun injuries without an overview on trauma centers because of their significance in the gun violence field and their prominence in curtailing and treating injuries.

The following brief description of a trauma center's response to a gunshot victim helps capture the importance of this institution. Paul Carillo (executive director of Southern California's Crossroads) describes the immediate moment a gunshot victim arrives at an emergency room, highlighting the complexity of treating a survivor and the actions required that go beyond immediate health care (Rodriguez, 2021):

> Responding in the trauma bay when a gunshot victim arrives, Carillo's first job is to make sure that everyone, including the patient, their family, and the medical staff, is safe. This can be a balancing act when often two rival gang members arrive simultaneously. By consulting with the responding paramedics, Carillo can ascertain what occurred and then, by examining any tattoos the victim might have, figure out what gang or groups might have been involved. From there, finding the right connections is critical to deescalating the situation and finding the proper recourse towards a restorative solution that steers all parties away from further violence. "I start texting and calling the street intervention workers, some of whom are my employees and some that work for other organizations, so that we can create a coordinated response," said Carillo. "The most important thing is the collaboration."

Close and trusting collaboration between trauma centers and communities requires a shared mission and effective communication between trauma centers and community organizations.

Urban trauma centers fill a unique niche in treating gun injuries. Clusters of urban victims presenting at centers are common and not due to mass shooting incidents (Beard et al., 2019). The following description of a night at Miami's Ryder Trauma Center highlights the intensity of work in these settings (Korten, 2016):

> MIAMI—The man on the table at Ryder Trauma Center in Jackson Memorial Hospital has been shot four times, in the forearm, chest, flank, and thigh. Dr. George Garcia spends two hours operating on him, removing bullet fragments, closing blood vessels and repairing tissue. By the time he's tied the last suture it's 2 a.m. Garcia tosses his blood-stained gloves in the trash and goes out to tell the victim's father his son will be fine. Then Garcia's pager signals that an ambulance is on its way with another shooting victim. And so, he heads to the observation bays to greet his next patient.
>
> There is no "typical" when dealing with blood spilling out of a body and scared families waiting for news. But this night is not unusual. During a recent eight-hour shift, Garcia worked on four victims from different parts of Miami (in addition to two stabbings and a brutal machete attack). As soon as staff stabilized and moved one patient to the operating room, another would come through the door. "It was like a war zone," recalls one nurse. Miami so often

resembles a war zone that the U.S. Army chose Ryder Trauma as its training center for all surgical teams that will treat soldiers in battle zones.

Again, an analogy to a war zone can easily focus on gun victims, but those who tend to their lives, too, are victims in a unique way, calling attention to how this affects them.

Trauma centers have aided our understanding of gun violence consequences and been an essential part of any comprehensive solution by enhancing the probabilities of injury survival (Choi et al., 2021):

> The recognition of traumatic injuries as an addressable public health epidemic rather than unavoidable accidents has led to the birth and expansion of trauma systems. Trauma systems represent comprehensive infrastructures to provide optimal care for injured patients, encompassing a wide spectrum from injury prevention efforts and an integrated network of trauma centers to concerted research agendas. The United States has led the development of this systematic infrastructure to tackle the burden of trauma.

Trauma systems offer three distinct, but interrelated, levels of care: (1) prehospitalization; (2) hospitalization; and (3) posthospitalization (Kang & Swaroop, 2020). Gun violence necessitates a comprehensive approach built on partnerships. It is important to think of these collaborative partnerships together rather than as a list with individual components. Hospitalizations will stand out in importance and be the focus of this section. These settings offer immense resources for crafting and delivering services. For instance, trauma centers generally have ready access to all blood components and are able to carry out massive transfusions, a key element in gun injuries (DeMario et al., 2018).

Trauma centers are hospitals with unique capabilities for handling traumatic injuries. They have met criteria developed by the American College of Surgeons and successfully passed a review by its Verification Review Committee. Not surprisingly, trauma centers are often in the point position in data gathering and initiating innovative approaches, and they are often regarded as key members in coalitions for gun violence prevention and treatment initiatives. It is not surprising that much can be expected of them in treating gun violence. For instance, trauma centers are strategically positioned to provide alternatives to gang involvement, a prime source of gun injuries, in partnerships with community organizations in providing education, conflict resolution, and access to employment (Halimeh et al., 2021).

Trauma centers are located across the country and identified by a specific designation, with a verification process referring to the types of resources and number of patients admitted on an annual basis, with level 1 being the highest level of surgical care and increasing the chances of patient survival. Patients experiencing gun injuries to the torso, for example, encounter differing morality risk depending on the hospital at admittance, with level 1 having a higher success rate than level 2 centers (Grigorian et al., 2019).

The time from trauma center arrival to an operating room, too, influences outcomes, with 10 minutes considered the maximum period, and if more, it translates into a threefold increase in fatality for hypotensive patients with gun injuries (Meizoso et al., 2016). These few minutes translate into reducing the time of potential intervention of neurosurgical emergencies (Anderson & Kryzanski, 2020). Prehospital triage of gun victims is a critical period in maximizing survivability, and it is a stage in the treatment process. Unfortunately, this period presents tremendous challenges for medical personnel and is worthy of deeper understanding; the development of triage guidelines and interventions that increase the likelihood of survival prior to trauma center admission is needed.

There are attempts to develop reliable prehospitalization trauma and injury severity scores for creating prehospital trauma triage and predicting in-hospital morality for patients, with MGAP score (mechanism of injury), Glasgow Coma Scale, age, blood pressure, and the anatomically based Injury Severity Score (ISS) proving to be higher in predictability (Galvagno Jr. et al., 2019). These scores help trauma centers evaluate their rates of success by developing much-needed baselines for comparisons.

Level 1 centers have a full range of specialists, with 24-hour access to necessary medical equipment. Trauma center designation is a process outlined and developed at the state or local level. Trauma centers are also designed according to patient age groups (adult, pediatric, or both). In the case of adult-pediatric-focused designations, there may be different designations according to grouping. Clearly, urban trauma centers are well positioned to save lives, with gun victims as the beneficiaries of their resources and expertise. Again, it is important that programs providing trauma prevention require trauma center verification. However, effective programs targeting children are rare (Willard et al., 2020) and take on added importance in communities of color because of their high percentage of children and youth.

NEIGHBORHOOD CONFLICTS AND GUNS

Differences between this country's safest and deadliest cities are largely due to gun violence (Everytown for Gun Surgery, 2016). Youth and young adults in high-violence neighborhoods cannot avoid contact or relationships with violence-prone peers (Dill & Ozer, 2016, p. 553): "The threat of neighborhood violence, actual violent acts, and the consequences of violent acts preoccupy youth's thoughts, interpersonal relationships, and the routes and modes they use to move around in their neighborhoods."

Eschewing engaging in violent behavior, while maintaining a relationship, is a skill that taps resiliency and coping mechanisms (Patton & Roth, 2016). Navigating violent environments is a survival technique that has started receiving research attention (Culyba et al., 2021). Community interventions on youth peer influence and group norms have the potential to act as protective factors against the risk of violence exposure (Nebbitt et al., 2021).

Coping strategies have gender differences between early-adolescent African American girls and boys exposed to community violence, with the former relying on fewer coping strategies and the latter requiring more resources to support their coping (DiClemente & Richards, 2021). Sociodemographic factors must be taken into account to create a picture of daily life based on the interplay of key factors, with gender a key social force. These insights necessitate research that uncovers nuanced factors influencing attitudes and behaviors on outcomes.

Interpersonal conflicts present youth with choices, with one urban study in two high-violence Baltimore neighborhoods finding 17% of participants expressing a willingness to use guns to resolve a personal conflict (Milam et al., 2018). Tapping their voices in how gun violence is conceptualized and, more importantly, how best to address it is a point of entry into developing community interventions targeting them (Beck, Zusevics, & Dorsey, 2019; Outland, 2019). A qualitative study of Chicago African American adolescent males (N = 5) and females (N = 3) uncovered several major themes in high- and low-violence neighborhoods, with all participants exposed either directly or indirectly to crime in their daily lives; most indicated feeling worried about becoming a victim, influencing their ability to socially navigate violence in their communities (Ray, 2020).

Dong et al. (2020) addressed dimensions of the daily navigation of violent urban environments by focusing on contextual and temporal factors, bringing nuance to understanding social navigation. They found that (1) activity triggers increasing risk of assault during particular times of the day or fleeting moments; (2) the daily act of setting paths affects risk for violent assault; and (3) situational triggers influence violence during the day or night. The interplay among individuals, group, and environmental factors triggers or mitigates the potential of gun violence. Dangerous urban environments are not the only barriers to engaging in physical activities (Payán et al., 2019). Nevertheless, severely restricting environments, including fear of parks and playgrounds, ascend in importance in high-violence neighborhoods.

Being at an outdoor/public space and engaging in unstructured activities, with an absence of guardians, increases violent victimization at a specific spatial-temporal scale at daytime and nighttime. Nevertheless, the presence of friends and environmental characteristics has differential consequences on violent victimization during daytime versus nighttime. Understanding how youth navigate daily activities is worthy of greater attention by elucidating how they navigate activity spaces, increasing our knowledge on how they understand the threat of violence.

Parental voices, too, must be sought. Parents in high-violence communities are often required to assume roles in preparing their children for navigating their surroundings if they are to be successful in protecting them from violence (Preston, 2020). "Violence management strategies" is a construct that captures a vast array of navigational skills to eschew becoming victims of gun violence (Dill & Ozer, 2016). These strategies are learned on the street and the home. When parents have experienced gun violence, these lessons take on even greater meaning for youth.

BOOK GOALS

Books facilitate storytelling, and even academic ones. This book is no different. However, it seeks to tell many stories with implications for different audiences. Writing a book starts with an intense curiosity of what new sources of knowledge will be uncovered or validated. The writing of a book on a controversial topic is challenging, and gun violence falls into that category, bringing unexpected twists and turns on this journey. For some of us, gun violence is very personal; for others, it is a subject of interest from a professional standpoint. The hope, regardless of motivation, is to arrive at the end of the journey better for having undertaken it.

This book does not diminish the importance of gun fatalities. Rather, it seeks to uplift the importance of injuries because they are more frequent in number, bringing an immense emotional and physical toll on victims and their social networks. Further, they extract a huge economic toll on communities, cities, and the nation. Anyone who lost a loved one or suffered a gun injury will attest to the emotional toll it has taken on them and their family. Bereavement is an emotion associated with gun death, but it does not have to be limited as such, with applicability to loss due to injuries, with religion and spiritual assistance helping in this process (Lee et al., 2020). Posttraumatic growth can arise from an adverse life gun experience. Experiencing violence trauma, however, does not automatically result in a life of despair.

Why a book on gun injuries? There are an abundance of books on gun violence, with a specific focus on gun fatalities too numerous to list. Unfortunately, the same is not the case for books on gun injuries. As this book goes to press, I can only find two books on this subject, which is disturbing considering the magnitude of this problem. Durando's (2018) *Under the Gun: A Children's Hospital on the Front Line of an American Crisis* is based on how one hospital (St. Louis's Children's Hospital) was challenged to be responsive to gun fatalities and injuries, and it was self-published. The other book by Louis Anatole Lagarde (1914), *Gunshot Injuries: How They Are Inflicted, Their Complications and Treatment*, is a photocopy of a manual written after the US Civil War and largely based on gunshot causalities during that period.

There is a serious gap in comprehensive knowledge on a subject not met by academic articles. To provide readers with a central publication that serves as a foundation for practice and further research and study advances this field. Such a body of work as represented by this book should be appealing to fellow academics. This book highlights potential roles in addressing gun injuries and presents arenas where we should be playing an active role within the community to prevent gun incidents as well as to propose projects and interventions for dealing with their aftermath.

This book addresses five goals: (1) It provides a picture of the extent and nature of gun injuries among children/youth and adults, with a special emphasis on those of color and cities; (2) it provides a series of concepts for conceptualizing urban-focused interventions; (3) it provides case illustrations of innovative interventions; (4) it highlights recommendations for practice, education, and research; and (5) it identifies crosscutting themes to move this field forward.

It is important that readers understand why this book has such a heavy focus on the nation's urban centers, particularly since gun violence can be found throughout the nation. Yes, gun violence occurs in rural America (Choi et al., 2020; Rowhani-Rahbar, Oesterle, & Skinner, 2020). However, urban population density, combined with concentration of poverty, educational issues, food insecurity, housing insecurity, and ready access to guns, is the formula for a combustible situation, and it raises key racial and structural questions.

Finally, my upbringing and practice experience have taken place in urban neighborhoods. This book's focus on urban neighborhoods of color is due to personal experiences, professional practice, and scholarship. My hope is that this book spurs interest in writing other books on gun violence on subjects that have been neglected, such as rural gun violence. Readers may have wished to see greater attention paid to domestic gun violence and suicides. The broad nature of gun violence necessitated that I focus as much as possible. Covering extensive material on domestic violence and suicides would have necessitated setting the requisite foundation to do justice to the topics. According to US estimates, guns murder 52 women every month, with intimate partners playing a significant role in these shootings (Everytown USA, 2019b; Van Brocklin, 2018a).

This book does not address gun injuries from failed suicides. Unfortunately, these efforts are usually very successful (90%); suicides account for over 60% of gun fatalities but only 3.6% of injuries (National Centers for Disease Control, 2019), and they are overlooked in this country (Kaplan & Mueller-Williams, 2019). The complexity and importance of domestic gun violence and suicides require multiple books, and ones written by authors with specific expertise on these subjects, which I do not have.

My hope is that this book spurs other books with foci on multiple sociodemographic population groups, such as children, women, specific groups of color, and immigrants, to list a few. These and other groups may share commonalities in response to gun injuries but also bring unique considerations that must be taken into account in developing policies, research, and interventions. Use of an intersectionality paradigm enhances our knowledge of the nuances influencing help seeking.

I have endeavored to make this book reader-friendly. A book on medical conditions, procedures, and outcomes brings terminology that many of us have trouble pronouncing, let alone understanding. Readers will encounter terms that are foreign to them, but I have used them out of necessity. When used, I have attempted to explain in a reader-friendly manner. Also, there are abundant citations providing options for further study.

BOOK OUTLINE

This book has four sections. Section I ("Grounding Gun Injuries") consists of three chapters. The initial chapter provides a contextual backdrop, introducing a variety of dimensions on gun injuries, including a detailed rendering of the immense types and consequences of these injuries. The next two chapters provide gun statistics and injury descriptions. Section II ("Social Science Perspectives on Gun Injuries") starts with a chapter on social perspectives, followed by cultural and economic perspectives, setting a broad foundation for viewing gun injuries. Section III ("Community-Centered Approaches and Urban Case Illustrations") provides two cases on gun injuries and innovative programs. Section IV ("Recommendations for Advancing the Field") has three chapters on education, research, and community practice, bringing together distinct, yet overlapping, approaches on urban gun violence.

CONCLUSION

This chapter did not minimize gun death consequences. Rather, it uplifted gun injury consequences because they deserve attention and corresponding resources by providing a broad canvas upon which to understand injuries. Gun injuries must fight to get the attention because of how they consequentiality alter victim lives, their social networks, and neighborhoods. Gun violence is a major public health and social problem, with immense consequences manifesting themselves in marginalized communities.

2

Gun Injury Research and Statistics

Recovering from a gunshot wound is not a vacation. You need to, like, write that on your hand or something.

—Suzanne Brockmann

INTRODUCTION

The impact of gun violence on the nation is understood (Zimring, 2020), and more so in its urban communities. Nevertheless, research over the past 50 years has not made dramatic progress, particularly as it relates to treating injuries and the rehabilitation process. This chapter builds upon earlier coverage and addresses aspects not well known and significant in the life of communities confronting this violence, setting the stage for interventions to prevent and treat victims.

Qualitative dimensions require a nuanced perspective on injuries, allowing a multidimensional picture to emerge to guide interventions. Injuries from a beating with a gun, for example, may not elevate to what we typically consider a gun injury. Nevertheless, this assault should be part of any comprehensive analysis of how gun violence takes shape.

Multifaceted dimensions illustrate the challenges in developing a clearer portrait of this public health issue, as well as better preparing us for expanding the universe on subjects requiring in-depth attention to present a comprehensive understanding of the rewards and challenges that await us. We must keep in mind that data are usually presented in a manner that focuses on the primary injury. Gun victims may suffer multiple injuries, making recovery more challenging, expensive, time-intensive, and dramatic in altering a life path.

Introduction of race and urban context in research must guard against racialized assumptions. Racism in "penetrative trauma research" is an important backdrop and reintroduces the history of racism in medicine and why gun-injured people of color must not be subject to research that further stereotypes them in the name of advancing knowledge (Fast, 2020). This chapter introduces research that serves the greatest purpose within urban communities. Very often, this research has benefited from community input throughout the endeavor. Greater

The Silent Epidemic of Gun Injuries. Melvin Delgado, Oxford University Press. © Oxford University Press 2022. DOI: 10.1093/oso/9780197609767.003.0002

attention to gun injuries translates into policies and initiatives that aid victims to lead more productive and satisfying lives, as well as prevention efforts (Gastineau & Andrews, 2020).

Gun injures are constant in many areas of the country and are spiking in others. One study of urban gunshot pediatric (0–18 years old) injuries over a 20-year period (1996–2016), for instance, found that it went unchanged over this period. In response, researchers called for the introduction of alternative community-based prevention efforts centered on neighborhood capacity building and economic strengthening as a way of reducing gun violence (Bayouth et al., 2019). This is an example of research informing practice.

Research ascends in importance for policymakers who heavily rely upon findings to inform policies and programmatic initiatives, and more so when addressing complex social problems. Gun violence clearly qualifies. Research produces a powerful "evidential" supportive role, in close association with character attributions of those pushing antigun policies (Smith-Walter et al., 2016). However, we must remember that public emotions may supersede evidence in creating responses to gun violence.

To plan is human and to implement is divine. Fidelity ascends in importance for the gun injury field (Lyons et al., 2020a) and prevention programs (Cirone, Bendix, & An, 2020), regardless of setting and strategy employed. Evaluations play a critical role in identifying cost-efficient and effective approaches in generating desired outcomes, making it essential to introduce emerging research concepts, which bring a host of challenges. Successfully confronting these challenges moves the field forward, particularly when the political environment actively undermines a rational and systematic approach to one of this nation's most pressing public health problems.

DATA REPORTING SYSTEM CHALLENGES

Gun violence reporting systems are flawed. Readers interested in research should read a report by the University of Chicago's National Opinion Research Center (NORC). NORC convened a panel of gun violence experts and issued a series of recommendations on gun data infrastructure that put a spotlight on their limitations (Roman, 2020):

> The recommendations here include steps to improve existing data collections that can be implemented immediately and long-term changes in strategy to build a more robust and scalable infrastructure. The development of a rigorous empirical research base to inform both citizens and policymakers requires a robust and sustainable data infrastructure. The most enduring data infrastructure is one that is comprehensive, flexible, and nonpartisan. (p. 1)

Two concerns stand out in their report: (1) existing data are not well organized and highly segmented, limiting their effectiveness for informing national policy and local decision-making; and (2) gun data are often arduous to access, narrow in scope, and dated when released, with few data sources and systems integrated. I would add a third: an overall lack of data on injuries (Thiels et al., 2018). This chapter addresses these and other recommendations to provide a comprehensive understanding research in data-driven solutions.

NONFATAL GUN INJURIES

Grasping an understanding of gun violence is complicated because use of a gun does not have to translate into shots being fired. A victim can be accosted with a gun as a means to scare or subdue them, and this does not entail shooting them. The Center for American Progress estimates that between 2009 and 2018 there were over 3.6 million, or 400,000 per year, nonfatal violent crimes involving a gun in the United States (Vargas & Bhatia, 2020). This figure translates to over 1,100 incidents per day or almost 46 per hour. Comparing this statistic to the 210 daily gun assaults resulting in physical injuries only magnifies the problem of gun violence in this country.

The Center on American Progress provided a race/ethnicity breakdown highlighting why communities of color are disproportionate impacted: "Black Americans faced the highest rate, with 209 crime victimizations per every 100,000 people from 2009 through 2018. Hispanics followed with a rate of 129 crime victimizations per every 100,000 people, while white Americans had a rate of only 91 crime victimizations per every 100,000 people." These statistics do not take into account exposure to gun violence.

Finally, gun threats during robberies and assaults primarily affect young people under the age of 30 years. When considering all of these subgroups together, young Black men have a rate of 470 crime victimizations per 100,000 people, over 4 times higher than the overall national rate. These statistics are best thought of as one portion of the underwater part of a massive iceberg.

Nonfatal gun injuries may not leave physical signs of injury and do not immediately require medical attention as their physical injury counterparts. However, invisible psychological scars often do occur, and as with physical scars, they, too, remain throughout a lifetime. The help we provide a gun violence victim necessitates we understand how these scars transform their life's trajectory, requiring research specifically on these victims, but first needing recognition as victims or survivors, taking their place alongside those physically injured by guns.

EXPOSURE TO GUNS AS A FORM OF INJURY

When one hears the phrase "gun injuries," it will no doubt conjure up images of physical wounds. Yet invisible wounds are often present and manifest in mental health and behavioral outcomes that are detrimental, as in the case of adolescents (Leibbrand et al., 2020). Further, gun violence must eschew a narrow focus on fatalities because it limits our creation of potential strategies, as previously addressed; injuries, too, need to assume a prominent stance using a disparities lens and broadening classification as a gun injury beyond a physical manifestation.

Gun exposure has many manifestations. Increasing visibility of who is at increased likelihood of injuries and the services they require allows for a comprehensive discussion on gun violence to occur.

It is rare when a high-gun-violence neighborhood only witnesses one shooting in a day. Sergio Hill, a gunshot victim in Washington, DC, for example, has personally known 70 gunshot victims and seen 30 shot, and he has been shot himself (Turner & Wise, 2019). Gun trauma exposure, for instance, creates a tolerance of guns by offering potential protection from further assaults (Wamser-Nanney, Nanney, & Constans, 2020). Witnessing, knowing, and physically experiencing gun violence are often not mutually exclusive. This subject requires further research and documentation, bringing implications assessment and the development of interventions.

Could individuals consider themselves victims of gun violence if a police officer pointed a gun at them? This dimension of gun violence has been addressed by requiring officer documentation to prevent gun shootings, and this is an important step toward curbing gun violence (Shjarback, White, & Bishopp, 2021). However, approaching this experience from a trauma standpoint, too, is important. I would consider myself a victim if someone pointed a gun at me, regardless if in uniform or not.

Community violence exposure, with guns playing a significant role, has a strong association with externalizing behavior that is violent, desensitizing, and normalizing, finding victims engaging in violent behavior, including use of guns (Phan, 2019). Effective gun measures need to look at cumulative exposure because it is rare that one exposure episode is the totality of exposure in high-gun-violence neighborhoods, and we need multiple dimensions to capture this phenomenon.

Many urban gun crimes go unreported, making it difficult for professionals to assess the experience of a gun assault, being threatened with a gun, or hearing gunshots. For instance, in Trenton, New Jersey, estimates show that 60% of all such incidents are not reported to the police (Mazeika & Uriarte, 2019). Residents are at an increased likelihood of calling the police in cases involving homicide rather than gun injuries (Huebner, Lentz, & Schafer, 2020). This hesitancy complicates the help-seeking and healing process for victims and their families because the community may not validate their pain. Determining the extent of exposure and the trauma associated with it will be underreported and remain invisible, compromising our understanding of the extent of this problem. When this trauma combines with other traumas, a disturbing picture emerges and compromises victims' lives.

A community trauma index must be broad and encompassing to be relevant in urban communities. This necessitates innovations in developing a more complete picture of the extent of exposure and strategies residents use to navigate their surroundings. The increased awareness of the deleterious consequences of exposure has spurred innovative approaches (mobile technology and geospatial analysis) to assess how it affects urban youth (e.g., in Chicago) and their ability to navigate and interact with their environment (Roy et al., 2021).

It is reasonable to associate gun injuries with physical outcomes. Nevertheless, this view is best symbolized by an iceberg metaphor, with the tip above the water capturing deaths and physical injuries and the part below the waterline representing emotional injuries (Vargas & Hemenway, 2021). For readers versed in mental health treatment, it broadens the view of gun injuries articulated by Sharkey (2018):

> The impact of an incident of violence is felt most directly by the individual victim, but it is not limited to that victim. The incident affects those who are present when it occurs and watch it unfold. The impact may extend further, to those who know the victim or perpetrator, to those connected through some affiliation to the actors involved (e.g., family, friend, classmate, ethnic group, gang, neighborhood), and to those in the community who become aware of the incident. That same incident may affect those who are completely unaware of what took place but who walk streets that have a greater police presence, enter schools through metal detectors, and look for jobs in places where business owners are reluctant to open shop. (p. 85)

An all-encompassing injury perspective is open to criticism as "too broad." We can focus on a subset, but the context upon which it rests is comprehensive, introducing interventions in areas that have not been a focus because of narrow views of injuries.

Witnessing a shooting is an injury because of resulting trauma. One study of exposure to gun fatality on mental health outcomes in four urban US settings found that people of color, young people, and those with low socioeconomic status (SES) and formal education reporting this exposure suffered greater mental health outcomes. African American/Black, Latinx, young people, and those of lower income and formal education were disproportionately exposed, translating into higher levels of psychological distress, depression, suicidal ideation, and/or psychotic experiences compared to those not exposed (Smith et al., 2020).

Physical symptoms among children and youth exposed to gun violence can manifest when the source is being exposed to gun violence, rather than direct physical injuries (Turner et al., 2019). Screening for injury prevention, including exposure to violence, has ascended in significance in pediatric emergency rooms (Attridge et al., 2020). Witnessing a shooting leaves deep psychological scars that are just as consequential as physical ones (Ali, 2019a):

> When Antonio Magitt was 9, he witnessed a shooting at a park near his home on Chicago's South Side. Afterward, he was afraid to leave his room. "I just felt like the more I stay in my room, the safer I will be," Magitt, now 21, said. "I didn't want to die at a young age." It took him several years of therapy to manage the fear of being shot. Other than going to school, he avoided all interactions with people, including at home, where he was also dealing with verbal abuse from his mother's boyfriend, who was living with them at the time. "But staying in my room, isolated, also took a mental toll on me," he said.

"I didn't know how to build relationships with people. It really caused a lot of emotional and mental trauma on me."

Antonio's experience casts a wide shadow from a consequential standpoint, making it relatively easy to overlook psychological symptoms when a focus is on physical injuries.

How does gun exposure influence educational attainment? Educational inequity is associated with gun violence, with increases in the former translating to decreases in the latter, with the latter translating into poor educational achievement (Stevenson, 2019). Low educational attainment translates into limited mobility, trapping a generation into a life where dreams of a better existence do not take place. It is arduous to focus on educational attainment when most of the time and energy needs to be devoted to health care and rehabilitation. Survival is a good way to think about this state of being, and anyone who has ever been in such a state knows how it shapes worldviews.

Exposure, with early adolescent bullying perpetration and victimization positively influencing attitudes toward use of guns to resolve conflicts, raises an alarm for communities and the nation (Nickerson et al., 2020). Violent childhood victimization increases the likelihood of committing a crime in early adulthood across all racial/ethnic groups (Lo et al., 2020). Adolescent fears of community violence moderate on social, psychological, and physical well-being dimensions throughout by the influential role provided by family support (Drinkard, Schell, & Adams, 2019).

Chicago's African American males have benefited from high parental involvement in school education and by early childhood intervention, which promotes adaptive behaviors in adolescence, reinforcing the importance of early family engagement and comprehensive initiatives (Giovanelli et al., 2018). We may question the importance of these findings when applied to gun violence. These social factors illustrate the interrelationship between engagement, wellbeing, and education starting at an early age to counter conditions leading to community violence. Constant gun violence fear is a form of victimization, with exposure in public places to all age groups, including young children (Beardslee et al., 2018; Mitchell et al., 2019; Wamser-Nanney et al., 2019).

We can expand the concept of gun exposure beyond witnessing this violence to include having a friend or relative being a direct victim of gun violence (McGrea et al., 2019). Such a move impacts on how wide a net we use to capture gun injury outcomes, and how we train providers to recognize symptoms overlooked in the course of assessing traumatic events. Bre'Anna Jones, aged 27, may count a loved one as a fatality but she, too, is a victim. She is illustrative of the wide circle of impact a violent act has both immediately and over an extended period of time, possibly lasting a lifetime (Van Brocklin, 2018a):

We both grew up in violent parts of Chicago, but we really felt like we beat it. We had moved to the suburbs and were planning our wedding. On the day it happened, I was at work when I started getting texts from his mother. I called her, and she was hysterical. I remember her saying, "He's dead, he's dead." I asked, "Who's dead?" She responded: "Jonathan. They killed him." I said, this can't be true. I checked Facebook, and the first thing that popped up was a video of him lying on the ground.

We've heard so many stories that I don't know what to believe, but it seems he was targeted. He was with friends, going to work out with his old high school coach, and he was the only one who was shot. At first, I was so angry. We went to speak with a woman about the funeral, and she said, "Give it some time. You'll find love again." My mind blacked out. Next thing I knew, I was asking my mom what happened. She said, "You went haywire. You flipped that lady's desk over." I had to apologize. That's not me. I couldn't go back to our condo, so my mom moved our stuff into storage. I know it sounds weird, but I kept some of his T-shirts and socks in a Ziploc bag to preserve his scent. Our 5-year-old daughter still looks at pictures of him on her iPad almost every day. I hear her talking to him in her sleep. I'm really trying to figure out how to deal with this in a positive way, because I know she's watching me.

Now, my anger has decreased, but I'm still sad all the time. Sometimes, I'll go into this dark space for days, and I won't talk to anybody, won't answer the phone. A year after his death, I started working on a book about my experience. It helps soothe my mind.

Bre'Anna's actions after this death are not extraordinary. Memories and corresponding pain can last a lifetime, bringing potential intergenerational trauma when her 5-year-old daughter grows up with this legacy of violence as part of their family history and life experience.

A literature review on youth exposure to community violence and physical health outcomes uncovered seven areas where there are consequences (Wright et al., 2017): (1) asthma/respiratory health; (2) cardiovascular health; (3) immune functioning; (4) hypothalamic-pituitary-adrenal axis functioning; (5) sleep problems; (6) weight; and (7) general health. The results were mixed with cardiovascular health and sleep showing the strongest consequences to this exposure.

Nightmares among treatment-seeking youth provide a window to better comprehend the extent of extreme stressors from traumatic experiences, with gun violence–related trauma being but one of many in their lives (Secrist et al., 2019). Thus, gunshot exposure necessitates age-appropriate assessment and programmatic responses, taking into account sociodemographic factors associated with health disparities (Stolbach & Anam, 2017).

Disparities exist in hospitals and health care (Chaudhary et al., 2018; Formica, 2021). Racial health disparities (Indianapolis and Wichita), not surprisingly, will find a way into how victims of police violence report injuries and access care by avoiding seeking of care (Lewis & de Mesquita, 2020). The American College of Physicians (2020) issued a policy statement on racism and health in the United States, shaping care and the importance of it becoming an antiracist organization (Serchen et al., 2020).

There is a relationship between race and outcomes in hospitalized gun-injured patients, even when African American/Black and White, non-Latinx patients share similar lengths of stay and in-hospital morbidity. African American/Black patients, however, had higher total hospitalization costs and charges (Peluso, Cull, & Abougergi, 2020): "This inequality was also seen after hospital discharge, as Black patients were less likely to receive rehabilitation resources compared with White patients with firearm injuries. Further research is needed, preferably at trauma centers regarding disparity in inpatient outcomes and disposition resource allocation among races."

The longitudinal pathway of adolescent violence exposure makes adolescents more likely to have difficulty envisioning positive future outcomes, and more likely to engage in gun carrying (Lee et al., 2020). Helping youth have positive educational experiences and develop a hopeful attitude toward the future will aid in countering violence exposure. Gun carrying increases the chances of an injury or a killing, whereas curtailing adolescent gun carrying lowers risk for engaging in gun violence (Gunn & Boxer, 2021). Violence exposure increases risk for a range of mental health conditions, such as emotion aggression toward peers, anxiety, depression, and emotion deregulation (Cromer et al., 2019).

Violence exposure (directly and indirectly), when combined with substance use, exposes adolescents causig negative outcomes. This increases the likelihood that they are threatened, further effecting substance use into adulthood (Beharie et al., 2019). The long-range behavioral consequences these youth exhibit will likely be viewed as separate incidents or behavior without regard to gun violence.

Gun fatalities have gotten the bulk of the attention in the popular and professional literature. Fatalities, however, need to be placed alongside injuries, because of the latter's greater numbers. The concept of "co-victim" has emerged to help capture the fallout of a close relative of a killing (Sachs, Veysey, & Rivera, 2020). Again, this concept applies to gun injuries.

We must examine the police on this matter. Police-induced fatalities are very much in the news today, too. However, there is a paucity of data on police-inflicted injuries, limiting understanding of violence and police–community relationships (Evans & Thompson, 2019; Pica et al., 2020) at the local and national levels. High-profile shootings such as that of Jacob Blake (Kenosha, Wisconsin), which was captured on video, led to his paralysis.

Urban African American/Black men's trauma exposure and loss because of police violence, injuries as well as death, introduce a dimension to this subject with increased saliency among this group. Police gun violence translates into state-sanctioned violence (Delgado, 2020c), which puts them in a vulnerable position, with limited recourse in seeking justice. Events in Minnesota, New York, Oregon, and Wisconsin during the 2020 spring and summer are illustrative of these concerns, with life and death being uncertain, and in the case of injury, so, too, the recovery process.

Not all gun trauma gets manifested equally in outcome(Smith Lee & Robinson, 2019):

Young men's narratives also revealed their efforts to cope with the chronic and traumatic stressor of police violence. Although young men's personal agency facilitated their application of strategies to try to keep themselves physically safe from police, they acknowledged their limitations to fully insulate themselves from the psychosocial toll of systematic police targeting. Therefore, equipping Black boys and men with strategies to protect their bodies and their psyches from the trauma perpetrated by state authorities is paramount. (pp. 175–176)

Hipple et al.'s (2020) call for law enforcement developing capacities for standardized data collection on gun violence to include nonfatal shootings increases our understanding of the nature and extent of this problem.

Understanding gun injuries is hampered by inadequate and underfunded data-gathering systems, making it arduous to develop a comprehensive understanding of this phenomenon (Everytown USA, 2019):

The [Centers for Disease Control and Prevention's] CDC's estimate for 2015 totaled 84,997 nonfatal firearm injuries, but this estimate was qualified by a confidence interval of 36,636 to 133,357 injuries. This is too large a range to estimate the true onus of gun injury with accuracy or to know whether year-to-year increases or decreases indicate true changes or, rather, are an artifice of data limitations. In the face of increasing scrutiny of the reliability of this data, the CDC recently removed 2016 and 2017 nonfatal injury data for firearms and all other injury types from its website.

There is a call from policymakers and social scientists to develop a single authoritative gun injury data source, although the National Violent Death Reporting System has made important improvements. This reporting system does not cover all 50 states or gather data on gun injuries (Post et al., 2019):

Instead, what is known about gun-related injuries comes from a patchwork of data sources such as administrative institutional data that suffer from clinical bias, undercounts, or sampling biases. There are numerous national sources of gun-related injury data including the National Crime Victimization Survey, the General Social Survey, Uniform Crime Reports, National Incident-Based Reporting System, and the Bureau of Alcohol Tobacco and Firearms. However, each source has limitations. The most current data on gun-related violence relies heavily on a single source: law enforcement reporting. However, survey, medical examiner, death certificate, police, and healthcare records may significantly undercount the incidence of gun-related violence. (p. 674)

Comprehensive injury understanding is undermined because data reporting limitations impede advances in services back in the community.

The location and severity of injuries, hidden or visible and sometimes used as a badge of courage, need recognition. There are various types of gunshot wounds, and sometimes it is impossible to remove the bullet. If impossible to remove, the bullet is a constant reminder of being lucky or seeking revenge. It might also result in a moment of clarity and potentially change the lifestyle of the victim or transform the victim into a helper.

Cori Romero raises a provocative but often overlooked point on what it means to be "lucky" to survive a gun injury (Shorr, 2017):

> People tell me all the time that I am a lucky girl. Which is true because I am. I am lucky to be alive. I'm lucky to have basically survived. But what nobody understands is that once you have gone through such a traumatic experience, it takes everything in you to make sure that your happiness and stability survives too. I'm not the same girl I was before this, but one day I know I will be better than I used to be. (p. 51)

The word "lucky" needs defining and operationalization. A victim's mother said it well on what luck really means (Dycus, 2020):

> People say to me all of the time, "You are so lucky he is alive." Yes, I am blessed my son was not killed. But there is nothing lucky about Dre's condition. He's not the same Dre . . . It's not just a shooting that is life changing, it's the adjustments and accommodations we have to make just to get through the day. For me, it's that I have to choose between having my son home and giving him around the clock care or being able to sustain a job and provide for him. That decision has been agonizing, but those are the moments that show our strength, those are moments that really matter.

A comprehensive understanding of gun injuries relies on gathering data on location, context, and type of injuries because no two gun injuries are the same (Powers & Rodriguez, 2020), including when age and other sociodemographic factors are taken into account.

The ripple effects of gun violence touch many individuals officially not counted as "victims" in official statistics. Those addressing gun violence are not immune from suffering from vicarious trauma, introducing another dimension to gun violence exposure injury (Free & MacDonald, 2019). Vicarious traumatization, or secondary trauma, is no small matter, and it is the "price of doing business" in our field. If we do not recognize it, we will not seek aid, therefore limiting our potential effectiveness and longevity in this field. This traumatization provides valuable insights into victim lives and humanizes us, grounding us in the seriousness of gun violence. Gun violence exposure has risen in importance and resulted in the creation of the Violence Exposure Assessment Tool, gun violence's association to trauma symptomatology (Beseler et al., 2020).

A variety of gun injuries cause readmittance once discharged, as addressed in the following section, complicating the healing process and increasing costs and demands on care systems and victim families. This outcome raises the importance of providers and victims having a greater understanding of how the journey to recovery is fraught with barriers and shortcomings, testing the resolve of victims, their families, and care providers. New Jersey Attorney General Grewal stated it well on the importance of healing gun wounds (Nieto & Mclively, 2020): "You can save lives by healing gunshot wounds, and you can also save lives by changing lives and turning them away from violence."

Finally, there is another aspect of exposure that warrants attention. Guns do not have to be fired to cause injuries, such as in the case of gun robberies, with an estimated 103 deaths, 210 physical injuries, and 1,100 threats involving a gun, resulting in a need to broaden the consequences of guns beyond a narrow confine of fatalities and injuries (Vargas & Bhatia, 2020):

> But not all acts of gun violence involve pulling a trigger. Often, perpetrators of crimes brandish firearms to intimidate and subdue victims, forcing them to comply out of fear of bodily harm. . . . While survivors of these crimes are left with no visible wounds, they often suffer from emotional distress and psychological trauma. It is vital that incidents such as these be considered in efforts to address the epidemic of gun violence that plagues the United States.

According to the Federal Bureau of Investigation (Center for American Progress, 2020), from 2009 through 2018, there were an estimated 160,000 assaults and 127,000 robberies involving a gun annually, with these figures being severely undercounted. Further, they do not take into account exposures discussed earlier. Gun assaults, again, fall disproportionately on urban communities of color, with young African American/Black males standing out as a subgroup of victims.

A CALL FOR A CIVILIAN GUNSHOT INJURY CLASSIFICATION SYSTEM

Significant advances have been made in identifying and classifying gunshot patterns (Savakar & Kannur, 2016), helping advance treatment strategies. However, despite the high prevalence of US civilian gunshot injuries, no universally accepted classification system currently exists, hindering our understanding and helping to shape our approaches to this public health issue. A call for such a system goes far beyond intellectual curiosity because of its real-life practical relevance for moving the field of gun violence forward both conceptually and practically. Some readers may argue that such a classification system is not possible in the United States.

Why do we need such a classification system? A gunshot injury classification system would aid in facilitating communication within and across professions, and serve a useful function in the development of data that different professionals could access. It would prove useful in developing treatment interventions and even predict the likelihood of outcomes, assisting in the allocation of resources. An overarching gun injury classification is possible whereby injuries are either acute or chronic. A simplified classification system, although attractive, is very limited in guiding the field because of the complexity of gun injuries. High-energy gun wounds cause unpredictable injuries (Lotfollahzadeh & Burns, 2020), compounding any classification process.

Readers may well favor a simplified framework on gun injuries: (1) penetration (enters but does not exist, and will either be removed or left if removal can prove consequential); (2) perforation (bullet enters and leaves the body); (3) re-entry (bullet enters one part of the body, such as a hand, exits, and then re-enters another part of the body); and (4) exposure (the victim is not hurt by the bullet but is exposed to a shooting, such as hearing, seeing, or learning about someone who is injured).

DiMaio's (2015) *Gunshot Wounds: Practical Aspects of Firearms, Ballistics, and Forensic Techniques (Practical Aspects of Criminal and Forensic Investigations)* (3rd ed.) presents a broad wound classification system of four categories: (1) contact, (2) near contact, (3) intermediate, and (4) distant. These categories identify and categorize challenges for guiding appropriate medical interventions.

Additional efforts have been advanced to develop a more detailed and comprehensive classification. Brito et al. (2013), for example, proposed a gunshot incidents (GSI) classification based on energy transfer, vital structural damage, wound characteristics, fracture, and degree of contamination. A GSI classification system must overcome challenges to have practicality:

> It should be comprehensive enough to include all relevant injury components and yet be simple if it is to gain universal acceptance. It must also identify a basic rationale by which an injury is accurately described, and categorized. Furthermore, it should demonstrate statistical merit characterized by adequate internal consistency and predictive accuracy. (p. 392)

Comprehensive yet simple at the same time may seem like a tall order, but accomplishing this goal aids using gun research in developing interventions.

I am hopeful that significant progress will be made in the next decade to create a gun injury classification system that bridges knowledge and service gaps in creative ways, bringing the field together with communities to achieve long-lasting solutions to what is arguably the nation's most pressing need. Hopefully, a comparable classification system can be developed on how recovery can be enhanced to aid national and local efforts in helping victims recover and leading lives that maximize their potential.

Gun injury outcomes interact with victim health status. In the case of victims of color, the physical and emotional trauma of a bullet compounds their compromised health status. Developing a gun injury classification system helps in resource allocations, interventions, and outcomes, and it must capture preexisting conditions. Such a classification system will prove worthwhile but not without its share of challenges. Developing such a classification system, with appropriate weights attached to wounds in combination with preexisting conditions, requires a multidisciplinary charge to capture nuances to make it worthwhile.

CRITIQUE OF EXISTING DATA AND CALL FOR RESEARCH

It is important to highlight one major challenge facing researchers, and that is the general absence of how to define commonly used terminology (Jordan & Harper, 2020). On the surface, readers may argue that this is not the case with gun injuries. True, a physical wound caused by a bullet lends itself to standardization on how to label and describe it. Emotional consequences do not. Health organizations must broaden their scope beyond physical manifestations to include social, cultural, and economic ramifications. That charge entails developing partnerships with community based social and civic organizations that provide these perspectives.

There is an unquestioned need to increase our knowledge of gun injuries within urban communities, and this entails increased research within this social and geographical context. A consensus exists that current funding for trauma research, with gun violence a prominent part, is not commensurate with the magnitude of the problem (Dowd et al., 2020). However, increased funding by itself is not the solution.

Winker, Rowhani-Rahbar, and Rivara (2020) argue that federal funding increases of gun violence research without corresponding laws to implement findings will solve very little. In fact, it adds insult to injury because when we have clear answers to this health problem, we will want to act upon it to prevent loss of life or a lifetime of serious medical needs.

CDC gun injury data are widely considered unreliable by social scientists (Campbell & Nass, 2019a, 2019b, 2020). The CDC sampling process changed when one hospital with low or no gun injuries was replaced by a hospital with many, calling for a broadening of sample size (Campbell & Nass, 2019). This sample size represents 2% of all hospitals across the country. There is no reason why there cannot be various nationally focused sampling approaches, with one focused on cities and hospitals with high concentrations of people of color and others focused on suburban or rural hospitals. Gun injury costs warrant multiple sampling approaches.

Historically, federal gun violence research support faced severe restrictions by the CDC (Maa & Darzi, 2018), with obvious and hidden ramifications. Estimates have this research receiving less than 1% of the funding spent on liver disease, an illness that results in very few deaths (Joint Economic Committee Democrats, 2017). Another unintended consequence of a two-decade absence of major funding resulted in lack of data to build a comprehensive national public health stance (Ahlin, Antunes, & Watts, 2021; Dzau & Leshner, 2018).

Guns are deeply ingrained in this nation's culture and highly controversial with the Second Amendment often uplifted as a right to bear arms. This controversy influenced the federal government, with the possible exception of the National Institutes of Health, to withhold funding research on gun violence (Rubin, 2016). The 1996 Dickey Amendment prohibited the federal government from advocating or promoting gun control, effectively freezing research (Cogan, 2019). The National Rifle Association played an influential role in this federal stance (Rostron, 2018).

Drawing on the collective experience and expertise of stakeholders helps to develop a multifaceted picture of gun violence, which is essential because of how local context shapes perception of this problem (Joseph, Bible, & Hanna, 2020). This process is never easy or fast. Reliance on local data for guiding interventions has merit but also brings serious limitations by not relying on a broader contextual understanding of gun violence (Manley et al., 2020):

> Gun violence remains a significant public health problem that is both understudied and underfunded, and plagued by inadequate or inaccessible data sources. Over the years, numerous trauma centers have attempted to use local registries to study single-institutional trends; however, this approach limits generalizability to our national epidemic. In fact, even easily accessible, health-centered data from the CDC lack national relevance because they are limited to those enrolled states only. (p. 475)

Clearly, a multifaceted funding research approach is in order, marshalling resources from national and local sources. This recommendation is revisited in Chapter 11.

Increased research funding must accompany efforts grounded within the sociocultural context of resident lives. Conventional research approaches have a propensity to emphasize singular forms of data to conceptualize and measure violence, such as place, particularly when it interacts with other forms of violence. This brings attention to the need for new approaches (Flynn, Mathias, & Wiebe, 2021).

A public health stance on urban gun injuries seeks data on geographical context to guide interventions. The gun violence field will argue that solutions are not achievable without data-driven research, and the complexity of this problem requires data-driven interventions (Rosenberg, 2019). Data, without contextual grounding, is not useful in appropriate community interventions. Powell and Sacks (2020) advance research recommendations: (1) creation of widely shared national research agenda on gun injuries and fatalities; (2) targeted research funds focused on gun injuries; (3) targeted support for development of a cadre of researchers to study gun injuries; and (4) enhanced access to comprehensive data sources. These recommendations will be discussed in this section, providing gain greater depth in operationalizing them.

I would add two more recommendations. There is a consensus that there are serious limitations in identifying gaps in data and quality issues, and we must fill these gaps to move the gun violence field forward (Sondik, 2021). Further, it is insufficient and unethical to call for increasing urban gun violence research without an equally strong call for research on strengths of the injured and community assets, countering gun-stigmatizing approaches (Hamby, 2020). A better understanding of the processes used to achieve a healthier recovery is needed, one requiring a resolve on the part of the survivor and an extensive support system based upon a cultural foundation.

Nationwide data-driven gun violence solutions depend on a research infrastructure that can help determine the success of interventions (Joseph, Bible, & Hanna, 2020). Further, there cannot be a narrow focus on gun fatalities (Educational Fund to Stop Gun Violence, 2020): "While increasing attention has shed light on the epidemic of gun deaths in the United States, far less is known about the millions of Americans who have survived gun violence. Unfortunately, publicly available and precise data on nonfatal gun injuries is limited and often not easily accessible." This conclusion should be alarming because research cannot be undertaken for the sake of research itself. Research must seek knowledge that can guide us to a better society and, in the case of gun injuries, not blame victims for their injuries. The majority of nonfatal gun injuries occur outside of the home, calling attention to developing new models that enhance predicting locations of shootings (Parker, 2020).

There is no reliable gun injury source for the United States (Kaufman et al., 2019), and that declarative statement sets the field marching. No research approaches should be ignored in increasing our knowledge base on gun violence. For example, there is recognition that translational research must assume a more prominent patient-centered approach to increase the effectiveness of violence reduction programs (Wical, Richardson, & Bullock, 2020). A patient-centered approach requires taking into account their sociodemographic characteristics, strengths, and lived experiences. No US trauma center collects long-term follow-up data 30 days after discharge, limiting understanding of long-term gun injury consequences (Lindsay, 2020).

Epidemiology, too, assumes prominence within the constellation of research approaches for advancing gun violence prevention, as it guides community-level public health services and actions. Epidemiologists bring unique interdisciplinary tools for identifying forces and for crafting interventions (Davis et al., 2018). Our knowledge of recidivism rates following gun injuries, a key aspect in any intervention project, remains incomplete, largely due to inadequate longitudinal follow-up and disparate databases (Marshall et al., 2020).

A consensus-driven gun injury agenda, in similar fashion to the nominal group research effort by emergency medicine (Ranney et al., 2017), can be conducted by other professions as a means of starting a comprehensive research agenda that enjoys a wide base of support. This technique is a group process involving problem identification, solution generation, and decision-making among professionals. Although the nominal group technique has limitations on specificity and number of participants, it is an excellent initial step in a lengthy process for developing sophisticated methodologies for understanding gun injuries. Further, it is not an expensive method compared to other more complex approaches.

The gun injury field will benefit from a standardized forensic recording form that enjoys wide appeal, in similar fashion to one developed for physicians in Thailand (Sripong, Samai, & Liabsuetrakul, 2019). These initiatives will prove labor intensive and politically difficult to embrace and navigate because of the decentralization of the US government, but they will help minimize differences across institutions and geographical boundaries in developing a comprehensive portrait of gun injuries.

The gun violence field finds itself advocating for a comprehensive cross-sector data collection (Hipple & Magee, 2017). Gross and colleagues (2017) strongly recommend a multidisciplinary and demographic-specific approach to preventing gun violence. However, prior to delving into gun injury statistics and how best to guide strategies, we must pause and point out key limitations of current data systems on gun violence, which provide difficulties for comprehensive and accurate understanding.

For example, a study of a police trauma registry of firearm assaults in Philadelphia (2005–2014), using the Pennsylvania Trauma Outcomes Study (PTOS) and the Philadelphia Police Department (PPD) database, found that PTOS data underestimated firearm assault incidence and overestimated mortality when compared to PPD data. This disparity between datasets reinforces the need for a comprehensive data collection system on the incidence, nature, and severity of gun injuries (Kaufman et al., 2019). These authors concluded that trauma registry data are by definition incomplete, calling for a combining of data sources for a more complete picture. When discussing city-level injuries, law enforcement data, too, are important for studying firearm injury.

According to Rowhani-Rahbar, Bellenger, and Rivara (2019), gun violence research can be conceptualized from three perspectives to better address data shortfalls: (1) availability, (2) accessibility, and (3) content. They note:

> Although much has been written about limitations in funding for firearm violence research, especially at the federal level, relatively less attention has been devoted to barriers in obtaining and accessing data needed to address important policy and practice questions. The availability of funds to collect data by researchers is imperative; nonetheless, there are circumstances in which the lack of access to pertinent information that is not readily collectible by investigators, regardless of research funding, can adversely affect specific areas of inquiry in firearm violence.

Research funding for the sake of funding is insufficient in helping to close important knowledge gaps, particularly as demographic changes are rapidly unfolding across the nation. Our sights must remain on relevance and not just amount of funding.

Gun violence research that systematically gathers data rather than an occasional survey offers greater relevance for injury data, but there are few researchers prepared to develop and improve these data systems (Hemenway, 2018):

> At Harvard, there are literally hundreds of courses on how to analyze data. These are important courses, but there does not seem to be a single course on how to create, manage, and improve data systems. I believe this is generally true at most other U.S. universities. Yet we all know the notion of "garbage in, garbage out." I believe we as a society should be devoting more resources to

training statisticians on the important topic of data system development and improvement, and paying more attention to the creation and maintenance of good surveillance systems. (p. 11)

Further, we need to create a new cadre of researchers willing to engage in participatory-driven research approaches.

A critique of case-control studies based on the National Violent Death Reporting System (NVDRS) identifies an inherent potential to present selection bias and misclassification through control selection (Lyons et al., 2020b). For example, studies using NVDRS compare groups of individuals who died by one mechanism, intent, or circumstance to individuals who died by another mechanism, intent, or circumstance, introducing selection bias; reliance on narrative summaries for exposure measurement has the potential to introduce misclassifications.

Williams, Bowman, and Jung (2019) concluded that government databases involving fatal officer-involved shootings (Supplementary Homicide Report, the National Vital Statistics System, and the Arrest-Related Death Program) are vulnerable to underreporting and classification errors. The authors are skeptical that these reporting systems are capable of sufficient improvement to make them viable. Rather, they recommend that an independent party, such as a university, receive funding to collect data from open sources and supplement it with data from public records and currently collected official government data.

Readers may ask what makes a reliable dataset. The National Emergency Department Sample (NEDS), which represents a nationally representative set of emergency departments (total of 950 and 20% of all the nation's emergency rooms), has this distinction (Educational Fund to Stop Firearm Violence, 2020). This database relies on Emergency Department gun injury visits, gathering information on the intent of injury, including basic demographics on victims, all essential elements in helping to develop a multidimensional portrait of victims. Insights derived from these data help shape how interventions get conceptualized and implemented.

Nevertheless, NEDS has limitations because it does not gather information on the extent of nonfatal gun injuries. Victims suffering minor physical wounds, for example, eschew treatment at an Emergency Department and escape detection. Nonphysical injuries, such as experiencing gun sounds, witnessing a shooting, threats, intimidation, or emotional abuse with a gun, are not part of this database. These data are not publicly available and considered arduous to analyze. Finally, this database records Emergency Department visits, not individual patients, making it impossible to follow the trajectory of patients seeking assistance for different encounters, which is very important because of the frequency they experience reoccurring gun violence. Databases as currently constituted underestimate the true costs of gun injuries on outcomes, care utilization, socioeconomic influences, and particularly over a lifetime (Rattan, Namias, & Zakrison, 2018).

A survey of barriers to gun injury (N = 113) and car accident researchers (N = 241) found those focused on the former face considerably more challenges in carrying out their research agendas because of limited funding and threats to personal safety, a vastly underreported subject (Donnelly et al., 2020). We can talk about personal sacrifice in working on gun violence prevention and safety concern for researchers and practitioners within these communities. However, personal safety from outsiders feeling threatened by the nature of the research conducted, including potential impact on policies, adds another dimension to impediments to gun research and is rarely discussed.

A number of strategies have ascended in importance to move the field of gun injuries forward. Santaella et al. (2016) advocate for sounder research on the association between implementation and/or repeal of gun legislation and gun injuries to inform interventions being crafted to local contexts. The field must not lose focus of the importance of an active research agenda to include evaluation of community initiatives to prevent gun violence (Dzau & Leshner, 2018).

Increased gun violence funding introduced advances for the field, such as in the CDC. The CDC's data science strategy enhanced the quality and breadth of current data-gathering efforts, filling critical gaps in the gun injury knowledge base (Ballesteros et al., 2020):

> This strategy includes goals on expanding the availability of more timely data systems, improving rapid identification of health threats and responses, increasing access to accurate health information and preventing misinformation, improving data linkages, expanding data visualization efforts, and increasing efficiency of analytic and scientific processes for injury and violence, among others. (p. 189)

The CDC's active role in gun violence research increases funds toward this cause, encouraging other federal and nonfederal sources to fund it as well.

No research process is without flaws, and once we accept this premise, whether focused on gun injuries or any other social issue, we become more open to critiques and the search for a better approach. This stance does not disregard important work already done on research methods. It means that we should be open to innovation in this journey. Openness to new ideas translates into an increased probability of research serving the public good, and in this instance, urban people of color who disproportionately bear the onus of this violence.

Increased federal funding must not be taken for granted, and constant vigilance is required to ensure that it continues (Denne, Baumberger, & Mariani, 2020, pp. 800–801):

> The CDC and the NIH are now tasked with moving ahead with plans to conduct this research with its new congressional appropriation. Public scrutiny of this research will undoubtedly be intense, and Congress will have to act anew every year to ensure this funding continues. Continued advocacy will be necessary to ensure that this funding is not eliminated again as it once was. Advocacy has long been a core value in pediatrics and all our academic societies have incorporated advocacy as a founding principle.

NATIONAL AND URBAN STATISTICS

National statistics serve the important function of painting an overall picture of gun violence, including injuries, in this country. Nevertheless, even though national statistics establish a broad context or foundation for gun injuries, they can obscure how gun violences manifests at the local level. More specifically, data are gathered at city and neighborhood levels in order to customize interventions to consider local circumstances and histories. This, however, does not mean that national urban statistics are irrelevant because they help us to understand regional differences.

The National Criminal Victimization Survey uncovered almost 80,000 gun injury victims in 2018, with the vast majority being people of color. These numbers, however, do not capture the physical and emotional tolls of victims and those entrusted to care for them after their injuries are treated at the hospital.

Although the gun injury epidemic can be cast as a national problem, there is no disputing that its urban centers experience the brunt of this violence. This focus brings rewards and accompanying challenges. Although there is a propensity to lump cities together, each is very different from historical, cultural, and social perspectives. True, we can standardize statistics per 100,000, for example, but no two cities are alike, even within the same state. It is essential we avoid an exclusive focus on major cities because top-ten cities in population are limited to only 10. Midsize cities, too, bring advantages and disadvantages in addressing gun violence because the federal government and big foundation funding initiatives often overlook them.

Most cities are small or midsize, and we rarely associate gun injuries with these cities. Rather, the cities of Chicago, Los Angeles, and New York garner their share of national attention. Nonetheless, with this caveat, we can turn to urban statistics on gun injuries, paying attention to areas rarely uplifted for national attention. Research in small to midsize cities, too, has helped to advance understanding of how gun violence has shaped neighborhood life.

Although gun violence is associated with youth and young adults, a New Haven, Connecticut, study examining gun injuries from 2003 to 2015 found differences. Victims were 27 years old on average and became older over the study time, from 23.9 to 27.6 years old across all racial groups, with differences among African Americans, Latinx, and White, non-Latinx groups (Law et al., 2017). If this is the case in other cities, it necessitates a nuanced understanding, taking into consideration longitudinal trends and racial disparities because present situations are best understood when grounded historically (Burrell et al., 2021). Other urban statistics will be provided in this book.

CONCLUSION

This chapter provided statistics on various aspects of gun injuries, including limitations in data-gathering approaches. Knowledge of gun injuries is still in its infancy and will change as more research is conducted from a multidisciplinary perspective. This chapter, in addition, presented quantitative and qualitative research approaches highlighting gun injury within urban communities.

Unfortunately, society and social scientists invariably emphasize "body counts," seriously narrowing discourse on gun violence and ignoring the social burden of this problem (Cook, 2020). Much discovery remains from a perspective that encapsulates community views of this violence, although this approach is not without challenges in bringing residents to the table as equal partners.

3

Gun Injuries

The emotional impact of being shot and left for dead on a quiet Berkeley street will be with me forever. I knew even then that I could become better or bitter. I decided to use my traumatic experience for the better.

—*Gabrielle Schang (Shorr, 2017, p. 53)*

INTRODUCTION

The quote that begins this chapter highlights survival after a gun injury and the path to "recovery" (Christensen, 2016):

> A patient can spend months at a hospital, during which time they may have several surgeries, including orthopedic, vascular and neurological. They will need further tests like CT scans and possibly surgeries after they get out to reconstruct bone or to repair internal organs. When a colon is severed for instance, victims may need to wait a year for the swelling to go down and only then can they have surgery to reattach it. Some may be permanently disabled. Then there is therapy.

Gun violence does not end with the injury; it is a recurrent, traumatic experience that affects survivors and their families for a lifetime.

There is no disputing the toll of nonfatal gun injuries in urban communities of color (Villarreal, 2020). No statistic captures the rush of emotions when the injury occurs and what follows in a victim's life, calling for efforts to capture these important narratives and corresponding insights to aid in shaping institutional responses. These thoughts and feelings flood victims' lives, including doubts about their future and those of their families and friends.

"Bullets are a vector of violence, a traumatic transfer of energy from one person to another" (Andrade et al., 2021, p. 255). What a simple description for such a dramatic and consequential act. Gun injuries are not part of life's rich pageantry; injuries assume lifelong markers or anniversaries of nightmares, rather than dreams waiting realization and celebration. Injuries are the majority of gun

The Silent Epidemic of Gun Injuries. Melvin Delgado, Oxford University Press. © Oxford University Press 2022.
DOI: 10.1093/oso/9780197609767.003.0003

violence outcomes but do not receive needed attention from research and practice standpoints (Theaker, 2020), including exposure to various aspects of guns. The subject of gun injuries benefits from greater attention, but it must be multidisciplinary and community centered.

The term "gun injuries" engenders a multitude of reactions and images, with many responses probably influenced by their portrayal in television and film (Nacasio, 2015):

> Penetrating trauma and tissue damage from projectiles are a bit different. They have the potential to cut through arteries and large veins without alerting the body's . . . muscles. With bullets, it all comes down to shot placement and passage—which, without the gift of surgical precision that no gunman will ever have, is another way of saying it comes down to luck. Aiming for limbs to create "flesh wounds" is a movie myth.

This is a glimpse into a victim's life on what happens when a bullet enters the body. Tissue effects of penetrating trauma can vary by the speed of the bullet entering the body and type of weapon used. Non-life-threatening wounds do not capture emotional consequences, which are largely invisible and traumatic.

For those new to gun injuries, we are undertaking a journey of discovery where we will hopefully never hear or see a statistic on this topic again without requisite grounding; that is, we will no longer read a statistic as just another numerical figure. Gun injuries cover a range of types and severity, making generalizing impossible, if not dangerous, and calling for increased injury research. Further, the historical paucity of gun violence research funding limits the creation of data that are consistent and comparable across sites and time periods (Krisberg, 2018).

According to the Centers for Disease Control and Prevention (CDC, 2020), males account for 85% of all gun deaths and 88% of nonfatal gun injuries. More specifically, gun homicide is the leading cause of death for African American/ Black males aged 15–24 years. It is no surprise that gun fatalities and injuries automatically are associated with males. However, an analysis of trends found an increase among females, as in Chicago (2005–2016), where they experienced a 74.7% increase in injuries (Fitzpatrick et al., 2019). Chicago has an estimated 1,500 gunshot injuries annually (Kieltyka, Kucybala, & Crandall, 2016).

Gun statistics on people of color often result in a dehumanized view of victims. We must humanize those killed if the nation is to make significant inroads on this problem. The names change, but their color remains the same, with recent national demonstrations highlighting the murders of George Floyd, Breonna Taylor, and other African American/Blacks by the police. We must also humanize victims regardless of their injuries to understand the consequences, and this necessitates media follow-up over an extended period.

This chapter introduces important narratives on injured victims and humanizes them in the process. We will not see the names of those injured by the police in public demonstrations, or the nature of their injuries, with some exceptions. However, those injured, too, have names, and often just go by a number in local media coverage, and they must be humanized. A book on those injured by guns at the hands of the police or fellow residents would provide a missing piece of the picture of the health ramifications of violence in urban communities. It is my hope that this book will be a good start.

It is important to begin with a discussion of aspects not normally part of the conversation on this subject because it illuminates the broad reach of injuries and the unintended, and invisible, ramifications. Readers will develop a broader appreciation of gun injuries and their various short- and long-term manifestations.

EXPOSURE REVISITED FOR PEOPLE OF COLOR

There is a propensity to think of wounds from a physical standpoint, but one does not have to be a direct victim of a gunshot to experience health repercussions and thus become a victim, too. Quimby et al.'s (2018) research of gun exposure among urban African American/Black youth found that almost half reported some type of gun exposure, indicative of a more widespread issue. Youth experiencing trauma because of direct gun violence often find themselves with other experiences of violence, introducing the role of context in incorporating gun violence exposure in a variety of ways (Turner et al., 2019) and understanding cumulative exposure.

Expected difference in outcomes can be present for African American adolescents when violence exposure is directed at a friend or family member, causing higher rates of anxiety than victimization of acquaintances or strangers. Exposure is also more likely to cause depressive symptoms, with high-exposure groups reporting more mental health symptoms (Sargent et al., 2020). Further, Chicago youths' exposure to community violence caused long-term chronic, pervasive, and spatially proximal vulnerabilites, causing higher levels of behavioral dysfunction (DaViera & Roy, 2020). Children and adolescents exposed to community violence are more prone to conduct disorder and health issues (Kersten et al., 2017), influencing their educational outcomes.

The correlation of asthma with exposure to gun violence, for example, has attracted research attention (Landeo-Gutierrez et al., 2020). One study of Puerto Rican children with African ancestry found increased risk of asthma (Rosas-Salazar et al., 2016), showing the interconnectedness of gun violence with other illnesses that rarely get associated. Measures of adverse childhood experiences (ACEs) consist of numerous types of violence, but not gun violence. Items included in ACEs are not systematically selected for inclusion based on their predictive power for negative consequences (Turner et al., 2020). This absence severely limits developing a comprehensive understanding of gun consequences in this nation's cities (Rajan et al., 2019):

> This systematic review of research over the course of two decades confirms the critical importance of classifying youth exposure to violence involving a gun as an ACE, of broadening the definition of gun violence exposure to include a broader spectrum of youth experiences with gun violence, and of expanding

the notion of who should conduct such screenings to increase the reach of both existing screening and intervention efforts. (p. 655)

Gun violence exposure is only now receiving the attention it deserves, with the promise that this attention will bring forth important breakthroughs.

PHYSICAL GUN INJURIES

Physical gun injuries can cause the following damages: (1) damage to vital organs, major blood vessels, and the nervous system; (2) severe bleeding; (3) broken bones; (4) infections (resulting from material and debris introduced into the wound by a bullet); (5) paralysis; and (6) scar tissue, which may result in ongoing pain (U.S. National Library of Medicine, 2018). These injuries can be seen through a different lens, presenting the problem with an understanding that a singular injury is rare: (1) body disfigurement; (2) mobility impairments; (3) blood toxicity because of lead in the body; (4) emotional trauma; (5) impaired senses (seeing and hearing); (6) pain; and (7) impaired verbal communication, for example.

The injuries listed here do not include gun violence exposure. The health consequences of this exposure need researching from perspectives we usually do not associate with this violence to develop a comprehensive understanding its range of ramifications, as noted in Chapter 2 and the previous section. According to the American Heart Association (Kuehn, 2019), those exposed to threats "to themselves, their family, or their communities . . . will experience an increased risk of myocardial infarction, stroke, ischemic heart disease, coronary heart disease, or death from one of these conditions."

Regardless of how we conceptualize injuries, there is no question about their significance for this nation's urban communities. Injuries are not mutually exclusive, complicating the treatment and recovery process. This complexity is addressed in this chapter, helping readers to appreciate the challenges of confronting gun injuries as a nation.

Hypervigilance, a state of heightened awareness and watchfulness, among urban residents exposed to community and police violence, leads to health consequences such as higher systolic blood pressure (Smith et al., 2019) and fear for one's life (Lipscomb et al., 2019). There is a call for greater research in this area because how we view the immediate world shapes mobility patterns and aspirations (Konstam & Konstam, 2019), and it has health consequences.

Gun injuries far exceed fatalities in frequency, with providers more likely to encounter injuries than fatalities, as covered in the previous section. Increasing survival rates of gunshots are now starting to be better understood (Zebib, Stoler, & Zakrison, 2017), introducing a set of challenges for providers. This quest for greater knowledge necessitates a sociocultural-socioeconomic context of injuries within urban communities of color, introducing a qualitative window to understand statistics.

Qualitative studies of survivors provide insights into their daily struggles, as shown in the data of African American/Black men in Philadelphia (Jacoby et al.,

2020b), uncovering four service intervention realms to consider in reaching this group:

1. Psychological distress, chronic pain, increased substance use, changes in sleep, and social isolation 12–36 months after injury
2. "Where it happened," which captures neighborhood recovery environments on factors such as traumatic stress, gun carrying, outdoor activity, and proximity to social networks
3. Postinjury influences on workforce participation, financial stability, and barriers to health care access
4. The need for outreach and service to address the previous three themes

Reinjury risk or even death is a phase of gunshot injuries that needs specific interventions (Pear et al., 2020); it can be considered as a distinct fifth phase on risk factors for further gun injuries and rehospitalizations.

High gun injury rates following an initial shooting are a reality, with a predictive rate of recidivism at 10 years of 16% for all patients, but for those at high risk (urban centers), it is almost one quarter (25%) (Marshall et al., 2020). A national study of reinjuries following penetrating trauma found that one in three patients presents to a different hospital for reinjury, increasing the chances of morality because of fragmentation of care (Parreco et al., 2020). A Baltimore level 1 trauma study of African American/Black men at high risk for recurrent violent injury found 58% becoming trauma recidivists, making their status a major public health issue (Richardson et al., 2016). Reducing victimization and perpetration as a goal stands out in importance in hospital emergency rooms (Brice & Boyle, 2020; Purtle et al., 2016). The likelihood of reinjury increases the significance of trauma centers preventing or minimizing trauma recidivism, and possibly saving lives and money.

Youth with assault-related trauma, for example, have a 25%–60% higher probability of suffering another assault, with another study finding those 10–24 years of age twice as likely to experience another assault within 2 years (Barna, 2020). Interventions after community violence episodes provide a point of entry to help youth build resiliency by introducing positive adult relationships or maintaining them where they already exist (Nunn, 2020).

Viewing gun injuries from an intentional or unintentional (accidental) perspective has great significance for victims, their families, communities, the nation, and for providers in the recovery process, both physically and emotionally. Injuries from gunfire fall within the broad category of penetrating injuries. There are efforts to develop scores on gun injuries to guide interventions. The Urban Injury Severity Score (UISS), for example, has shown promise for developing better predictors of eventual mortality from gunshot wounds, thus shaping services and increasing understanding of gun violence consequences (Tobon, Ledgerwood, & Lucas, 2019).

Not all injuries are life-threatening, but they still wield social and physical consequences, such as facial injuries. The face is often the first physical feature that people notice on a person, and thus it shapes first impressions (Jamrozik et al., 2019). Facial gunshot wounds can cause extensive damages to soft and hard tissue, making repair challenging for maxillofacial surgeons and necessitating extensive and long-term follow-up (Roochi & Razmara, 2020). I was unsuccessful in finding prevalence rates of facial gunshot injuries, limiting a comprehensive understanding and development of best practices.

Reducing gun injury lethality requires drawing upon multiple factors. One factor is advancement in surgery, calling attention to the importance of gun injuries rather than the conventional focus on fatalities. Even though deaths often determine the reach and depth of gun violence, injuries must be part of any equation in determining the full extent of this epidemic (Zakrison, Puyana, & Britt, 2017).

Gunshot injuries to the head bring greater morbidity than other penetrating traumatic brain injuries (Turco, Cornell, & Phillips, 2017; Vakil & Singh, 2017). On a positive note, a rare study of racial inequities in treating inpatient gunshot injuries to the head found no disparities (Chiu, Fuentes, & Mehta, 2019). Wounds to the thoracic aorta have a high probability (92% to 100%) of being deadly, with those fortunate to survive because of low-caliber and low-velocity bullets, which restrict hemorrhage to the aorta wall (Kuo et al., 2019).

Adverse Childhood Experiences

These measures consist of numerous types of violence but not gun violence, which is severely limiting in developing a comprehensive understanding of consequences with guns in this nation's cities (Rajan et al., 2019, p. 655). A systematic review of research over the course of two decades confirms the critical importance of classifying youth exposure to violence involving a gun as an ACE, of expanding the definition of gun violence exposure to include a broader spectrum of youth experiences with gun violence, and of expanding the notion of who should conduct such screenings to increase the reach of both existing screening and intervention efforts.

Traumatic Brain Injury

Traumatic brain injury (TBI) is a major cause of disability and death (Miller et al., 2020) and is widely associated with increased medical comorbidity and significant economic health costs (Holzer et al., 2019). TBI is the impairment of typical brain function, which can be temporary or permanent, resulting from brunt force trauma such as a bullet damaging brain tissue (American Medical Association Resident and Fellow Section, 2020). Level of brain damage functioning can be classified as mild, moderate, or severe, based on the extent of brain impact against the skull, with a corresponding brain functioning prognosis and implications for treatment and rehabilitation.

Gunshot TBIs, also called craniocerebral gunshot injuries (CGIs), are no longer restricted to military engagement and are now found in civilian and urban settings. These injuries are less frequently presented in trauma centers than closed head trauma; penetrating brain injury brings a worse prognosis and is considered the most lethal of all firearm injuries, with victims often dying before admittance to a trauma center (Alvis-Miranda et al., 2016). Estimates indicate that TBIs cost the nation $64 billion per year, making it a major health cost (Scarboro, Massetti, & Aresco, 2019). Duda et al.'s (2020) literature review on outcomes of pediatric CGIs found research on this condition to be limited.

Gunshot TBIs accelerate the brain death cascade by causing lesions, which result in a higher death rate when compared to brain injuries from accidents (Cacciatori, Godino, & Mizraji, 2018). TBI brings cognitive challenges with long-lasting consequences, and again, as argued throughout this book, requiring further research that takes into account the unique rehabilitation methods for this population. These methods can enhance brain plasticity and the process of recovery for victims, bringing potential economic benefits for the nation (Said et al., 2018). TBI caregivers must contend with emotional needs of their family and their own needs, such as sadness, anger, and guilt, as well as altering their own lives to carryout caregiving activities they are often not trained to undertake (Degeneffe, 2019).

Retained Bullet Fragments

General opinion may be that bullets need extraction to aid in recovery, but this may be easier said than done. Retained bullet fragments (RBFs) can prevent the use of magnetic resonance imaging (MRI), making locating and removal that much more difficult, with potential for triaging of victims and with long-term consequences in cases where fragments remain after treatment (Fountain et al., 2021).

RBFs may stay in place because they are too dangerous to remove due to their proximity to key organs, serving as a constant reminder of the traumatic episode that caused them. This does not mean these fragments do not migrate elsewhere in the body, creating more serious medical conditions years after the initial injury, up to almost two decades after the initial injury (Marantidis & Biggs, 2020).

We often associate lead poisoning with paint chips and water, but it can also be associated with gun injuries. Lead toxicity due to bullet fragments left in the body can cause recurring visits to an emergency room and persistent weight loss, vomiting, and nausea, symptoms that can be easily confused with a stomach virus (James, Fitzgibbon, & Blackford, 2016; Weiss et al., 2017a). There is a case of ankle lead arthropathy and systemic lead toxicity secondary to a gunshot injury after 49 years (Ramji & Laflamme, 2017). One Chicago hospital estimated 1,000 gunshot victims were treated every year, with 75% discharged with bullet fragments still in their bodies (Chan, 2019).

The gravity of lead toxicity from bullet fragments is starting to get recognition as evidenced by the CDC releasing the nation's first report linking these two together (Weiss et al., 2017b). There are no time limits regarding lead toxicity. For example, lead poisoning 9 years after a gunshot wound caused by RBFs can even result in quadriparesis (Nally, Jelinek, & Bunning, 2017). Unfortunately, our understanding of this issue is challenged because the CDC stopped collecting sufficient data due to a drop in the number of states participating in this program.

Remarkably, considering the extent and significance of gun injuries, there are no standard medical guidelines on bullet removal and therefore no comprehensive understanding of the consequences of RBFs (Apte et al., 2019). Their management remains poorly defined (Smith et al., 2020), limiting comprehensive understanding of them and development of standard procedures with a high likelihood of success. This is a problem of great significance in communities of color due to the extent of gun violence and the prodigious amount of resources needed to support victims.

When a bullet fragment remains in a body, it increases the probability of posttraumatic stress disorder (PTSD) and depression (Smith et al., 2018):

> The most important finding of our study is that the presence of a retained bullet was associated with more severe depressive symptoms in Black men who sustained firearm injuries. Specifically, men with retained bullets had depressive severity scores 3.5 points higher than those without retained bullets when controlling for other variables. Our results suggest that there may be value in removing retained bullets. (p. 138)

Health professionals argue that a gun injury has greater consequences than a physical injury from any other cause (Rowhani-Rahbar, Zatzick, & Rivara, 2019; Tasigiorgos et al., 2015). This injury stigmatizes, even when victims are innocent bystanders.

Injuries: Lower and Upper Extremity

Lower-extremity injuries, in similar fashion to other forms of gun injuries, are underresearched (Maqungo et al., 2020). Lower-extremity arterial injuries presented at a level 1 urban trauma center can complicate injuries and result in limb amputation and even death, with increased prognosis dependent on a multidisciplinary approach, and initial trauma work-up facilitating rapid diagnosis and early engagement of multiple specialists (Tanga et al., 2018). As noted in the section on gun injuries in children, upper-extremity injuries have similar consequences for adults.

Lower-extremity vascular injuries cause a higher risk of amputation and death when compared with non-gun-penetrating trauma, a common outcome when comparing gun and non-gun injuries (Siracuse et al., 2020). Limb amputation changes lifestyle, and depending upon which limb(s), it may necessitate a new living situation, not to mention social implications and possible stigma.

Upper-extremity gun injuries bring notable morbidity for adult victims, with critical neurovascular structures posing heightened risk, proving challenging for orthopedic surgeons. The vast majority of gun injuries involve extremities but with concomitant orthopedic injuries (Congiusta et al., 2021). Bullets can strike a bone and break into multiple pieces, causing many fractures, or simply ricochet and hit other body parts. Treatment for bone defects varies by location and severity of injury, typically requiring staged treatment. Nerve injuries after wounds are common, too, with spontaneous nerve recovery expected to occur in most cases (Omid et al., 2019).

Injuries Post Discharge

A victim's life is altered with many surprises, as experienced by Joshua, who was 19 years old when he was shot (Parnell, 2020):

> Joshua Arrington, 28, was shot in the arm and abdomen in Dec. 2011. It ended up costing him more than a stint in the intensive care unit. "It was bad. I did the initial amputation and then I had to get a revision. After that, I thought I was healed. I thought everything was good to go," he said. "Six months later, my leg was not healed. I had an infection on the inside of the leg. So they ended up having to cut me five inches shorter." Arrington, a high school basketball star, was 19 years old when he was shot in Hollis, Queens. He never thought the two bullets would cost him a leg and require years of recovery. He's also surprised that some people consider gunshot victims to be akin to heroes. "Unfortunately in today's community, you get shot and you walk away from that, you're the man, you know? They don't understand how traumatic it is. Yeah, yeah, I'm the man—from the outside," Arrington said of how he was viewed on the street. But on the inside, it's different. "I didn't know what I was dealing with," he said. "What am I going to do at home?" . . . Despite the hardship, Arrington rose above. He put the mechanical skills he learned in high school to work and eventually opened his own body shop in Freeport, L.I. He also visits high schools to talk about gun violence, and he works to mentor other amputees. He says something has to be done about the guns.

The pressures survivors with amputations face upon discharge are multifaceted, including contending with new identities and trauma. Turning a tragedy into an opportunity to give back cannot be overly stressed in the recovery process.

The 30-day period after initial discharge finds over half of gun victims being readmitted, with 16% having more than one readmission, with costs exceeding $500 million during a 5-year (2010–2015) period (Fransdottir & Butts, 2020). An exhaustive study of 11,294 gunshot wounds (1996–2016) in one center found mortality rates decreasing and those with multicompartmental injuries increasing (Manley et al., 2018). This mortality rate decrease speaks to advances in treating victims and postdischarge needs (Raja & Zane, 2016). Youth victims experience greater quality of life decline when compared to youth with chronic illnesses (Levas et al., 2020), raising a need for resources in making adjustments to their lives, including trauma help.

Vascular Injuries

Gun injuries requiring vascular repair are associated with higher injury severity scores and mortality, with the majority of vascular operations performed in the abdomen/pelvis and extremity area (Siracuse et al., 2019). Imaging plays an important role in management of abdominopelvic gunshot trauma (Sodagari et al., 2020). Pain and movement impairments compromise abilities to readjust to life.

Pain medication will likely be prescribed for gunshot injuries, thus opening the door to abuse (Lee, 2013). It is impossible to examine this topic without grounding it within factors or forces shaping provider views of pain management among African/Black and others of color treated for gun injuries (Aronowitz et al., 2020).

"Minor" Gun Injuries

There is no such thing as a "minor" gun injury, although descriptions in the popular press may give this impression. The same can be said about a "minor" surgery. If you are the one having a "minor" surgery, it is not minor. If it is someone else, then it is. One has only to talk to someone experiencing a "minor" gun injury to note this. Countless injuries are not addressed in this section because of a lack of space, such as facial reconstruction in gunshot trauma (Benateau et al., 2016). More research is needed on victim demographic characteristics because of the costs of these procedures. Bone fractures are often a given with gun violence, bringing increased demands on Emergency Departments (Arceo et al., 2018).

Hand injuries often require surgery, leave scars, and impair hand mobility (Hutchinson et al., 2019). Gun injuries to hands may appear as relatively minor ("nonthreatening") when placed alongside some of the other injuries addressed in this section. However, we must not lose sight of trauma associated with any gunshot injury. Understanding of gun hearing impairment is in its infancy. Hearing disorders can occur over an extended time and be a result of one explosive episode (Colson, 2019; Skrodzka & Wicher, 2019). Multiple exposures have not been explored and likely result in cumulative damage.

Finally, even what is considered a "superficial" wound can be much more. A superficial wound belies the true extent of an internal injury (Antonucci, 2019): "On clinical examination, superficial skin laceration and hematoma around a scalp entry wound (and, if present, exit wound) belie the more substantial intracranial findings seen on medical imaging. These can include skull and metallic fragments within brain tissue, resultant hemorrhage in and damage to eloquent cortex and white-matter tracts, edema, and herniation" (p. 304).

Injuries Requiring Rehospitalization

A gun injury can result in an increased rate of rehospitalization over many years or even a lifetime (Oakes, 2019). One study found 40% being released with complications requiring prolonged rehospitalization, with 30% dying when shot and the remaining 30% making it to an emergency room and being released (Kalesan et al., 2018).

Gun injuries must not be limited to a narrow temporal perspective of a few months post injury. A long-term mortality view reveals that injury exposure is an important predictor of death within 5 years, and most significant in the first year after injury, which most likely will be the result of gunshots rather than a medical complication (Fahimi et al., 2016). Longitudinal studies of life trajectory require a holistic view of fatalities and injuries resulting from adjusting to life, including falls and other accidents. This temporal perspective, taking into account socioeconomic and sociodemographic factors, shapes services and supports.

Victims who are rehospitalized have a higher likelihood of dying, and their care is more costly than when they were initially hospitalized (Bonne et al., 2020). Interventions targeting survivors represent a group worthy of special attention to both save lives as well as money. Surprisingly, using a gunshot scar as a focal point for reflection or connectedness with others with similar trauma experience has not received the scholarly attention it deserves, bringing a dimension to addressing trauma that lends itself to creating group interventions, as addressed later in this book.

Posttraumatic Stress Disorder

Surviving a gun injury often requires extensive medical attention, including mental health services for both the victim if he or she survives and the loved ones. It is important to note that a catch-22 exists for victims of color: Johnston (2020) argues that traumatized victims must be labeled as having a "mental disorder" in order to get treatment, thus pathologizing them. The "pathologizing of the wounded" is used by *DSM-5* to capture the phenomena of shaming gun survivors and commodifying their pain. Labels take on added significance among people of color because of how they assault self-worth and further marginalize them.

Gun injuries have more severe short- and long-term consequences compared to other injuries (Foran et al., 2019), often not conforming to typical expectations associated with physical symptoms. Their physical, social, psychological, and economic manifestations get shaped by their nature, as do our responses to them. PTSD is high among violently injured patients (Coles, Tufariello, & Bonne, 2020). PTSD for relatives of those killed by guns must not be overlooked, either. Life post death creates victims beyond the fatality, as with Jarren Peterson's boyfriend (Moms Demand Action, 2021): "Gun violence has been devastating for me in so many ways. Living with grief, fear and anxiety and, on top of that, having to sort through logistics of planning a funeral and dealing with trials was incredibly difficult. Even getting a new car after James was killed in mine felt like an impossible and unbearable task." PTSD coexists with other challenges.

Other Physical Injuries

This section highlights other physical injuries for illustrative purposes. Certain injuries receive considerable attention in the literature, as evidenced in this book. Other injuries, however, although potentially deadly, have not benefitted from increased focus, such as injury to the thorax. The thorax is a common site for fatalities because it contains the chief organs of circulation and respiration, with upper and lower extremities following in severity (Pallin et al., 2019).

Where in the body these injuries occur brings significant ramifications (Testa & Legome, 2017): "Gunshot wounds to the head are the most lethal of all firearm injuries. Estimates have a fatality rate over 90%. Those to the myocardium have fatality rates reaching 80%. Intra-abdominal injuries from gunshot wounds tend to involve the small bowel (50%), colon (40%), liver (30%) and abdominal vascular structures (25%)" (p. 1). On the surface, this seems self-evident. All gunshots can be dangerous with lifelong implications. The degree of severity rests on the type of wound. Those with severe acute kidney injury (SAKI) from gunshots, for example, are twice more likely to be at risk for death than those without SAKI (Athavale et al., 2019).

Bone fractures are usually associated with youth, but rarely associated with gunshot injuries. Bone injuries are comorbid injuries associated with firearm-related fractures among male youth of color (Blumberg et al., 2018). How successfully and rapidly bones heal determines long-term adjustments, which if unsuccessful, may influence self-esteem, limit career options, and can even result in further injuries. Hip-region injuries are likely to cause complex peritrochanteric fracture and fracture patterns associated with local structures and abdominal viscera, making treatment challenging and resulting in long-term follow-up (Maqungo et al., 2020).

Deng et al. (2019b) argue that knowledge of adult brain injury resulting from gun violence is in desperate need of updating and systematic characterization because of the dangerousness of this injury. Cranial gunshot wounds, for instance, have the highest mortality of all gun injuries (Crutcher, Fannin, & Wilson, 2016), bringing severe long-term consequences for survivors (Menger et al., 2017). Head gun injuries also bring higher likelihoods of encountering complications and even death when there is a prolonged period between the gunshot and surgery (Sertbaş & Karatay, 2020). This finding increases the importance of reducing time from injury onset to emergency surgery. Gunshot injuries to the chest, too, have a high death probability with a small percentage of victims arriving at a hospital alive (Truesdell et al., 2020).

Gun eye injuries are relatively common with head and neck wounds. The complexity of this part of the body, including the functional relationships of the periocular region, brings profound challenges and involves a wide range of providers (Erickson et al., 2020). Eyesight is too important to compromise because it can cause other injuries. Eye injuries, for example, account for 6.5% of all ocular trauma injuries, with the majority of open globe injuries secondary to gunshot injuries having poor outcomes (Castro et al., 2020). There is a dearth of research on visual outcomes from gun injury. One study of head gunshot victims showed 44% suffering long-term visual damage (Chopra et al., 2018).

INJURIES AND SUBGROUPS

We must view gun injury statistics with attention to subgroups to develop appropriate interventions, with gangs a case in point (Dierkhising, Sánchez, & Gutierrez, 2021; Maldonado, 2019; Tapia, 2019; Wiley, 2020). Clearing gang- and drug-involved nonfatal shootings are very low when compared to other gun incidents and circumstances, and this is particularly problematic when they constitute the majority of urban gun shootings (Barao et al., 2021). This calls for greater attention and deployment of resources to curb gun violence deaths and injuries. An intersectional perspective breaks down stereotypical views of ethnic and racial groups. Subgroups allow application of an intersectional lens to injuries stressing important considerations in shaping our understanding of this public health issue.

Why pay special attention to people of color? African American/Black people have a rate of 113.8 nonfatal injuries per 100,000 (highest in the country), which is 10 times greater than White, non-Latinx people; Latinx people have a rate twice that of White, non-Latinx people (Everytown USA, 2020c). When introducing race and the urban scene, gangs emerge as important subgroups for attention. Urban gang activity increases gun violence due to higher likelihood of access to guns in solving disputes. Concerted efforts on gangs and gun violence are not new. In the 1990s, for example, Boston's Operation Ceasefire addressed the violent mix of gang violence and the crack cocaine epidemic. Interventions can be crafted by taking into account special circumstances that create these conditions. Gun injuries among prominent subgroups require tailored outreach and engagement proposals.

Vigil (2020) argues that gang members suffer from multiple causes of social marginalization, when combined with a street identity, increasing the likelihood of gang membership. If this argument is correct, gang gun victims who become paralyzed are further marginalized, making rehabilitation more arduous to achieve, and worthy of a special group status.

There is a strong relationship between gang membership and gun violence (Huebner et al., 2016), including members shooting themselves accidently (Drayton, 2018). The pathway between gang membership and violence is high (Wu & Pyrooz, 2016). Gang members with gun wounds face particular challenges worthy of specialized outreach efforts (Watts, 2019). Gangs have traditionally supported their members, meeting a range of experiential (emotional) and instrumental (concrete gains) needs (Aspholm, 2020). Gang life is dangerous, bringing injuries as well as death, within and outside of prisons (Fahmy et al., 2020).

There is value in providing alternatives to gang membership prior to experiencing a gun injury (Berdychevsky, Stodolska, & Shinew, 2019). Alternatives are also necessitated when gang members suffer a life-altering gun injury. As with non-gang members, gang members have the emotional need to engage in meaningful social relationships, even though the relationships they choose inside gang life increase the probability of participating in violent acts. They often develop strong membership alliances that meet a range of needs, and finding substitutes in postgang life is a tremendous challenge for them (Bubolz & Lee, 2019; Cuevas, 2019). Having debilitating gun injuries complicates this search. Further, finding substitutes takes on significance in high-violence neighborhoods (Vlaszof, 2017):

> Mindsets of participants in the current study reflect the belief that they have no other choice than to join the gang. They live in low-income and high-crime neighborhoods. Many come from broken families with no father figure present. They have family members in the gang, and most of their childhood friends belong to their gang. Their enemies belong to rival gangs. Most of the participants reflected on having no opportunities for change as they have criminal records, they are in and out of jail or on probation, and the prospects for an optimistic future are few. (p. 103)

Adding severe gun injuries, and maybe even those resulting in paralysis, highlights the value of pursuing a constructive or less dangerous path in their lives, but also the challenges they face in achieving this goal.

Victims, depending upon injury seriousness, may be unable to perform their duties as gang members (Lauger, 2019), thereby diminishing their roles and worth, and relegating them to a world between gangs and their community. They are often ill prepared for this new existence. This translates to treating their injuries through other sources of support, which may involve their families (Ralph, 2012):

> Of course, some disabled gang members will prefer to resume their activities, and in such cases, they are not so much willfully ignored as forgotten about, marginalized, or neglected. Hence, in contrast to members who die in gang wars and become martyrs—those bygone affiliates often emblematized on graffiti'd R.I.P. t-shirts—the disabled gang member, who cannot contribute to the organization in the way that is most valued (that is, as a street-corner drug dealer) becomes like the presumably honored war veteran who begs for change by day, and is tucked beneath a highway underpass by night.

The shift in roles and change in social network provides the survivor a chance to assume new productive and contributing roles. Community service introduces the potential for innovative strategies, as addressed later in this book.

Many families, however, are not positioned to help gun violence victims, leaving health and social service systems to fill this void. Religion and houses of worship may not be support sources unless a close family member, such as a mother, is part of the congregation and is able to tap this network. This support (spiritual, experiential, and instrumental) can be powerful (Galiatsatos et al., 2021):

> Spiritual leaders and places of sanctuary that help build relationships can and do provide meaningful assistance to the mental health and wellness of

individuals identified as second victims. Future initiatives aimed at firearm violence should not only provide aid to second victims, but consider the faith-based personnel and organizations in an effort to assure all local resources are allocated and utilized, thus ensuring that all second victims have the opportunity to recover and heal in accordance with their beliefs and culture. (p. 1833)

Being wheelchair-bound limits former gang members from engaging in behavior that is conducive to being reinjured or killed, with this new existence opening the door for engagement in constructive pursuits. Houses of worship can provide and broker services for victims (Harper, 2020). Latinx youth witnessing community violence, for example, found religion and spirituality benefiting them, calling for religious institutions to partner in community interventions (Jocson et al., 2020).

Children and Youth

Comprehensive understanding of gun injuries requires attention to children and youth. Gun violence and injuries have prodigious consequences for urban children, not only because their adult relatives have been hurt or even killed by guns. They, too, are directly affected from physical and emotional standpoints. It says so much when children are considered fair targets for gun violence (Baltimore Sun Editorial Board, 2021):

It's devastating enough for children to have loved ones and friends who have ended up the latest gun fatality statistic; it is unfathomable that the child is the actual victim. When the victims of the city's gun violence are so young, it suggests a new level of desensitization to life among the perpetrators of violent crimes. Where children and the elderly were once hands off, everybody is now fair game: kids, teens, bystanders. Shootings are done in daylight, without care about who is around to witness it and oftentimes with everything caught on video. Not even a deadly pandemic can stop the gun violence.

Gun violence consequences are no longer the problem of a particular age group but best viewed across the lifespan.

We can compare children living in high-violence urban communities to those in war zones (Johnson, 2016), which says so much about stress, missed opportunities, PTSD, and life expectancy. Cities such as Chicago and Oakland have trauma 1 centers available for training military doctors to treat gun wounds for combat duty (BBC, 2014). A great deal has been learned on the battlefields, with this knowledge finding its way into medicine (King, 2021). It is a two-way street: Military doctors enhance their skills and those of trauma 1 doctors.

Pediatric gun fatalities are the leading cause of death for this group, and no one is safe in high-violence neighborhoods (Hatchimonji et al., 2020a). Gun violence naturally equates with loss, either physical, emotional, economical, or life itself. Motifa Akosua (2019) expresses the meaning of loss quite eloquently when sharing the lesson her stepfather taught her about street violence and friends:

> Loss is something that we all experience at some point, but in urban communities we experience it more often. When I was 10, my stepdad told me, "You need to start taking pictures with all your friends cause one day you're gonna look at them and say 'He's gone, he's gone, and he's gone. He's in jail, and he's in jail.'" At the time, I didn't know what he was talking about. Now, when I think of loss, I think about my friend in eighth grade. . . . Lamar's funeral took place near our homes in Richmond [California]. It was my first time going to a friend's funeral. From there, the shooting and deaths kept coming. I can't count the number of people I know who've been shot or died on my hands. There are definitely more than 10. If you count the adults that I knew as a kid, more than 20.

Motifa, although commenting on those killed, suggests that the number of those injured would be a multiple of those killed. When adding trauma, the numbers are even higher for victims and their loved ones. Motifa, too, is a victim.

The increasing number of children who become gun victims has increased attention to their critical care use, hospital costs, and mortality in pediatric intensive care units (PICUs) (Kamat et al., 2020). When assault-injured urban youth at an Emergency Department were questioned about life expectancy, one third stated they were uncertain they would live to age 35, raising questions on perceptions of risk (Lennon, 2020). When one has a lifespan that is half that of a comparable victim who is White and non-Latinx, this fosters increased engagement in risky behavior and a dismal view of the future.

The belief that one has a dramatically shortened lifespan can translate into an attitude that life must be lived to its fullest while you can. The concept of living to old age and retiring is outdated in this population. Pediatric gun injuries necessitate an increased knowledge base to aid trauma surgeons to treat and prevent these injuries (Swendiman et al., 2020). The same can be said about other helping professions. One Chicago study of gun-victimized adolescents found they were at increased likelihood (2 to 3 times) to perpetuate gun violence in adulthood, spotlighting long-term consequences of creating a group of perpetrators over an extended lifespan (Telpin et al., 2021). Early interventions target youth with gun violence to prevent them from aging into adult perpetrators of this violence.

Since the start of summer 2020, Chicago has seen a spike in shootings with children bearing a brunt of this violence, including killings (MacFArquahar & Charito, 2020). Estimates have a child or adolescent wounded by gunfire every 40 minutes in the United States (Everytown USA, 2019). Nevertheless, gun injuries are a silent epidemic (Beard et al., 2019), translating into silent victims (Furman, 2018), with multifaceted consequences for different groups, including children and youth (Brewer et al., 2019; Lee & Schaechter, 2019; Ngo et al., 2019; Parikh et al., 2017; Veenstra et al., 2015).

Virtually no city escapes gun violence, particularly for youth of color. Big cities such as Chicago, Detroit, Los Angeles, Miami, Philadelphia, and New York are not the only cities confronting gun violence. Denver, for example, experienced 700 youth gun violence acts each year, with 74 dying (2012–2017), with those injured at a rate of 7 per every death requiring hospitalization; the period 2012–2018 witnessed 175 being admitted due to gun injuries (Denver Public Health, 2019). The type of gun injury was not recorded.

Pediatric gun injuries at a St. Louis trauma center found the majority of injuries were categorized as assaults and intentional (65%), and occurred between 6:00 p.m. and midnight (Choi et al., 2016), introducing a temporal dimension. This profile can be found in this country's urban centers, calling for violence initiatives focused on time period, population groups, and the geographical areas that are prone to having gun activity. Nighttime is a dangerous period, making trauma centers particularly busy during that time period (Lenart et al., 2020), increasing the need for intervention strategies, and more so during summer months.

An Indianapolis study of adolescent gun carrying and gun arrests over an 11-year period found the majority of repeat offenders were first arrested for gun carrying, raising the importance of an intervention after that initial arrest to prevent escalation (Magee et al., 2020). Interventions effectively delaying the onset of gun carrying bring the benefit of reducing gun violence and gang membership, strongly supporting early intervention as young as 9 years old, which may seem very young, but connects well with the demographics of communities of color because of their youthfulness (Bonne et al., 2020). Peer networks influence adolescent gun-carrying risks and gun use in commission of a crime (Robertson et al., 2020).

A population-based analysis of gun injuries among children aged 5–19 years presenting at Emergency Departments (2010–2015) found younger children having a higher likelihood of fatalities when compared to older youth (Cook et al., 2019), raising the importance for targeting this age group. Pediatric gun injuries, in similar fashion to adult counterparts, carry different morbidity outcomes depending on the age, race, income, location, and other sociodemographic factors (Diebel et al., 2018).

Urban substance use and abuse is often a section of any gun violence portrait. It, for example, is strong among urban Latinx youth (Sanchez et al., 2020), raising their probability of becoming victims. Gaining greater understanding of guns and substance use intersection is imperative (Culyba & Sigel, 2020). Once injured, the likelihood of subsequent gun injuries and even death goes up dramatically. Gun injury interventions not only provide a service to victims but can also prevent future fatalities. Gunshot wound recidivism is not restricted to adults, with pediatric recidivism also a serious problem, with incidents occurring within the first (32%), second (53%), and third years (66%), with nine youth shot on three separate instances with a 22% mortality rate (Gibson et al., 2016).

Nordin et al. (2018), using findings from the National Trauma Data Bank (2007 to 2014), found unintentional gun injuries represent 33% and almost 10% of mortality in young children. Rhine et al. (2020) note a disturbing uptrend, finding children (0–4 years old) in PICUs following gun injuries, highlighting their particular vulnerability and increasing need of more rapid transport to PICUs.

Potential children fatalities require creating a classification system that separates gun and non-gun injuries (McGaha et al., 2021). A review of the incidence and outcomes of gun-injured adolescents, taking trauma center levels into account and whether adult or child focused, found they have a higher survival rate at children trauma centers, with data supporting victims treated at the closest trauma center regardless of age focus (Swendiman et al., 2020).

Racial inequities are a major concern, and these are not limited to adult gun injuries and fatalities but are found among children as well, with White, non-Latinx children having worse outcomes from suicide injuries. This is also the case with suicide victims regardless of their age (Sakran et al., 2020). Pediatric gun injury racial inequities are a strong predictor of health care outcomes, with White, non-Latinx, and African American/Blacks more likely to be gun injury victims (Hughes et al., 2020). This disparity calls for a nuanced understanding of strategies to address gun injuries (Carter et al., 2020; Carter & Cunningham 2021).

Children with traumatic injuries requiring extensive and comprehensive rehabilitation will also experience racial/ethnic inequities in accessing postdischarge rehabilitation, with those receiving services at children urban teaching hospitals accessing better discharge outcomes and services (Shah et al., 2019). These settings bring the inherent benefits of trauma 1 settings for adult counterparts, discussed later in this book, because activity volume translates into increased efficacy of interventions and ability to innovate, which is a sad commentary.

Gunshot fatalities and injuries are a common reality in many urban communities, with research lagging far behind other areas of injury prevention (Cunningham et al., 2019). Gun injuries among children and adolescents, as with adult counterparts, stand out because they suffer from a lack of a national surveillance system. This hinders the development of a comprehensive picture of the issue (Zeoli et al., 2019) and the planning of long-range interventions (Brooks, 2019).

Interventions require an active research agenda that taps community perspectives in construction of purpose and questions, involves youth in conducting research, and minimizes professional biases (J.W. Collins, 2019). Research democratization opens the door for inclusion of questions, including wording to elicit desired responses, to ensure that findings address community conceptions of gun violence. This approach is a paradigm shift in how conventional research gets conducted with professionals in control of all facets.

Gun injuries among children and adolescents presenting in emergency rooms account for over 20,000 visits per year in the United States (Flaherty & Klig, 2020). These visits transpire with victims' families in emotional turmoil, hoping that their child does not die or suffer lifelong compromises. These youth invariably suffer a variety of physical and emotional consequences necessitating a coordinated system of care for an extended period, with numerous appointments, interrupting daily life in often unforeseen ways. Their families, too, worry whether the next assault will result in a death.

Developing care systems requires research and a cadre of professionals attuned to their unique needs and the challenges they face in maintaining a gun-free existence (Delgado, 2020b; Fischer et al., 2020). We face a challenge in researching gun injuries while concomitantly not losing sight of protective factors, a vastly underfunded research focus, because not all youth carry and engage in gun use (Schmidt et al., 2019). Youth risk and resiliency is a delicate balancing act of need without losing sight of potential (Nelson-Arrington, 2020). This dilemma is not unique to gun violence.

Grasping the extent of gun injuries is dependent on how it is conceptualized, with distinctions going far beyond semantics. Gun injuries can result from penetration of a bullet and/or the emotional consequences of this act. Injury can be the result of trauma from witnessing a shooting, which is invisible from a physical standpoint but has life-altering outcomes, including untreated childhood trauma (Leasy, O'Gurek, & Savoy, 2019; Soyer, 2018). Early childhood trauma as a driver of urban youth (12 years or younger) violence and gang involvement is an outgrowth of toxic stress (Ross & Arsenault, 2018).

Being shot at and missed leaves an invisible scar, bringing its own set of invisible symptoms. Sharing this experience may elicit a response about being so lucky, which is not what is typically in the mind of victims. Further, they were lucky, but will they be so next time?

Gun injuries occupy the second most frequent cause of death among children appearing at a trauma center (Petty et al., 2019). Statistically, pediatric gun injuries account for 5% of all pediatric injuries but represent the highest fatality rate (Feldman et al., 2017). Upper-proximity pediatric injuries often cause severe health consequences (Dabash et al., 2018). Gun injuries do not have to be by powder guns and can involve other forms of guns, with air and BB guns prime examples (Apelt et al., 2020; Hyak et al., 2020). One study found pediatric and air gun missile injuries resulting in 16 children visiting an emergency room daily (Burnham & Lee, 2020).

Emergency rooms are prominent places for gun violence interventions as primary points of entry into health systems (Betz et al., 2019), particularly in pediatric cases (Abaya, 2019). Emergency room utilization for gun injury is an opportunity point for postdischarge services to prevent gun retraumatization (de Anda et al., 2018). Follow-up appointment adherence ascends in importance in pediatric cases, and undoubtedly involves multiple family members taking the child to appointments and altering daily schedules.

Urban trauma centers have increased in prominence in creating successful approaches for pediatric relief, recovery, and rehabilitation (Olufajo et al., 2020; Romo, 2020; Rosenfeld & Cooper, 2017; Wolf, 2019). These centers have also created prevention strategies and longitudinal studies (Mikhail & Nemeth, 2016). Injuries are a critical intervention point for preventing violence escalation, and more so for youth victims (Carter et al., 2015), so enlisting them in creating and implementing interventions is paramount.

Children and adolescents surviving gunshots are lucky. However, how lucky are they when survival means high hospital operative and readmission rates, along with sustained long-term mobility impairment, and suffering from mental health sequelae (Phillips et al., 2020)? Not surprisingly, African American adolescent males carry the heavy burden of gun injuries (Bachier-Rodriguez, Freeman, & Feliz, 2017). Black/African American children are at greater risk for gun injuries when compared to White, non-Latinxs, regardless of socioeconomic class (Kalesan et al., 2016). Low-socioeconomic-status neighborhoods are associated with gun injuries and are heavily of color, bringing together class and race (Abaza et al., 2020).

A gunshot injury can translate into a lifetime of health-seeking behavior, compromising educational attainment, which limits career options. Perpetrators apprehended and sentenced to long sentences or life in prison often leave their victims with a lifetime in a compromised state worrying about finances and health care (Kalra, 2019): "In Dolores's case, the financial and health care burdens are astronomical. Although her shooters received 45 years to life in prison, Dolores believes that the punishment was too light. For the past 20-plus years, she has lived in her own cage of immobility, hunger, and debt" (p. 1705). Helping the Doloreses of the country meet concrete and emotional needs is critical in any rehabilitation process.

The economics of this trajectory spill over to their families, neighborhoods, and friends, too, suffering trauma that can become part of a family's legacy. When a family member is shot, it brings an increased probability of a history of intergenerational gun injuries and other fatalities touching their social network (Van Brocklin, 2018): "When a bullet tears a jagged path through a body, it may end one life . . . and dramatically alter others: the friend who was standing right there, the neighbor who called for help, the girlfriend who rushed to the scene . . . victims represent a little acknowledged or studied diaspora of trauma."

When gunshots cause pediatric brain injuries, their mortality rate is 45%, with hypotension, cranial, overall injury severity, and suicidal intent resulting in poor prognoses (Deng et al., 2019a). Naturally, speed in operating on injuries is always a significant factor, and of greater significance with craniospinal gunshot wounds, with the increased passage of time causing a corresponding increased likelihood of death (Sertbaş & Karatay, 2020).

Children subgroups encounter different chances of survival, just like adults. Pediatric gun injuries have a higher probability of death among young versus older children (Esparaz et al., 2021). If not death, they suffer wounds (physical and emotional) with lifelong consequences. The young (5 years and under) are 2.7 times more likely to become fatalities when compared with older children (15.3% vs 5.6%), and those under the age of 1 have the highest hospital mortality (33.1%) (Cook et al., 2019).

Even when injuries are not life threatening, they still bring prodigious implications. A literature review of comprehensive treatment of orthopedic trauma injuries found a need for a portrait of the global gun orthopedic injury burden (Held et al., 2017). An 8-year study of orthopedic gun injuries of 43 children and adolescents (average age of 12.7 years) at a major urban trauma center found half required surgery with long hospital stays secondary to the complexity of the injury, highlighting long-term consequences (Perkins et al., 2016). Difficulty in walking and running takes on added significance in children when compared to adult counterparts.

Finally, pediatric palliative care is an unimaginable dimension to gun injuries that will eventually result in traumatic outcomes for a parent, and an intervention point for helping them find meaning and purpose from the eventual death of their children (Vente, 2020). This aspect of gun injuries is largely devoid of systematic interventions, calling for community efforts involving trusted local institutions, such as self-help organizations and religious institutions (Cook, 2021; Delgado, 2021).

Gun injuries, even when comparing them to injuries from sharp instruments and involving male African American children, are more likely to require intensive care unit (ICU) admissions, indicating high injury severity among children and adolescents (Wolf et al., 2019). Prevention initiatives on gun initiation focused on youth victims, their peer influences and social norms, decreases the likelihood of engaging in gun violence (Goldstick et al., 2019). Consequently, the death of children requires specialized efforts for grieving, as does supporting them in their recovery (Delgado, 2021).

Adults

Adults endure the majority of gun deaths and injuries. When focusing on injuries, the portrait that emerges covers a range of outcomes for different age groups. Young adults with violent injuries, for example (ages 18–25 years), which can include weapons beyond guns, experience trauma post discharge and significantly higher mortality rates post hospital release (Kao et al., 2019). Over 25% present with a second unrelated trauma or death, thus calling for disrupting a cycle of violent injury. Other challenges follow victims as expressed by Dr. Marie Crandall, Professor of Surgery, University of Florida College of Medicine Jacksonville (Amnesty International, 2019): "Even basic follow up care after being shot is challenging . . . if a patient is unfunded and uninsured, they have to rely on charity care for rehabilitation, wound care, etc." (p. 7).

The postdischarge period plays a critical role in treating and preventing gun violence. The 90-day period after a wound is critical due to a substantial increased probability of experiencing another gun injury (Kalesan et al., 2019). A higher probability of engaging in gun violence following an injury calls for special interventions, with tools such as the Violent Offender Identification Directive (VOID) taking on greater prominence in shaping research and programmatic initiatives (Wheeler, Worden, & Silver, 2019).

New Orleans' *CeaseFire* hospital-based response team, for example, established within the University Medical Center (UMC), a level 1 trauma center, targets victims aged 16 to 25 years, to prevent retaliatory violence post discharge from the hospital (Hammack, 2019). Helping youth understand heightened risk for violence, and providing them with conflict resolution skill sets, can find a home in hospital violence prevention initiatives (Snyder et al., 2020).

Paralyzed Victims as a Special Subgroup

Spinal cords injuries go by many different terms. Loggini et al.'s (2020) litera-ture review on cervical spine immobilization, seizure incidence and prophy-laxis, infection incidence and antibiotic prophylaxis, and coagulopathy vascular complications concluded that there is a paucity of literature and research on this condition for surgeons to draw upon. The same applies to professionals aiding in victim recovery.

Gun injuries are among the leading causes of penetrating spinal column injuries in the country (Gutierrez et al., 2020; Su et al., 2020), requiring multiprofessional collaboration to facilitate community reentry and extensive rehabilitation (January et al., 2018; Moore et al., 2016). These collaborations are enhanced with special funding initiatives supporting service provision innovations. Estimates show that between 5% and 20% of gunshot victims experience a permanent dis-ability (Fowler et al., 2015). Further, victims have a 400% greater likelihood of being disabled rather than dead, with African Americans/Blacks and Latinxs the most likely to suffer from a disability (Ralph, 2012).

This section focuses on a subgroup: namely, those paralyzed by guns. Few gun injuries elicit a greater reaction than paralysis because of the emotional and phys-ical consequences that it brings. Chapter 8 and 9 are devoted to case illustrations on how lives were severely altered because of a gun injury, but those individuals turned tragedy into a positive life-altering experience by helping others. These illustrations show the power that rehabilitation can bring for helping others de-velop hope. Rehabilitation can take multiple forms, with continuing education often forming a cornerstone of this journey and programs specifically tailored to meet their needs, as with victims paralyzed getting their GEDs (Washington, 2017). Opportunities for engaging in peer-supported activities bring an added group dimension to this process.

This section introduces case illustrations on challenges worthy of special attention. However, prior to presenting these cases, it is worthwhile to examine how the body responds to trauma causing paralysis. Readers may come across the term "violently acquired impairments" (VAIs), one of countless acronyms, with disability being one of the most prominent injuries (Green, 2019). We usually associate gunshot paralysis to a body's extremities, but paralysis can also occur to a face, bringing consequences for rehabilitation and social adjustments (Owusu, Stewart, & Boahene, 2018; Tripathi, Floriolli, & Caughlin, 2017).

Spinal cord injuries (SCIs) have witnessed treatment improvements over the past decade, but there is a paucity of effective treatments (Hachem, Ahuja, & Fehlings, 2017). There have been significant advances (diagnosis, stabilization, and sur-vival rates) in treating SCIs, translating into changing how medical management gets practiced (Alizadeh, Dyck, & Karimi-Abdolrezaee, 2019). Penetrating spinal trauma generally affects a younger group, necessitating greater reliance on public funds, with surgery limiting progression of neurological damage, increasing sta-bilization, and controlling infections (Morrow et al., 2019).

SCIs call for greater research on spinal cord regeneration, requiring a longitudinal view. Victims facing long-term rehabilitation have struggles unique to these injuries, including creation of new self-identities and self-concepts (Cooper, 2016). New self-identities provide a window for significant changes in attitudes and lifestyles to carry out new roles as community contributors. When these identities are affirming, they bring a transformative potential.

A bullet can cause a cut, shear, tear, or crush in the spinal cord, permanently impeding its function. SCI outcomes are shaped by severity and location of the injury, manifesting themselves in partial or complete loss of sensory and/or motor function below an injury location. Most attention to gun SCIs focuses on the lower span, causing paralysis of the legs. Upper spine injuries may ultimately cause dependence on a breathing machine, bringing unique challenges that are just as devastating from a quality-of-life standpoint. Gun SCIs are not rare (Casey, 2018):

> Spinal cord gunshot wounds are more common than you think and more devastating than you can imagine. What can we learn from victims? Hollywood depictions of gun violence would have you believe that gunshot wounds result in either instantaneous death or mere "flesh wounds." The reality is much more nuanced as gunshot wound survivors comprise a spectrum that range from complete recovery to profound disability. Many face lengthy hospital stays, multiple medical complications and financial burdens that often go unnoticed after the initial spotlight on the tragic shooting fades away.

Gunshot injury falls into the severe category, accounting for 17% of all SCIs, with under 1% surviving and eventually discharged without permanent neurological damage.

It is natural to think of those paralyzed by focusing on their lack of mobility. Those are the initial and most obvious impressions. The consequences, however, are far reaching, as described by Harris (Van Brocklin, 2019c): " 'A lot of people just think you sitting in a wheelchair, but they don't know we get diarrhea, we get constipated because of medications we take, we get urinary tract infections. We're constantly in pain, I'd say 80 percent of our life,' said Harris. 'Can you imagine being in pain every time it rains?' " Impaired mobility is but one of the challenges victims face. Advances in treating SCIs have resulted in victims living longer. That translates into other challenges.

The United States has 1,000 new SCI cases every month, including all age groups and causes, not just gun violence. In 2018, 78% to 80% of SCIs occurred among men (Perkins, 2020). Individuals with SCIs experience high rates of rehospitalization, with costs estimated in the hundreds of thousands of dollars, causing profound financial burdens. These costs present a narrow financial view, particularly when we expand costs to include rehabilitation of those with secondary comorbidities and compromised employment prospects (Merritt et al., 2019).

SCIs present a lifelong condition, translating into dramatic lifestyle changes, with life never being the same for the victim or the victim's family. SCIs bring both well-recognized and hidden consequences due to ripping and shattering

trauma of the spinal canal and cord; bullet or bone fragments are often left behind after surgery, which can manifest as uncontrollable itching and the sensation
of bugs crawling on the skin (Van Brocklin & Fernandez, 2018). The following
classification is illustrative of a range of outcomes of what can happen with SCIs
(Perkins, 2020):

> SCIs [spinal cord injuries] can be complete or incomplete. With a complete
> SCI, communication between the brain and spinal cord doesn't occur and the
> patient will have a complete lack of sensory and motor function. With an in-
> complete SCI, communication between the brain and spinal cord is altered
> but not completely lost. SCIs can be further classified as sensory incomplete or
> motor incomplete. With a sensory incomplete injury, sensory function is pre-
> served below the injury but not motor function. With a motor incomplete in-
> jury, motor function is preserved below the injury but not sensory function. In
> addition to being classified as complete or incomplete, SCIs are also classified
> as primary or secondary. Primary SCI represents the injury when it initially
> occurs; after the injury occurrence, it's referred to as secondary SCI. . . . The
> signs and symptoms associated with an SCI are dependent on a variety of
> factors, including severity and injury location. In patients with an SCI, the
> functional loss that occurs is below the area of injury. A loss of function to most
> of the body, including the arms and legs, is called tetraplegia. An older term
> for this is quadriplegia, which is still commonly used today. Loss of function to
> the trunk and lower extremities is called paraplegia. Broadly speaking, the pa-
> tient with an SCI will have partial or complete loss of the following: sensory or
> motor function; the ability to regulate bowel or bladder; and the ability to reg-
> ulate heart rate, breathing, and BP [blood pressure]. Approximately one-third
> of patients with neck injuries will require assistance with breathing. Injuries
> at the C1 to C4 level affect the phrenic nerve, which goes to the diaphragm
> and aids in breathing. These patients may also have difficulty regulating their
> temperature.

This description offers a partial glimpse of what happens to a person who is
experiencing this condition. Media accounts of paralyzed gun victims never go
into this level of detail, yet these physical consequences still do not capture the
emotional reactions of the victims.

A study of all patients with penetrating head or neck wounds over a 4-year
period at an urban level 1 trauma center found that it is of tremendous benefit
to have prehospital spinal immobilization, raising implications for treatment and
procedures for transporting victims to trauma centers (Schubl et al., 2016). This
takes on added significance in minimizing further damage at a shooting scene,
calling for specialized training of emergency responders.

Jeremey Posey's story highlights part of the journey to recovery and how it
keeps dreams of survivors alive; this is even more essential than just keeping
them biologically alive. This story provides a glimpse into the immediate period
following surgery (Clayton, 2018):

Once the surgery drugs and pain medicine wore off, Posey started having nightmares that lasted into his stay at the Fairmont rehabilitation and wellness center in San Leandro, right outside of Oakland. "It was tough. The constant pain, the magnitude of losing my ability to walk and having to relearn stuff," Posey recalls of his time in rehabilitation. "I had to learn how to close my hand so I could write. I had to learn how to get dressed by myself, use the bathroom, shower by myself and take care of myself." Posey spent six months in Fairmont's acute rehabilitation and another three months in the facility's long-term nursing department. Meanwhile, a local not-for-profit that helps people with disabilities found him a wheelchair-accessible apartment where he could live on his own. Eventually he moved into an apartment in San Leandro, where he's been for the past three years. Although he was denied victim's compensation funds by the state board, he receives social security, and disability benefits from the state. Before the shooting that left him paralyzed, Posey envisioned "a life of grandeur." "My expectation was being able to make a lot of money playing sports or being a performer," he said. "I was really into the materialistic side of what life offers."

Posey dreams of a new life and obtaining a degree in broadcast or mass communication. Our primary job once saving a life is to save the dreams. Dreams are the cornerstone of recovery plans by engendering purpose and potential actualization of life post injury.

SCIs secondary to gunshot wounds require services be structured to take into consideration living circumstances to maximize rehabilitation outcomes (DiZazzo-Miller, 2015). Adolescents and young adults with SCIs, although certainly not limited to this injury, require informal caregivers be involved as part of a provider team, taking into account the needs of patients as well as caregivers. Community caregiver support groups are of equal importance as victim support groups because those supporting survivors are victims themselves, although they are rarely conceptualized in this manner.

SCIs require caregivers to assume therapeutic roles in helping victims address anger directed at perpetrators (Haywood et al., 2019), with caregivers also needing support with this mental health role. When examining gun-inflicted SCIs among Latinxs, a group that will increasingly be the focus of gun injury research, for example, we find that their quality of life is compromised, and this is further hampered by a lack of staff with linguistic and cultural competences (Balcazar, Magaña, & Suarez-Balcazar, 2020). Latinx youth need targeting by violence intervention programs (Barton, McLaney, & Stephens, 2020).

Jalil Frazier's postinjury life shows how circumspect he became after a paralyzing gun injury and why posthospital care and rehabilitation are so demanding and in need of special attention (Gambacorta & Ubiñas, 2018):

Whenever Jalil Frazier opened his eyes, he found himself on an island barely wide enough to contain his frame, boxed in on both sides by unforgiving metal bars. A familiar landscape surrounded him: four bare walls, a beige tile floor, and a tiered chandelier that hung above what used to be his family dining room. He was stuck on this uncomfortable hospital bed because he'd had the misfortune of being inside a North Philadelphia barbershop on an unseasonably warm January night at the same moment two men barged in, looking to rob the place. Frazier, whose round face is framed by a scraggly beard and short dark hair, glanced at three children who happened to be in the shop. He was a father, with two kids at home, and felt an instinctive urge to protect them. He hurled himself at the would-be thieves. One had a handgun, and fired two shots. The bullets punched through Frazier's midsection and leg, ricocheting off his insides, tearing through tissue and bone before exiting his body. At age 28, he was paralyzed from the waist down. Frazier was a hero by anyone's definition of the word, but he was also a victim, one of the estimated 116,255 people who are shot in the U.S. every year. He belongs to an often-overlooked fraternity of gun-violence survivors who are left with lifelong disabilities, whose ranks include schoolchildren, movie-theater patrons, politicians, and grandmothers . . . hospital bills are just a small portion of the financial burden that's shouldered by survivors, no matter if they're injured in a nationally prominent mass shooting or a spurt of inner-city violence that attracts only glancing media attention. Many struggle to navigate a confusing web of local, state, and federal assistance programs, which are plagued by steep backlogs and in some cases can award as little as $1,500 to victims whose injuries require expensive lifelong care. Some, in their desperation, turn to Kickstarter or GoFundMe campaigns to help them obtain such basic needs as handicapped-accessible housing, transportation, and even functional wheelchairs.

The financial barriers of debilitating injuries complicate a recovery that is physically burdensome and fraught with emotional consequences. Jalil Frazier typifies the challenges of paralysis in a world not responsive to these needs, much less because of a stigmatizing gun injury.

Penetrating spinal trauma is an epidemic within a broader gun injury epidemic (Hurlbert, 2019):

For every 100 patients with a spinal fracture at their institution, 10 are from a gunshot wound. Almost half of those have an associated neurological deficit, most likely to be a permanent motor- and sensory-complete spinal cord injury. Patients with penetrating trauma have more serious associated injuries, consume more acute- and long-term care, and are less likely to be discharged home than those with blunt spine trauma.

Labeling spinal trauma an epidemic highlights how it alters victim lives and those around them, as well as the importance of injury-specific treatment/rehabilitation.

In 2018, Philadelphia's Department of Public Health (2020) studied a subgroup of gun injuries focused on paralysis. That city had 1,403 nonfatal shootings and 351 deaths from guns, with four survivors for every death, with paralysis being one long-term health consequence. In 2016–2017, there were 1,132 individuals hospitalized, of which 156 (about 14%) had a paralysis. Partial paralysis is losing use of any of the following: a limb, a side of the body, or the lower body. A complete body paralysis is loss of all four limbs. Significantly, 74% were African American/Blacks and over 50% were under 35 years old.

Unfortunately, clinical medical trials and research foundations, key in the advancement of medical science, have a propensity to ignore gunshot paralysis victims in favor of other forms of injuries, and the presence of medical complications associated with these injuries has proven to be more attractive (Gambacorta & Ubiñas, 2018):

> "It has to do with ballistics, with the transfer of energy," Schuster explained. "Even if the bullet doesn't actually go through the spinal cord or the spinal canal, there's a blast effect, there's a percussive effect. The injuries are much worse than they appear." Some of Schuster's patients end up at Magee Rehabilitation Hospital, where Mary Schmidt, the spinal cord injury program director, prepares survivors and their families for the maladies that can arise from paralysis: bowel and bladder issues, pressure ulcers, urinary tract infections, chronic respiratory problems. The spinal cord can't regenerate, but many of Schmidt's patients struggle to accept the permanence of their injury. "Every day, people come in here and say, 'I don't need to talk about that wheelchair, because I'm going to walk out of here,'" she said.

Gun spinal injury complexities magnify the stigma of these injuries, bringing a tendency to blame victims for engaging in lawless acts or simply being at the wrong place at the wrong time. Regardless, SCIs bring unique challenges for providers and family members (Orthopedist, 2016). Other injuries can prove to be more hopeful and worthy of attention.

Invariably, gun victim homes are not physically equipped to handle a wheelchair or have family members with the requisite skills and time to aid in advancing rehabilitation goals. This translates into often necessitating admittance to a nursing home or having someone with expertize to locate support services, which requires very specialized knowledge. Few homes and families can provide aid, and this is even more challenging in low-resource families with limited local options to secure services.

The social isolation victims experience post injury calls for facilitating creation of a social network that supports postinjury life adjustments. Wise (2019) quotes Dr. Samuel Gordon, who runs a weekly Urban Reentry Group at MedStar, National Rehabilitation Hospital, in Washington, DC, another program that shows the promise of group gatherings of victims who are paralyzed:

> In the more than two decades since Gordon helped establish the group, patients have cycled in and out of the program. Some patients defy their initial prognoses and regain significant use of their lower bodies. Others simply move away. Several, when faced with the prospect of living the rest of their lives in a chair, have taken their own lives or intentionally neglected their own health to

die. "Getting a spinal cord injury or a serious brain injury puts you in a sort of catch-22, where you're both still alive, but not fully alive. Somewhat in prison. You're not in a typical prison, but you're kind of imprisoned by your disability," Gordon said. Despite the obvious difficulties of losing most of your mobility in the prime of your life, Gordon says that for many in the group, ending up in a wheelchair marked a positive turning point.

Victims can assume new identities if provided with supports, becoming vital assets for a community. Extolling these interventions does not diminish the challenges faced in mounting them. The logistics of transporting group participants to and from meetings are considerable, for example.

Assuming a new identity can involve participating in a range of activities, including a sporting outlet. Former gang members disabled by gun violence bring a high level of experiential legitimacy to antigang forums when hosted by disabled former gang members. This service not only helps others, it also empowers them. Those victims can easily withdraw from their communities and society, feel sorry for themselves, and assume invisible and disempowered roles, and many members of society would not argue with them. With the necessary support, however, they can assume advocacy and public service roles (Ralph, 2012):

On the face of it, the violent event associated with injury allows the disabled gang member to rise in social stature and moral standing, similar to the war veteran in contemporary American society. And like the war veteran in contemporary society, the rhetorical effect of this patriotism stands in sharp relief to reality. Unlike the gang member who has been labeled as a police informant (or "snitch"), disabled gang members in Eastwood are not given a "dishonorable discharge"—rather, they are released from service. An "honorable discharge" would be the appropriate analogy here.

Life post gang membership is dramatically changed by paralysis, translating into a moment of reckoning for victims. They can live out their lives without a positive purpose or find a new purpose and sense of belonging. This "come to Jesus moment" requires support from within and outside of the family and recognition of challenges in making this transition.

GUN INJURIES AND RACIAL INEQUITIES

Health inequities require attension because of their importance in helping us understand gun violence's manifestation in urban communities of color, setting the stage for research and intervention initiatives. A wide range of illnesses in the literature have dominated the subject of health inequities (Bohan, 2018; Hagan & Foster, 2020; Stein & Galea, 2020), and more recently, COVID-19 (Braithwaite & Warren, 2020). Inequities, too, apply to gun injuries (Boeck, Strong, & Campbell, 2020; Motley & Banks, 2018; Thomas et al., 2020; Wamser-Nanney et al., 2019).

Guns and other violence disproportionately affect low-resource communities of color (Papachristos, Brazil, & Cheng, 2018). People of color bear a disproportionate burden of gun violence, with an estimated 26 African American/Blacks killed and 104 having nonfatal injuries, and every other day at the hands of police (Everytown USA, 2020d). The violence epidemic within the African American community has gotten attention from urban researchers and practitioners, taking a public health approach to building a knowledge base (Frazer et al., 2018).

Critics of this country's tolerance of urban gun violence argue that the reluctance of federal government research on this epidemic is tantamount to fostering this slaughter (Carswell, 2019; Delgado, 2020c). This stance helps explain why this epidemic is silent. Public health contextually grounds gun violence, helping guide interventions, particularly when embracing social justice values (Bauchner et al., 2017). Leonard (2017) argues that gun violence has such strong racial disparity outcomes because "race and space overdetermine who is afforded the rights of safety and security, and where violence is normalized, expected, and therefore nothing to worry about. Race, space, and class affect the legality and illegibility of gun violence" (p. 101).

A health equity approach translates into differing institutional responses based on sociodemographic characteristics and circumstances surrounding the incident, with the concentration of resources on distinct groups experiencing an increased risk. The emotional responses to traumatic injuries among urban African American/Black men, for instance, differ according to intentionality and recovery support available to aid them (Jiang et al., 2018), weighing heavily upon their distrust of providers.

Further, African American/Black male trauma survivors (witnessing or experiencing direct victimization) are less likely to seek mental health services compared to other gender-ethnic groups (Motley & Banks, 2018), introducing a gun violence dimension that remains largely invisible but wields tremendous consequences. Urban male youth have a higher likelihood of gun carrying when being a victim or witness to violence, with each additional exposure translating into a higher probability of doing so (Reid et al., 2017).

Understanding gun violence exposure is expanding as injuries draw greater scholarly interest. A Chicago study, for example, found a relationship between low birth outcomes and gun violence exposure, further highlighting racial inequities (Matoba et al., 2019). There is a call for universal health insurance to ensure equal access for all with traumatic injuries (Chaudhary et al., 2018).

Health equity interventions rely on research to inform practices building on a community's assets (Kim et al., 2020). On the prevention front, inequities require considering geographic regions with stricter gun laws experiencing fewer emergency admittances for children, calling for legislation limiting access to guns by better understanding regional gun culture in shaping interventions (Patel et al., 2018). There is a call for gun violence using zip codes stratified by age and circumstances (Borg et al., 2020). This approach targets areas prone to gun violence, allowing quick deployment of resources.

"Hot spots" rely on a spatial and spatiotemporal cluster analysis (Urrechaga et al., 2021). Interventions have emerged in geographical areas labeled as such (clusters of gun violence), with research showing its effectiveness at suppressing gun violence (Braga et al., 2019; Loeffler & Flaxman, 2016). However, existing hot spot research suggests modest outcomes, with violence increasing in areas close to high-crime places targeted by these efforts (McManus et al., 2020). Hot spots are also strongly associated with pediatric gun injuries, introducing victim age clustering (Slye et al., 2019).

Police-led gun violence initiatives, as in St. Louis, have been attempted but require sustained support to achieve desired outcomes, particularly for reducing violence in hot spots (Koper, Woods, & Isom, 2016). There is nothing wrong with seeking short-term results. It becomes a problem when undertaken without a long-term strategy to maximize returns on funds. Gun violence is a health problem that has taken decades to create and may take as long to eradicate.

Gun injuries, too, bring an ethical necessity for prevention education taking into account inequities (Ahiagbe, 2020), bringing yet another portal to social justice. Fletcher (2020), for example, advocates for a racial equity framework for gun violence prevention initiatives considered a key element of public health, which benefits from a robust funding of research. Finally, trauma inequities, too, are subject to social contextual factors influencing manifestations (Mikhail et al., 2018). If going unrecorded, they officially do not exist.

Ezeonu (2008), although referring to gun violence in Toronto's Black community but applicable to the United States, notes that "social policies are often products of contestations among different claims makers trying to dominate the definition of particular social problems and the policy initiatives for their control" (p. 193). Contestations have winners and losers, with communities of color losing. Losing on gun violence translates into the salient epidemic category, with winners reaping the benefits, making their cause national with appropriate publicity and funding (Gallagher & Hodge Sr., 2018):

> The current gun situation in America is an issue that needs to be understood and responded to by care ethicists. It is an issue whereby some groups are less visible and less valued than others. It is our view that this is contributed to by a lack of empathy. This is an issue that ethicists need to care about as it impacts negatively on marginalized groups and communities and on caregivers who have to respond to the suffering of victims of gun violence and their families. (p. 3)

Race and class fuel lacking empathy for victims, their families, and communities, spotlighting the initial step in finding solutions.

MULTIPLICITY OF PERSPECTIVES ON GUN INJURIES

Not all gun violence and injuries are the same, nor are the circumstances leading to violence and treatment. Too often media accounts of gun injuries fail to provide any details other than "life-sustaining wounds" or "non-life-threatening injuries." These descriptions do not capture the extent of the injuries and how the lives of victims were altered. There is good reason why an entire section of this chapter has been devoted to this. Dolores's case (*New England Journal of Medicine*) provides insights into how a gun injury alters a victim's life, and in this case she was 19 years old when a bullet shattered her spinal cord (Kalra, 2019):

> After a year of surgery and intensive rehabilitation, Dolores regained her ability to walk with the assistance of a cane. But that was merely the first step on a long road to recovery. She struggled to move her bowels despite taking motility agents, and her bladder remained flaccid, devoid of nerve stimulation. For the past two decades, Dolores has had to push a plastic catheter into her urethra every 4 hours to relieve her body of urine. She carries her catheter bag everywhere she goes — to the store, to social events, and to her favorite event of the year, the Houston Rodeo. "Sometimes, my catheter beeps in security check," she smiles for a moment. Then the gravity returns to her face. "I've been destroyed by this — mentally, physically. Sometimes I feel like I'm going crazy." In addition to the physical disability, the financial costs are crippling. Even with health insurance, Dolores spends more than $200 per month on medications and catheters. Living off a $700 disability check leaves little money for rent and food. "I've had the same bed for 15 years. I can't afford a new one. Sometimes, I put pillows where the mattress sinks in," Dolores says. "Once a week, I go to church for free food. I get there at 5 a.m. or they run out. With food stamps, I buy eggs, beans, tortillas." (pp. 1704–1705)

No statistic does justice to how Dolores's life changed. Her family's lives were altered and their dreams for her changed because she was at the wrong place at the wrong time.

The long-term functional, psychological, emotional, and social outcomes for gun injury victims benefit from early identification and a database on long-term and multidisciplinary longitudinal care (Vella et al., 2020). These victims have unemployment rates (14%) and substance use (13%) increasing 5 to 8 years after their injury. These conditions impede recovery, increasing the likelihood of injuries reoccurring or resulting in premature death.

Successful recovery is dependent on how well we understand and deliver services (Amnesty International, 2019):

> A lot of people don't understand how long it takes to recover from a gunshot wound. There is the physical aspect, but then the mental and emotional aspect takes much longer . . . this impacts how I think about the future. . . . Will I be able to dance at my wedding? What if I have kids, can I keep them safe, can I play with them? . . . All these questions come from the shooting, it happens because I was shot (William "Tipper" Thomas, gunshot survivor). (p. 17)

It is no surprise that there are no standard timelines assigned to gun injuries, with each injury dependent upon a host of factors and influenced by race and socio-economic class.

The percentages of urban gun murders solved varies depending upon the victim's race, and if White, non-Latinx, the probabilities go up dramatically. Gender inequities interact with racial inequities (Vaughn, 2020). If a person is of color, they are much lower (Delgado, 2021). If murders go unsolved, one can only imagine what happens with an injury. Gun killings have a higher clearance rate than gun injuries, as in Boston, with a significant gap in clearance (43% versus 19% for nonfatal gun assaults), largely due to investment in intensive efforts within the first 48 hours to solve murders (Cook et al., 2019). A recent es-timate has San Francisco (15%) and Boston (10%) with some of the nation's worst clearance levels, making it a mistake to equate percentages of solved crimes with big high-profile cities.

The Trace (Ryley, Singer-Vine, & Campbell, 2020) examined shooting clearance rates with this startling observation:

> The odds of an arrest are particularly low when victims survive, in part because those crimes tend to be assigned to detectives whose caseloads are exponen-tially higher compared to their colleagues in the homicide department, who are often overburdened themselves. The chances are even lower if the victims, like Little, are people of color. When a black or Hispanic person is fatally shot, the likelihood that local detectives will catch the culprit is 35 percent—18 per-centage points fewer than when the victim is white. For gun assaults, the arrest rate is 21 percent if the victim is black or Hispanic, versus 37 percent for white victims.

Gun assault increases, including injuries, have led to development of local law enforcement models, as in Kansas City, Missouri, helping solve gun crimes as a future deterrent. In 2018, the Kansas City Police Department (KCPD) initiated the Crime Gun Intelligence Center (CGIC) to address the impact of gun vio-lence in that city. This model has six components (Novak & King, 2020): "(1) the comprehensive collection of ballistic evidence, (2) timely entry and correlation and crime gun tracing, (3) ATF analysis, (4) identification of NIBIN [National Integrated Ballistics Information Network] leads, (5) collaboration between local and federal law enforcement, and (6) prosecution of offenders who commit gun crimes." This model focuses on law enforcement (local and federal), identifies in-vestigative barriers, and proposes strategies to improve outcomes.

A multidecade gun injury analysis found hospitalized injuries increasing sig-nificantly from 1993 to 2014, reflecting hospital propensity to use inpatient serv-ices for injuries that are more serious and outpatient services for less, with the health care system incurring a greater financial burden (Kalesan et al., 2018). However, an epidemiologic review of 13 years (2003–2015) of Pennsylvania gun hospitalizations found them declining but with outcomes remaining unchanged (Gross et al., 2017).

Taxpayers invariably pay higher taxes to cover costs of hospital care and corresponding services because costs are disproportionately concentrated among young men of color who have a higher likelihood of being uninsured or covered by Medicaid (Howell & Gangopadhyaya, 2017). The scope and costs of this nation's gun violence problem far exceed hospital expenses (Hemenway & Nelson, 2020). A focus on achieving rehabilitation, for instance, is rarely possible because of uncovered costs associated with this phase of recovery (Amnesty International, 2019):

Hobson told Amnesty International that she was still in debt because of the medical bills she incurred for treatment after she was shot: "I was a victim, I had nothing to do with my crime. I was just in the wrong place at the wrong time according to detectives. But today, I cannot tell bill collectors I was in the wrong place at the wrong time and expect my debt to disappear." Although Hobson had health insurance, she still incurred costs associated with emergency health care (around US$50,000) and her recovery in hospital (around US$35,000). The injuries caused by the shooting were severe and Hobson continues to need regular health care for which she has to pay. For example, she has a leg brace to aid with walking, which cost US$800. She needs to visit a podiatrist regularly because of the calluses on her feet linked to her use of the leg brace. The most conveniently located podiatrist does not take her insurance and she needs to pay him US$50 per session. . . . "If I could go every week, it would be US$200 a month, but because of budget constraints I try to stretch it to as much as once every 2–3 months." Hobson visited a psychiatrist briefly, but her insurance did not cover these sessions and the cost was prohibitive. Hobson's case was declared inactive because there were no leads. Today, she is an activist working with gunshot survivors nationally, providing them with trainings and safe spaces to heal. She is also the Miami coordinator of the national network crime survivors, Crime Survivors for Safety and Justice. (p. 23)

Megan Hobson's recovery is lifelong, bringing financial hardships and compromising lifestyle and living standards associated with severe gun injuries.

Greater aid to victims and the organizations supporting them is needed. One such fund is the Victims of Crimes Act (VOCA) designed to compensate victims of violence and assist organizations. Nevertheless, these funds were severely underutilized by gun victims. These federal funds can be more accessible by easing eligibility requirements and provision of technical assistance (Educational Fund to Stop Gun Violence, 2020). Increased technical assistance moves the field, with research findings guiding best practice models.

Street crime victims have help-seeking options, including mental health services. However, very few access or receive this service, with an overwhelming majority without any knowledge of their rights or eligibility for compensation and services (Roman et al., 2019). From 1993 to 2009, 9% of serious violent crime victims received compensation from a victim services agency, illustrating barriers to this aid (Nieto & McIively, 2020). The Office of Victims of Crime share many characteristics with governmental bureaucracies (Gambacorta & Ubiñas, 2018):

> Pennsylvania has an Office for Victims of Crime, which is supposed to help with everything from funeral costs to recouping lost wages. In a recent year, the office denied nearly 100 applicants for not filing their paperwork on time or failing to cooperate with law enforcement, records show. The fund fields 8,600 claims a year, and pays out about $13 million—an average of $1,511— with awards topping out at $35,000. The largest chunk of funding, more than $4 million, was spent on forensic exams for sexual-assault victims. "If a victim is successful in obtaining funds, it is simply inadequate to meet the incredible needs that they face." Bureaucratic red tape is the last thing a gun victim needs to encounter while struggling to regain some degree of equilibrium in their new life as a gun statistic.

A lack of success, as in Pennsylvania, offers a dismal picture for the victim assistance field, with those of color and non-English speakers encountering greater barriers because of discrimination. The Gifford Law Center's (Nieto & McIively, 2020) report "America at the Cross-roads: Reimagining Federal Funding to End Community Violence" provides an in-depth critique of the Victims Crime Act.

Sodhi et al. (2021), in discussing gun survivors, draw upon a historical analogy on service barriers being reminiscent of barriers to reparation by the US government. This is manifested by how bureaucracy, administrative requirements involved in seeking health care, and the failure of victim compensation funds to aid those of color. It takes a marshalling of immense resources to aid gun victims, and we can ill afford to leave any stone unturned in seeking these resources.

CONCLUSION

This chapter addressed the extent and complexities of gun injuries, taking them beyond a statistic. Injuries are complex, with no such thing as a minor injury. Further, we can never relegate them to a prescribed time period; they often extend to a lifetime, especially when considering trauma. Multiple injuries were discussed. Those paralyzed by gun injuries received additional attention in this chapter because of the multiple resources required to aid them.

Section II provides a context to understand injuries by introducing social, cultural, and economic dimensions.

SECTION II

Social Science Perspectives on Gun Injuries

Social science shapes gun violence views and strategies to prevent and treat the injured. This section breaks out the social sciences along three perspectives (social, cultural, and economic), with each providing a valuable snapshot that, when combined with the others, provides a composite picture of this public health epidemic.

Social Science Perspectives on Gun Injuries

4

Social Perspective

If you live in DC, you are automatically considered a survivor of gun violence. Our city is way too small to not feel compassion for people in front of and behind the gun.

—Tia Bell, founder of the antigun violence project T.R.I.G.G.E.R.
(Spearman, 2020)

INTRODUCTION

Social sciences play an important role in demonstrating the effectiveness of interventions of those at highest risk of perpetrating gun violence or being gun violence victims (Abt, 2019). This chapter's opening quote saying that if you are alive in this neighborhood, you are a gun violence survivor is quite telling; in the case of injuries, you are survivor with ever-persistent trauma. Social contextual forces (cities versus rural settings) influence behaviors across an entire spectrum. The concept of "urban penetrating trauma" highlights gun violence as an urban public health problem (Gurney, Menaker, & Springer, 2020). A "rural penetrating trauma" concept will probably elicit an expectation of farming accidents.

Gun injuries are both a metaphor and a lived experience, underscoring their profound significance in a survivor's life (Ralph, 2014), and they are an integral part of this country's social fabric. Academics interested in this violence are challenged in writing about social context, and I include myself in that group. This challenge is not because of a paucity of content. The broad expanse social context brings facilitates inclusion of numerous gun violence topics, making understanding the subject arduous.

A social structural view introduces the importance of social interactions and relationships that examine options, or perceived or actual lack of them, in behavior. Community brings this construct to a local level for use in empirically and theoretically studying gun violence, and more so regarding urban gun injuries and responses once they have occurred. Furthermore, this broad canvas is sufficiently expansive to include different professions and academic disciplines to help us understand gun violence. Further, how the injured fare back in their communities needs examination. We can find memorial T-shirt businesses in the

The Silent Epidemic of Gun Injuries. Melvin Delgado, Oxford University Press. © Oxford University Press 2022.
DOI: 10.1093/oso/9780197609767.003.0004

nation's urban centers, but no such outlets exist for gun injuries. Victims cannot wear these shirts, nor can their social network. Injuries supposedly pale compared to death, complicating the victim grieving process.

An intersectional lens brings a nuanced view of gun violence on race, developing an understanding of injury manifestations (Delgado, 2020a). This information informs community interventions, bringing a practical dimension to this coverage. Although a social perspective is associated with a "big picture," there are subareas such as micro-sociology violence where people are directly threatening violence but may not carry out this act (Collins, R., 2019). The fear these threats engender leaves invisible scars.

There is value in having a strong social perspective on any subject, particularly gun violence. A social perspective does not exclude micro-macro connection viewpoints (Newman, 2020). The data provided in the previous chapter offered various perspectives on gun injury research. Grasping the sociocultural context of gun injuries helps achieve "clarity of purpose," providing a northern star guiding us through the maze that major social problems present to the public, social scientists, and care providers.

As noted in Chapter 1, this book does not cover gun control in depth, although my views are obvious. Readers interested in this realm will find numerous reports and scholarly material. I will, nevertheless, cover gun control in this chapter but in a manner that introduces broad dimensions rather than entering into the "weeds" of detail.

WORD OF CAUTION

A social perspective on gun injuries enhances our understanding. However, a word of caution is needed because an aerial view is of limited use in creating interventions without an on-the-ground view connecting these perspectives. We must guard against broad conceptualizations because they mask significant differences within and between subgroups. Gun injuries require understanding broad conceptualizations but attending to how differences exist within neighborhoods and groups. A social perspective guards against this occurring and the damage that stereotypes create.

Conversely, a simplistic singular gun violence stance would be a serious barrier at preventing and treating injuries (Cukier & Eagen, 2018): "Socioeconomic disparity, instability, inequality, lack of democratic processes, health, social and educational policies, and the availability of drugs, alcohol, and weapons are all contributing societal factors to violence rates" (p. 11). Gun violence encompasses multitudes of activities and ties many social issues, highlighting the complexity of addressing its root causes (House, 2018), and how certain neighborhoods bear a disproportionate "disadvantage." Neighborhood disadvantage is a construct that captures the interplay of social forces; it is structural in nature, and it positions these geographical settings as having high rates of gun violence (Dalve et al., 2021).

Grasping who are victims and perpetrators, assuming the latter are victims, too, and the circumstances that make them so and how can we enhance positive outcomes, is what social science is all about. Statistics shape understanding of gun violence, but they miss the stories behind these numbers, calling attention to qualitative data bringing in-depth insights into the struggles and actions needed to aid victims. Framing gun violence from a health perspective brings a broader social grounding using a multidisciplinary approach to studying and responding to it (Ransford & Slutkin 2017):

> This health framing is important because it recognizes that violence is a threat to the health of populations, that exposure to violence causes serious health problems, and that violent behaviour is contagious and can be treated as a contagious process. Relatively standard and highly effective health approaches to changing behaviours and norms are increasingly being applied to the problem of violence and are showing strong evidence of impact among individuals and communities. (p. 4)

Ransford and Slutkin highlighted how a health stance for understanding violence does not stigmatize a subject that is inherently stigmatizing. This stance is not without controversy due to the highly racialized aspects of this problem.

Preventing gun carriage among high-risk urban youth must be a part of any comprehensive strategy at preventing gun injuries and death, particularly since communities of color have a high percentage of residents in this age category. A social contextual perspective finds peer networks, history of victimization, perceptions of community violence, and retaliatory attitudes combining to shape firearm carriage. Strategies for reducing carriage must operate at the nexus of individual and social determinants (Sokol et al., 2020). Concentration of other social problems such as substance use/misuse increases the likelihood of violence related to the drug industry. Thus, much depends upon the gun lens we use.

Gun injuries in a Miami-Dade County (2002–2012) level 1 trauma center, for example, witnessed dramatic increases in admittance disproportionately affecting young African American/Black males, supporting use of targeted community engagement and prevention initiatives (Zebib, Stoler, & Zakrison, 2017). Outcome inequities exist along a variety of dimensions, including gun violence and children (Irizarry et al., 2017), with no age group escaping, although certain groups bear a disproportionate impact.

WORD OF CAUTION: PLACE OF TRAUMATIC GUN INCIDENT

I have taken the liberty of lodging place of traumatic gun incident within a social perspective category. Some readers may argue it is best situated within a cultural context. I have gone the social perspective route by emphasizing factors that are easily identified and measured. Remarkably, there is a paucity of information on the social, physical, and neighborhood location of gun injury events (Newgard et al., 2016). Various dimensions of place are covered throughout this book, but the topic is of sufficient importance to revisit with additional information.

Understanding gun injuries requires in-depth knowledge of how context shapes interventions. Labeling gun violence from an urban perspective can limit guiding interventions because "urban" is a catchall term, representing the first step in a lengthy process of contextually describing where this violence transpires and is treated. When an injury occurs within a victim's neighborhood, it garners even greater significance because the victim must either move out postinjury or stay and recover while facing an environment associated with violence. Some locations are easier to avoid. Suffering a wound in a local playground, for example, may make it easier to avoid. If the injury occurs in a major community intersection, within their residence or just outside, avoiding this area becomes challenging. Moving may be impossible due to finances or because there is a support system in the home neighborhood.

The term "hot spots," as discussed earlier, captures the need to focus resources on areas prone for gun use. Although this perspective represents an important view, there are other dimensions from a victim's standpoint. Place is but one aspect, with time of day, season, weather, and proximity to home, for example, adding to this context.

The emotional aspects attached to these factors are as important as the physical injuries. Further, they need to be separated if a victim is to launch a meaningful recovery. We know relatively little on the relationship among gun violence, adolescent mental health and behavioral outcomes, and proximity to school or home, for instance (Leibbrand et al., 2020). There is a call for increased research on the association between living near a neighborhood shooting and utilization of a local Emergency Department (ED) for stress-responsive complaints (South et al., 2021), taking into account age and the importance of gun violence exposure.

GUN VIOLENCE: EVERYDAY OCCURRENCE
IN URBAN CENTERS

The United States has almost 38,000 (over 100 per day) gun deaths annually, of which 13,000 are homicides and 25,000 are suicides, with almost 85,000 injuries, or over 230 per day (Team Trace, 2018). According to the Centers for Disease Control and Prevention (CDC), between 2015 and 2016, firearm injuries increased by 37% (85,000 to more than 116,000), and the biggest single-year increase in over 15 years (Campbell, Nass, & Nguyen, 2018). This problem is not limited to adults. The consequences for children are dramatic and lifelong, with 1,300 of them dying and 5,790 wounded each year, with males, older children, and those of color overrepresented and with the worst outcomes (Fowler et al., 2017).

Gun injuries are considered "middle spaces" by Cruz, whose poem is included in a book written by those disabled by gun injuries; the program sponsoring this work is discussed later in this book (Nishimura & Robledo, 2019):

Why do you think the "middle space" often gets ignored? It's funny that you say that. One time I went to a show, where they were giving out guns as toys, and in that event, there was a whole bunch of gangsters from LA with masks and all of that, who were talking about guns. And you know most people, they talk about people that died, when it comes to gun violence, and they don't talk about people who still have to suffer the rest of their life. You are absolutely right, but I guess, this is our program—where we show people the "reality," and I guess this is what we are trying to do. Right now we are doing community events, but we want to expand. Violence is everywhere right now, in every country. It's crazy.

We can enhance understanding of violence through a variety of mediums, and poetry is just one of many. Poetry, composed and spoken by gun violence survivors, resonates among audiences with similar painful experiences.

For every gunshot death, there are six people with wounds of varying physical severity (Fairchild, 2016), not taking into account emotional ramifications, making gunshot outcomes worthy of greater reliability of injury-related data (Hink et al., 2019). This statistic does not take into account trauma from witnessing a shooting visually or hearing gunshots (Mitchell et al., 2019). Further, it does not provide details on their nature or long-term consequences. This, too, is recognizable as a silent epidemic, and it needs to be part of any comprehensive portrait in high-violence neighborhoods, reinforcing a public health stance (Dicker, 2016; Santilli et al., 2017).

Again, we must acknowledge resiliency and eschew a deficit approach to gun violence statistics. I do not want readers to think that the urban organizations highlighted in this book are few in number in this country. There are many, and they are increasing in number and size. For example, Wheelchairs Against Violence (WAG), a New York City organization founded in 2013 by Kareem Nelson, paralyzed by a gunshot wound, has a mission that rings familiar to readers:

Wheelchairs Against Guns is an organization of exceptional individuals living with physical disabilities due to gun violence. Our mission is simple—to protect children from the dangers of bullying, gangs, and gun violence. Through workshops at inner city schools and churches, we share our stories of personal tragedy to serve as a warning to students and teach them ways to avoid toxic situations through conflict resolution, critical thinking, and self-esteem building techniques.

Kareem's background shares much with countless other youth of color, and it serves as a testament to human will (resiliency) under adverse circumstances (Wheelchairs Against Gun Violence, 2020):

Kareem Nelson was born in Harlem New York to Charisse Jackson & Michael Nelson. Raised as an only child by his mother, Kareem was a bright student and exceptionally talented in the sports of boxing and football. As a teenager in the late 80's, he witnessed the traumatic effects of the crack epidemic on his neighborhood and soon found himself involved in the streets. While in Baltimore Maryland on Father's Day 1995, the street life gave Kareem his first serious warning. He was shot in the back over a petty dispute. The attack left him paralyzed and confined to a wheelchair. After taking some time for reflection, Kareem became determined to change his life for the better. In 2013 he founded Wheelchairs Against Guns to teach inner city kids positive ways to deal with the bullying, gangs, and gun violence they are exposed to in their neighborhoods.

WAG-type organizations vary in size and mission; they take a personal tragedy and transform victims into community assets; these people require support and, where that does not exist, WAG develops it.

WEAPON AND CALIBER USED

The weapon used in gun injuries introduces a dimension worthy of attention, and one rarely mentioned in media coverage. For instance, Schellenberg et al.'s (2020) research of national trends, injury patterns, and shotgun victims presenting at a trauma center found that they represent 9% of gun injuries and are primarily the result of White, Non-Latinx men in their 20s in the southern parts of the country. The saying "God and guns" is not a New York City expression; it is a southern expression (Claiborne & Martin, 2019). Shotguns are not an urban weapon of choice, introducing a sociocultural dimension. However, other gun types are weapons of choice or availability.

Gun type significantly influences fatalities and injuries. One does not have to be a forensic expert to conclude that gun type and ammunition shape injury outcomes. There are major differences between a single gunshot and multiple wounds, challenging provision of trauma care at the scene and the recovery process (Zeineddin et al., 2021). Multiple-wound cases are often due to automatic weapon assaults, with a wide range of outcomes depending on bullet entry point.

There is good reason why semiautomatic assault weapons and large-capacity ammunition magazines were banned by the federal government from 1994 to 2004. Data on high-volume gunfire (HVG), which involves greater than 10 shot episodes, is limited (Koper et al., 2019). However, Minneapolis gun violence (January through August 2014) numbered 135–167 incidents and resulted in 20%–28% of all victims (Koper et al., 2019). High-capacity semiautomatics account for 22% to 36% of guns used in crimes, with estimates having them involved in 40% of serious violence incidents (Koper et al., 2018). Wounds caused by AK-47s and AR-15s do not leave small bullet entry wounds; they simply blow off body parts, leaving victims as paraplegics or even quadriplegics (Korten, 2016). Further, semiautomatic rifles are convertible to fully automatic using a $50 kit (Roth, 2019).

Further, handgun radically invasive projectile (RIP) bullets are fragmenting or expanding bullets, legal in the United States, causing a high number of fragments when hitting the body, increasing the severity of injuries and further challenging treatment and healing (Hakki et al., 2019). These bullets have a primary intent of causing greater tissue damage than conventional bullets, with victims having multiple injuries hampering their recovery process.

Reducing deaths and injuries by high-capacity weapons necessitates states introducing restrictions on high-capacity magazines (Koper, 2020). Not surprisingly, gun violence lends itself to researching various aspects. Better understanding assault weapons, as a subset of gun violence, is required because of the immense damage these guns cause (Danner et al., 2020). Increase in gun instrumentality (size and technology) has made them more lethal (Braga et al., 2020). Military-style weapons have increased in use over the past two decades, making them significant in causing fatalities and injuries.

Guns found on the streets are increasing in caliber, with a higher probability of severe victim health consequences, as with semiautomatic weapons (Braga & Cook, 2018; Koper et al., 2018; Manley et al., 2019). The following description provides a glimpse on why these weapons are so destructive (Brooks, 2019): "An AR-15 assault rifle and a .223 round is designed to tumble and create maximum soft tissue damage. . . . The entry wound from an AR-15 is about a half-inch in diameter. The exit wound is like 8 to 10 inches in diameter. So bullets leave holes in people's bodies, in their lives and in their communities." Semiautomatic weapons make it easier to shoot bystanders, increasing indiscriminate body counts and injuries.

AVAILABILITY OF GUNS/GUN CONTROL

This nation would not have a gun problem if we did not have guns. This simplistic declarative sentence belies its profound significance to gun injuries and fatalities, yet gun control generates intense reactions across the nation. Everyone seems to have an opinion on gun control—pro and con. The National Rifle Association (founded in 1871) is the nation's oldest gun organization and a prime interest group in advocating for guns. It has been a powerful lobbying force, although its influence has waned due to recent controversies involving leadership and financial misdeeds.

A public health approach does not take an either/or position on this contentious issue. Rather, it approaches this problem by providing solutions that reduce deaths and injuries from suicides and accidents, in addition to reducing gun availability for those intending to do harm to others. Nevertheless, the highly politicized nature of gun control has challenged the public health field in carrying out actions to address this epidemic (Branas, Reeping, & Rudolph, 2020; Crosby & Salazar, 2020), with formidable forces fighting against passing legislation limiting access to guns and ammunition.

Merry (2019, 2020) argues how warped narratives and interest groups shape US gun policies. Narratives originate at the local level and work their way to the national stage, attracting local and national interest groups, becoming talking points and themes in campaigns for and against gun control. There is such a thing as "gun culture" (Mencken & Froese, 2019). Gun culture serves symbolic and practical purposes in our society, empowering owners, which influences gun policies at state and national levels.

The US gun industry gets overlooked in gun control discourse and public health (Parsons, Vargas, & Bhatia, 2020). This industry is generally unregulated.

The Bureau of Alcohol, Tobacco, Firearms and Explosives (ATF), the federal agency charged for overseeing this industry, is widely regarded as underfunded. This lack of funding, when combined with numerous restrictions on carrying out its mission effectively, leaves the federal government impotent in protecting the public.

We can say it is OK to own a gun but not make ammunition available (Center for American Progress, 2019): "But often missing from the conversation about firearms are questions on ammunition—namely, the role of easy access to ammunition and ammunition accessories in the epidemic of gun violence in the United States." Access to ammunition must be part of any solution on gun violence. There may be a constitutional right to own a gun, but there is no right to own ammunition. The successes and failures of silencers and ammo magazine restrictions offer many lessons on the quest to introduce future controls (Spitzer, 2020a).

Gun control generates its share of scholarship; although summarizing this literature is beyond the scope of this book, it will be briefly touched upon here. Rand Corporation's report presents a synthesis of over 200 research findings on the effects of gun policies (Smart et al., 2020). The research literature review is very powerful in its conclusion—namely, limiting gun access decreases gun violence along a variety of spheres (Sivaraman, Marshall, & Ranapurwala, 2020). Mind you, that conclusion is not worthy of a scholarly prize. The saying "Guns do not kill people, people kill people" comes to mind. This argument is challenged because guns do not kill people on their own, requiring a human pulling the trigger (Shammas, 2019). People with guns cause injuries. Violent impulses find outlets, and guns are an efficient way of carrying them out.

Is there a difference between access, possession, and carrying a gun? This difference may see semantic, but the consequences are not. The differences among youth, for instance, has much to do with their social network and level of criminal engagement, such as having friends in gangs, drug business, violence, and other delinquency acts (Mattson, Sigel, & Mercado, 2020; Sweeten & Fine, 2021). These activities are at high risk for violence that escalates to the use of guns, increasing the significance of altering social networks.

Illegal firearm availability (IFA) is associated with gun violence within neighborhoods such as Newark, New Jersey (Yu, Lee, & Pizarro, 2020). We cannot acknowledge youth carrying demographic characteristics while concomitantly ignoring the sociocultural context, with state gun control laws playing an influential role in these outcomes (Preidt, 2021).

Baltimore's underground gun market illustrates how they assume a position of currency to buy drugs or pay debts, an indicator of how guns become part of a community's economy (Crifasi et al., 2020). The relationship between illegal gun trafficking and injuries is very strong, calling for preventing gun transports into high-violence urban communities (Ciomek, Braga, & Papachristos, 2020). Gun violence prevention is impossible without communities trusting law enforcement, with illegal gun possession a precursor to shootings, calling attention to enforcement efforts at curtailing gun availability (Webster et al., 2020).

Perceptions of gun access and carrying (risk factors) among male adolescent offenders require recognition in addressing this group (Keil et al., 2020). Gun availability in these neighborhoods, although difficult to determine, must not be overlooked in understanding gun attitudes and behaviors. Understanding this social network and targeting strategies to reach it take on great significance to mitigate gun violence among this age group.

Obtaining weapons is not difficult, even in states with strict gun laws. Bordering states with lax gun laws are problematic as well. An underground weapons pipeline, as in the Bronx and Brooklyn, New York, calls attention to the need for a national approach to gun control measures (Braga et al., 2020a). One does not have to own a gun because availability is easy, such as "borrowing" or even "renting" one, further uplifting the importance of gun control (Cunningham, 2016). On a final note, Dr. Carmona, a former US Surgeon General, poses policy-related questions to the field of public health that take center stage with gun violence consequences (2020):

> Many questions legitimately arise based on the scientific public health approach to numerous societal challenges, including guns. Examples would include, but not be limited to, why then do states allow gun purchases without any training and certification of competence? Why do we not mandate comprehensive background checks to ensure that only "appropriate" persons are given the privilege of gun ownership? In fact, the public, in a bipartisan manner, is overwhelmingly supportive of this concept but leadership fails to act. Why do some states allow "concealed carry" without evidence of competency? Certainly, an armed citizen without training or demonstrated knowledge of firearm safety and competency could arguably be construed as a societal risk just like a person driving a car without training or licensure or a surgeon wielding a scalpel without training or competency certification. (p. 3)

The Surgeon General's questions are often overlooked in gun control debates, and public health must weigh-in with answers. The answers do not address illegal gun ownership, which poses different challenges.

URBAN FORCES

Urban violence discussions must tackle the definitions of "urban" and "violence" (Pavoni & Tulumello, 2020). Gun violence is an urban phenomenon with half of gun assaults transpiring in 127 cities, comprising 25% of the nation's population (National Institute for Criminal Justice Reform, 2020a).

Understanding why gun injuries assumed such prominence among urban residents, and the barriers that need to be surmounted to prevent and treat them, benefits from social science views. Identifying barriers that keep gun injuries from reaching a prominent position that is worthy of major national and local initiatives is not achievable by waving a magic wand and wishing it so. Rather, it requires national political will and understanding of major structural barriers in need of dismantling. This necessitates major funding initiatives and, more importantly, a recognition that funding is well spent when viewed through a long-term rather than short-term lens.

Population concentrations in urban centers translate into higher chances of guns causing widespread victimization. High population density exacerbates any epidemic, as we have seen with COVID-19 and gun violence. News generated by these events travels quickly throughout a neighborhood, with the message shaped by the methods and characteristics of the sender—Twitter, word of mouth, and so on.

VICTIM PURITY AS A SOCIAL CONCEPT

The public's ability to connect emotionally with victims humanizes them and their circumstances, including the nature of injury. This engenders empathy, which is critical in the creation of a public outcry for investing in finding gun violence solutions. However, this outcry and corresponding rally do not occur when there is a perception that the victims got what they deserved; this has far-reaching implications for solving this problem and investing resources in aiding the victims.

Victim purity translates into victim worthiness (Whalen, 2020). Media bias in minimally or ignoring gun violence in communities of color (racism, news-reporting practices, and territorial stigma) is evident across the country (The Joyce Foundation, 2020), as in Chicago (White, Stuart, & Morrissey, 2021), one manifestation of victim purity. In contrast, one study of mass public shootings found that young, White, non-Latinx women victims receive disproportionately greater media coverage (RAND Corporation, 2018).

The life choices made by some gun victims put them on a violence-prone path, making it difficult for them to elicit public sympathy. As noted by a gun victim services provider (Korten, 2016): "'I do not look at our patients through rose-colored glasses,' Joseph says. 'Some of our patients, no matter what you try to do to help them, they have decided on a course, and you can only do so much. But it's hard to make the argument that somebody who is 15 years old has chosen the path that they are on.'" Placing age within this context illustrates the unfairness of labeling these victims as "impure."

The concept of purity is heavily laden with social meaning, with sexual behaviors standing out. However, purity is rarely associated with gun violence and who is a victim is shaped by race and ethnicity. The "nullification" of caring about victims, such as those of color injured by the police, is influenced by racial characteristics (Johnson & Lecci, 2020). Gun victim purity can assume various manifestations, including shaming. A literature review (in the past 10 years) found one article on victim purity and gun violence, highlighting the challenges gun victims face.

The concept of "ill eligible black death" is an example of victim purity (Leonard, 2017):

All gun violence is not created equally. All victims of gun violence are not created, seen, or treated equally—no "all lives" don't matter. Those assailants, those mostly men who hold, pull the trigger, and use their guns to inflict pain and death on others, are most certainly not created or treated equally. Race, gender, zip code, and class all matter. (p. 101)

Deserving versus undeserving victimhood is deeply rooted in social forces and values, influencing how we address victims, and in our case, gun victims. This observation applies historically and in the present day.

Victim purity rests on race and socioeconomic status, as seen in how school mass shootings get media coverage (McCoy, 2020):

Shootings in predominantly white schools repeatedly receive an outpouring of support, such as donations to pay for funerals and the almost immediate provision of trauma-based services. Yet the regular shootings in urban communities get little help or attention, in large part because many Americans view urban violence as normal and someone else's problem.

Adding vulnerability to this discussion introduces factors such as age and gender (Niemi & Young, 2016):

If you are mugged on a midnight stroll through the park, some people will feel compassion for you, while others will admonish you for being there in the first place. If raped by an acquaintance after getting drunk at a party, some of us are impacted by your misfortune, while others will ask why you put yourself in such a situation.

Gushue and Wong (2018) note, for most of the public, being a gang member victim makes one deserving of the consequences of this lifestyle.

What social factor determines if someone feels indifference, sympathy, or scorn for a victim? Is it political affiliation? Race/ethnicity? Gender? Age? Social class? The nature of the crime? Geographical context of urban versus rural? These and other factors create a context that reinforces victim views and influences self-perceptions, both of which are factors in seeking rehabilitation. Niemi and Young (2016) observe that the interaction of one's stance on privilege, loyalty, obedience, and purity, when compared to the values of caring and fairness, increases being prone to blaming the victim. Race is undeniable in victim blaming (Dukes & Gaither, 2017; Wright & Washington, 2018). If gun mortality in communities of color is racialized (Walker, Collingwood, & Bunyasi, 2020), so are gun injuries (Niemi & Young, 2016):

Victim blaming appears to be deep-seated, rooted in core moral values, but also somewhat malleable, susceptible to subtle changes in language. For those looking to increase sympathy for victims, a practical first step may be to change how we talk: Focusing less on victims and more on perpetrators —"Why did he think he had license to rape?" rather than "Imagine what she must be going through"— may be a more effective way of serving justice.

Increasing victim worthiness is essential in constructing interventions and obtaining funding for the cause of gun injuries, preventing and treating it.

"Victim purity" is a concept with particular relevance to gun injuries because it shapes public perceptions of the injured, how the media characterize them and their suffering, and how it translates into decisions on resources for their treatment, including victim compensation. Degree of victim purity elicits divergent public and official responses, such as when a young infant is injured when compared to a teenager who shares all of the demographics except for age. Arguably, nowhere are these disparate victim views more salient than with gun fatalities, and by extension, injuries, among White, non-Latinx, and people of color. The former is viewed as tragic and the latter as ordinary (White, Stuart, & Morrissey, 2021).

Is it possible to be an ideal victim? The ideal victim construct is one worthy of sympathy and possesses specific characteristics. One linguistic study analyzed victim descriptions to explore how people conceptualize this status. They project vulnerability, innocence, and helplessness, and they have experienced harm (Lewis, Hamilton, & Elmore, 2019). Gun victims of color rarely elicit this purity reaction, impeding their recovery.

One study of the relationship between media reporting gun injuries and victim characteristics in Cincinnati, Ohio; Philadelphia, Pennsylvania; and Rochester, New York, found 1,801 victims (intentional non-self-inflicted), with 900 covered by local media. Eighty-three percent were African American/Black, with 48% receiving some form of coverage. However, if the victim was male, less than 40% received coverage compared to news coverage if the victim was a woman (Kaufman et al., 2020).

Victim blaming is a social-political-psychological-cultural process that considers a bullet a natural outcome of engaging in criminal activity, regardless of whether the violence centered on selling drugs, guns, or other contraband items. These victims will not elicit a positive societal response or a willingness to support them in their recovery; this translates into limited support services and resources.

Victim purity is rarely applied to perpetrators. A deeper understanding of how perpetrators view causalities is an overlooked dimension to gun violence, and when obtained, it provides a profound understanding of motivation and feelings on the harm they cause (Topalli, Dickinson, & Jacques, 2020). If we embrace the view that gun violence is two sides of the equation, with victims and perpetrators being victims, both perspectives introduce purity standpoints.

PLACE OF TRAUMATIC GUN INCIDENT

Neighborhoods often form an integral part of an individual's identity, helping in the self-discovery process, particularly when there is contact with those from other neighborhoods allowing comparisons. This neighborhood influence, I am sure in the case of readers born and raised in urban neighborhoods, is an integral part of their identity, and that is certainly my case. What does this have to do with gun injuries? Social context grounds lived experiences, and in this case, where the injury occurred is vital to understanding.

Where a gun injury occurs becomes an integral part of a traumatic incident, although more research is called for in understanding how this relationship influences behavior. I have spoken with victims that avoid the place where the injury transpired, even if it entails walking a circuitous route that adds considerable time to their day. I have also spoken with others that make it a point *not* to eschew the place where they were injured. The latter viewed the spot as a reminder of their grit and survival.

This neighborhood (South Bronx, NYC) had, and continues to have, the nation's poorest zip code. The moment I come across someone born and raised there, it creates an instant recognition and need to compare notes on where they lived, went to school, recreated, and so on. Not all neighborhoods wield a strong influence on self-identity. Understanding the power and influence of community identity in shaping the meaning of a gun injury, such as stigma, determines local representation on coalitions and task forces, for example.

Viewing neighborhood from a social perspective lens increases our understanding of the practical and symbolic meaning of urban gun violence. To this day, I still remember witnessing gun violence, where it took place and the time of day it occurred. When I think about it, I have a visceral response all these decades later. Those gunshots are forever associated with those locations. These feelings are not unique to me; undoubtedly they are shared by countless other urban residents who have experienced gun violence.

Communities with distinctions of being labeled "Chiraq" (Chicago), "Little Mexico" (Kirkwood, Atlanta), "Chopper City" (New Orleans), and where I grew up in The Bronx (Little Korea), which dates me, capture war-like images that stigmatize residents when venturing outside of their neighborhoods. Educators and providers of services cannot overlook children and youth identity development within neighborhood context.

DRIVE-BY SHOOTINGS

It is appropriate to follow up the previous section on place with attention to drive-by-shootings, a social phenomenon that brings another layer to this act and is associated with the urban scene. Drive-by shootings are defined as follows (Dedel, 2007): "A drive-by shooting refers to an incident when someone fires a gun from a vehicle at another vehicle, a person, a structure, or another stationary object." Drive-bys are not a recent occurrence as shown by the research undertaken by Hutson, Anglin, and Eckstein (1996); in Los Angeles drive-by shootings in the early 1990s, they found that 38% to 59% of those shot were innocent bystanders.

The mere mention of a drive-by shooting, a frequent type of mass shooting, elicits intense reactions, assuming a vivid and visceral reaction in urban centers. Not surprisingly, it is associated with gangs, further stressing the role of guns. Gang violence triggers other violence, calling for early intervention to prevent or limit retaliations, which often result in multiple injuries and deaths (Brantingham, Yuan, & Herz, 2020), with innocent bystanders injured or killed. Sadly, surviving is an excellent predictor of a victim perpetrating a gun injury or fatality.

Drive-by causalities warrant more attention because of their prominence in urban gun violence, although there is much to learn on this subject. Dedel (2007) noted that there were no national data on the volume of drive-by shootings, which is the case almost 15 years later. The lack of attention and data must not inhibit raising this topic because it affects daily life routines often taken for granted in non-violence-prone neighborhoods.

Few gun episodes generate the violence drive-by shootings engender because of its randomness, degree of causalities, and the terror it causes, with no location or events exempt, including funerals, as noted by a Chicago funeral worker (Eng, 2020): "The disrespect has reached a different level. It's very scary for us, because what's going to happen at the next funeral? A lot of times people don't realize that they've set a precedent like, 'If I got away with this, then someone else is going to think they can get away with it.' And it puts their minds to doing dishonorable things."

Drive-by shootings occur in high-gun-violence communities and often involve automatic weapons. Solemn or sacred rituals are not exempt, furthering eroding community connectedness. Contextually grounding these shootings takes on greater significance than shootings in other public places, such as playgrounds or street corners, for example. Funeral shootings are social-community indicators that can help researchers understand how gun violence has escalated to a high level within a neighborhood, introducing a new dynamics when sacred events no longer are sacred.

Those killed in such shootings are but one dimension of the problem, but nonfatal injuries to those attending and bystanders are more numerous, casting a wider net on causalities and engendering widespread fear about bringing at the wrong place at the wrong time. Stray bullets broaden the sense of risk because of the indiscriminate nature of this violence (Wintemute et al., 2012), including injuring and even killing very young children and infants, and tearing at the social fabric of these communities because these causalities are unwarranted and preventable.

Drive-bys represent one gun violence episode in an escalating series of violent events (Dedel, 2016): "In these cities, an individual drive-by shooting is often one in a series of confrontations between street gangs with ongoing tensions. Attacks often are followed by reprisals, which are followed by counterattacks. The same individual may come to the attention of police as a perpetrator, victim, and witness" (p. 6). Merging these roles complicates having a simple reason for becoming a gun injury statistic, yet this puzzle needs completion to make a significant dent in this nation's gun epidemic. Drive-bys create a need for specialized researchers to study this act and the injuries resulting from it, aiding our understanding of act-specific interventions. These shootings open the door for innovative interventions, such as ecological changes impeding drive-bys.

CONCLUSION

Contextual grounding plays an instrumental role in social work and other professions in helping us understand urban gun violence occurrence and the circumstances leading to the use of guns. More importantly, this chapter approached gun violence interventions as establishing a foundation using the latest thinking and data. A social perspective on gun injuries allowed for casting a wide net in capturing this phenomenon, helping readers develop an appreciation for a wide gun injury lens.

Grasping the social meaning of guns is essential in coordinating public health campaigns on the outcomes they cause (Galea & Abdalla, 2019). Social workers, for example, are paying attention to injuries in the home when a gun goes off accidentally. Intentional gun injuries generally occur outside of the home, and in urban communities of color, in shootings to resolve disputes or in committing a robbery, for example. Other professions, too, seek collaborative partnerships to address this violence.

5

Cultural Perspective

Guns undoubtedly signify a public health crisis. Too many people die and too
many communities suffer lasting trauma, and often in patterned ways that
public health expertise is designed to address. But scratch the surface, and it
becomes increasingly clear that guns signal a social crisis as well.

—*Metzl (2019, p. 4)*

INTRODUCTION

Gun cultural symbolism has many meanings and can translate into a social cause.
The social context covered in earlier chapters focuses on gun violence as an epidemic, silent or public. However, is there such a thing as a gun culture? I think
readers agree that there is such a perspective, and it wields a prodigious degree of
influence in how we think about this topic and how deeply ingrained it is in this
country (Boine et al., 2020). Guns have many different purposes and meanings
in our society, practical and symbolic (Hunter-Pazzara, 2020), but there is no
denying their cultural significance.

Culture should not be restricted in such a narrow fashion because its power
shapes behavior and crosses all spheres of influence. I am also cognizant of how
culture can be used to marginalize groups of color, such as "the culture of poverty,"
with all of its destructive forces, or "blaming the victim" for their dire situations.
Victim purity captures this latter perspective. Guns have powerful symbolic cultural meanings of what constitutes manhood and power for men, often forming an
integral part of their identity (Green, 2019). It makes sense that the consequences,
too, bring similar meanings, including hiding the ramifications, in the case of
trauma, or eschewing services that aid in the rehabilitation process.

This chapter examines cultural aspects of how gun injuries are manifested,
perceived, and responded to. I have separated out cultural from social perspectives,
although readers may argue they are two sides of the same coin. Nevertheless, a
cultural lens allows prominent introduction of factors on values, which lead to
social ramifications. Spotlighting culture allows understanding deeply held beliefs

The Silent Epidemic of Gun Injuries. Melvin Delgado, Oxford University Press. © Oxford University Press 2022.
DOI: 10.1093/oso/9780197609767.003.0005

that shape human outlook and behaviors, influencing how gun injuries influence victim perceptions and responses and their social networks.

Viewing gun injuries through a cultural lens introduces deeply grounded belief systems, profoundly shaping personal values and experiences. "Southern culture," for example, is not limited to cuisine and celebrations. Louisiana, Alabama, Mississippi, Washington, DC, and Alaska lead the nation, with Hawaii, Colorado, Maine, New Hampshire, and Oregon rounding out the bottom five with the lowest rates of gun violence (Everytown USA, 2020c). The southern region stands out for dangerousness, with a greater propensity to own guns when compared to the rest of the country (Gonzalez, 2020), largely due to historical traditions carrying over to the present day.

Culture also opens the window to various perspectives, such as the meaning and consequences of gun violence and forgiveness. Forgiveness is a complex construct but one worthy of further investigation in healing and rehabilitation (Delgado, 2021; Odak, 2021). Understanding its cultural roots and manifestations will provide helping professions with a window through which to engage religious and spiritual leaders as partners in gun violence initiatives.

This chapter provides insights into a variety of gun violence cultural dimensions that enrich our understanding of how gun injuries are perceived and responded to in our society. Cultural factors, with an embrace of social justice values, introduce a sociopolitical view on gun violence and treatment of injuries. When grounded from a social justice stance, this subject could have been included within a social perspective, but I elected to use a cultural lens by emphasizing belief system roots.

PUBLIC HEALTH AND GUN CULTURE

Culture shapes views of ourselves and others, and shapes interactions, wielding tremendous influence on health behavior. Culture is not the exclusive domain of anthropologists. In fact, it would be a challenge to find a helping profession that does not include a segment of its education to culture in shaping help-seeking responses. Gun violence is but the latest chapter of this education. Social work, for instance, devotes considerable attention to its study. As with the social sciences covered in this section, cultural perspectives shape all aspects of gun possession, use, and its consequences.

Has public health played an instrumental role in advancing understanding of gun culture? This is an appropriate question to ask since this book has embraced a public health stance on gun injuries. Abdalla, Keyes, and Galea (2021) argue that public health has much ground to make up in taking gun culture into account:

> However, public health scholarship has lagged behind in efforts to understand gun culture, which may guide public health action on the gun violence epidemic. This paucity of scholarship concerning gun culture stands in contrast to other health outcomes such as alcohol, in which public health scholarship has focused on the intersection of alcohol policy and drinking culture to formulate recommendations to reduce alcohol-related harm. (p. 7)

Public health's knowledge of culture will be enhanced as other professions join forces, including how best to prevent and treat injuries.

GUN INJURY SCARS AS SYMBOLS

Gun injury literature has emphasized damage to internal organs, called attention to the importance of curtailing blood loss, and highlighted how lives have been dramatically altered. Scars, however, are part of this picture; they are a connection between internal physical damage and emotional responses to this injury, and they serve as a constant reminder of victimhood. In other words, scars bring a gun violence dimension that practitioners and researchers must take into account, highlighting what is probably a deep psychological manifestation pertaining to self-image and playing an instrumental role in any rehabilitation effort.

Scars are the remnants of an injury that can cause a variety of victim responses, in their immediate social network, community, and society. Interestingly, a literature review uncovered a paucity of scholarly material on gun injury scars, even though surviving a physical injury causes scars that may be hard to hide or disguise and therefore wield prodigious influence on self-image and the nature of interactions with others.

Gun injury scars are a distinctive form of scar, and their meaning needs cultural contextualization to grasp their significance. Discussion of gun injuries rarely addresses how one's skin is altered and its significance in society (Barret & Barret-Joly, 2020): "Skin is the largest organ in the human body. Burn injury to the skin can range from being relatively trivial to one of the most severe injuries the human body can sustain" (p. 459). The gender of those with facial injuries, too, brings forth social implications, with women facing far greater consequences when compared to their male counterparts.

A scar's location, size, coloring, and whether it is obvious or hidden, as in the stomach (Zhitny et al., 2020), for example, has social implications beyond physical consequences. This context is multifaceted and incorporates social, cultural, and physical dimensions. Facial disfigurement (scars), however, brings unique reactions (Jamrozik et al., 2019): "People's physical appearance can have a profound impact on their social interactions. Faces are often the first thing we notice about people and the basis upon which we form our first impressions of them. People with facial disfigurement are discriminated against throughout their lives" (p. 117). Although no literature could be located on differences between gender, differences emerged in my conversations with victims and providers, with women, for instance, experiencing greater consequences.

A facial scar is regarded as a severe form of injury, taking on even greater significance with children (Giran et al., 2019). When children suffer facial scars, particularly taking into account the developmental stage when the violence occurred, it translates into a possible lifetime reminder. Eyes, a rather small portion of the body, are the third most exposed part, increasing their subjectivity to gunshots, bringing forth increased likelihood to damage to vision and corresponding psychosocial problems. The complexities of these injuries require use of multidisciplinary teams (Silas & Akang, 2019).

Gun injuries cause gunpowder placement around the skin where a bullet enters, burning the area and causing a tattoo. These scars bring a degree of similarity with conventional tattoos, because depending upon where they are located, both have the capability of generating questions, discussions, and even stigma, and they are greatly influenced by cultural factors and consideration. One is intentional and the other is not. Both introduce a cultural standpoint on how they are viewed by victims, their social network, and society. Tattoos and gun scars, however, can be closely associated when artists perform work on scars to cover them, reducing the stigma associated of these wounds (Sanders, 2019). "Covering" these scars serves practical and symbolic goals.

Scars lend themselves to a multiplicity of health, cultural, and social perspectives. Megan Hobson's response to her gun scars was to see them as connecting with others sharing similar experiences (Shorr, 2017): "Scars are stories. I used to joke and say I wanted scar removal surgery but it's a miracle I'm alive and my scars tell that story. When I first discovered the SHOT project, it was the scars pictured that captured my attention. It's what continues to connect me to people all over" (p. 51). Scars bring a physical and symbolic artifact to the storytelling that can be quite powerful in making a key point for the survivor storyteller.

Turning a physical gun scar into a positive life-altering experience (story) should assume a prominent role in efforts to help victims, providing a course of action benefiting victims and their community. Scars, depending on location and severity, however, can hamper economic mobility because they may scare potential employers and require constant explanation in social situations.

NEIGHBORHOOD KNOWLEDGE

One does not have to be an academic interested in how culture influences human behavior to appreciate the concept of neighborhood knowledge. Neighborhood knowledge (also referred to as "street knowledge") is an essential component of any social work community practice course, for example. The "hood" is the canvas for any painting capturing the essence of a neighborhood. As in any picture, there are bright and not so bright spots, or shades.

We often consider cultural perspective from a highly individualized standpoint, but this concept wields influence when viewed through a wider lens. Neighborhood wisdom, a form of cultural knowledge, is such a lens and helps capture idiosyncrasies of place; seeking to account for distinct local threats can play an immensely important role in helping "outsiders." Calvin's response to noise heightens how this sensor is interpreted within an urban context (Berardi, 2021):

Growing up here, it just sharpens your senses. It makes you more aware of your surroundings—aware of who you're hanging out with, where you're hanging out, what time you're outside at. You always gotta be paying attention, cause a shooting can happen any time. Like if I hear a car skid off, I'm automatically ready for a drive-by. A tire pops, my mind automatically goes to a gunshot. A car's driving real slow, I'm watching it closely, you know. You really gotta be able to identify what's dangerous and what's not dangerous out here, just cause the stakes are so high. (p. 103)

Collective knowledge sets the context for a consequential understanding of gun victims, and when derived from victims, it must take into account their positionality within a neighborhood.

Neighborhood knowledge remains to be systematically utilized with how victims navigate their surroundings post gun injuries. Life changes for victims accessing services. Life within the neighborhood, too, changes and the daily activities are altered, either because of fears about another gun injury, limiting social-recreational activities, or how they fulfill daily requirements concerning shopping, worshiping, or attending school, as with victims still engaged in educational pursuits.

The following section addresses street behavior called "code of the streets," a close cousin of neighborhood knowledge. The latter, too, dictates behavior. However, the code's focus is on a narrow path that is respect centered. Neighborhood knowledge is broader, capturing perspectives and elements intended to help residents navigate their surroundings beyond violence. The code of the streets has a very specific scope and focus on reputation and behavior that can lead to violence, with the use of guns standing out. Neighborhood knowledge, when combined with a strong neighborhood identity, may be arduous to separate and captured by researchers and institutions developing appropriate interventions, including public messages to curb gun violence.

CODE OF THE STREETS

If readers ask me what would be an excellent example of urban culture and gun violence, it would be the code of the streets because of its relationship to reputation and respect. The code of the streets has multiple manifestations. Further, this code is not limited to the United States and has been found in other countries such as the largest Roma neighborhood (Bulgaria) in Europe, with differences taking into account local forces (Kurtenbach, 2021).

A cultural perspective on police cooperation, for example, referred to as "perceived procedural injustice," uplifts the role of a code (culture) of the street that discourages this interaction, introducing a cultural context to a social grounding (Kwak, Dierenfeldt, & McNeeley, 2019). Culture is learned and transpires within a social context. When that context is geographically limited, it reinforces the importance of the local scene.

The nexus of individual norms and environmental influences, such as those in risky urban neighborhoods, represents a potential point of entry for social scientists and helping professions. They can intervene to alter attitudes and behaviors and, with this book as a guide, gun violence leading to injuries (Kurtenbach & Rauf, 2019). One aspect of this code is hypervigilance and the degree of energy devoted to avoiding being a victim of street violence, as well as the internalized stigma from feeling that victimhood is deserved (Singletary, 2020a; Smith et al., 2019).

Anderson (2000) is credited for popularizing this concept with the publication of *Code of the Street: Decency, Violence, and the Moral Life of the Inner City*. This book dispelled the notion that urban African Americans experience violence as a random act—a simplified view of life for this population group. Violence is highly regulated (rules established and enforced) through an informal but well-established code of the street. This code is governed by an ability to project strength (possibly fear) and command respect, dictating behavior in public places and spaces. Violence is a mechanism through which to exercise power and enforce conduct in neighborhood streets.

A bifurcated ("decent" and "street") view of urban families simplifies a picture that is quite complex, with families consisting of law-abiding members alongside those who are not and willing to engage in violent behavior. Further, codes are not static and consist of different levels of beliefs and adherences, with high levels of code adherence displaying greater propensity to engage in criminal and violent behavior and bringing diminished conflict management skills (Erickson, Hochstetler, & Dorius, 2020). Although codes cover a wide range of behaviors, their impact is great when focused on violence (Moule Jr. & Fox, 2021). Further, when a code is discussed, it assumes everyone within that geographical area or entity understands what this code is, disregarding factors such as race, ethnicity, documented status, religious beliefs and practices, gender, and sexual identity, for instance. A code is not the law of an entire neighborhood because no neighborhood is monolithic.

Urban residents of color are cognizant that living by this code can reduce their potential for victimization and allow them to survive violent encounters (Burgason et al., 2020). Maintaining a strong street reputation, which can be enhanced or diminished by social media, becomes part of daily existence in high–gang violence urban neighborhoods, as addressed by Stuart (2020) in an article titled "Code of the Tweet: Urban Gang Violence in the Social Media Age." Gun violence can be the result of slights or revenge for prior violent acts, rather than the conventional gang fighting or warding off infringement on a drug market.

VICTIM COOPERATION WITH LAW ENFORCEMENT

Seeking justice for victims willing to bring charges against perpetrators is an important emotional step in the healing process. It is influenced by a host of factors, including the code of the street. Cultural manifestations of violence and help seeking are present in shaping gun violence in communities of color. The social sciences and helping professions need a grasp of the varied ways race-based secondary trauma shapes perceptions. This requires specific research identifying cultural and historical barriers to community policing.

Bifurcating police gun violence views on race/ethnicity casts them into distinctly different roles—"warrior" (toward African American/Black and Brown people) and "guardian" (White, non-Latinx), with this worldview dictating their behavior and subsequent views of the community they patrol (Carlson, 2020a). Police distrust runs deep in communities of color (Lipscomb et al., 2019), and that was my experience growing up in the South Bronx and remains so to this day. Further complicating the help-seeking process is distrust of institutions such as schools and hospitals. This is the perfect formula for further marginalizing victims—conditions creating high probabilities of gun injuries followed by inadequate care to treat them.

Community police distrust carries over to COVID-19, calling for better police–community relationships in addressing social distancing, providing facemasks, and offering disinfectants to reduce the likelihood of contracting the virus (Southall & MacFarquhar, 2020). This calls for "credible messengers" and "violence interrupters" to broaden their mission from violence prevention to COVID-19 prevention. The challenges of the epidemic highlight the importance of the "right people" intervening in communities.

Violence interrupters bring insights (knowledge), and the respect (experiences) they command in neighborhood streets makes them ideal messengers in antiviolence campaigns. Changing community norms and violence expectations by actively enlisting interrupters/navigators (often consisting of community health providers), and other providers with high levels of trust, is a widely embraced method in stemming gun and other forms of violence (Delgado, 2020a; Taxman, 2019).

Although victim cooperation with authorities can fall into a code of the street category, its significance warrants its own cultural attention. There is often a victim-offender overlap (Roman et al., 2019). The importance of this overlap requires more serious theoretical and empirical attention (Berg & Mulford, 2020). Sustaining a gun injury perpetrated by a known assailant translates into a range of opportunities for a victim response (Felson & Lantz, 2016):

> If victims are angry, aggrieved, or seeking protection, one would expect them to cooperate in the prosecution of their offenders. Their interest in seeing the offender punished is often stronger than anyone's. It is therefore not surprising that the most frequent complaint that victims make about the police is that they were too lenient toward the offender. . . . We know, however, that some victims are unwilling to cooperate with authorities. Victims who will not participate are a challenge for the criminal justice system since complainants and witnesses are often required for prosecution. . . . When victims refuse to participate, offenders are more likely to evade punishment and to think they can commit crimes with impunity. (p. 97)

Victim–perpetrator relationships play instrumental healing roles, with a potential for retribution as a critical act as counterproductive to the healing process.

Knowledge of victim–police cooperation is very limited, which is odd considering the role it can play in restoring confidence in the criminal justice system (McLean et al., 2019). Xie and Baumer's (2019) model helps explain victim help-seeking patterns: (1) understanding the myriad of help-seeking reactions of a victim to a crime incident; (2) considering victimization context, taking into account social networks, organizations, neighborhoods, local governments, and nation-state; and (3) understanding how help-seeking decisions at a given point in time influence future help-seeking actions. The interplay of these arenas shapes help-seeking behaviors.

Police may be first responders to a gun injury scene, rendering critical aid, the first step in a recovery process (McLean et al., 2019). Lack of gun injury cooperation may also mean that a victim wishes to seek revenge, turning one incident of violence and expanding it to others. Anger can be a key motivator for vengeance, with the potential for exposing them to another attack resulting in death.

Obtaining victim cooperation in apprehending and charging perpetrators is contingent on race, with White, non-Latinx victims with a high severity of injury more willing to cooperate with police when compared to victims of color (Hipple et al., 2019). No statistics are available on the percentage of injury cases solved. These statistics, when available, will no doubt parallel those for murders in urban communities, which are low in being solved when compared to White, non-Latinx victims (Cook et al., 2019; LoFaso, 2020; Paddock et al., 2019). Brunson and Wade (2019) discuss lack of community cooperation in police gun investigations and the trust challenges they present with initiatives relying upon a cross-section of collaborative agreements.

FAMILY INVOLVEMENT IN DISCHARGE AND TREATMENT PLANNING

How families view the death of a loved through the lens of gun violence, and when and who can do so, has received important attention (Cook, 2021; Delgado, 2021; Reny et al., 2020). Unsolved murders bring great anguish for families, drawing increased scholarly attention (Altholz, 2020). Similar attention has not been the case with unsolved gun injuries. How to make this painful process therapeutic helps advance the field for families of surviving victims and puts professionals in a better position to provide aid.

An emphasis on saving lives puts the spotlight on saving victims. However, what happens after a life is saved and they go home is a black hole, as noted by Dr. Sanchez, a trauma surgeon (Hersh, 2019): "We pat ourselves on the backs because we saved their lives after a horrible gunshot wound, but what happens after they go home?" Often the circumstances leading to a shooting have not changed. An active campaign involving family in the postdischarge period is self-evident and has heavy cultural overtones, particularly in defining a family. Similar attention to gun injuries is sadly absent, even though these two conditions share many similar aspects and often involve the same professions. Further, we may have a propensity to focus on a direct victim, but as noted, the term "victim" must not be narrowly defined because violence always claims many.

Does it make a significant difference in recovery to have loved ones at a hospital during the time a victim is receiving treatment? A society that is individual oriented means that the number of individuals with the patient will be limited at a time when there should be an extensive network present to aid in recovery and minimize the seeking of revenge shootings. Understanding how the recovery process unfolds provides insights into how families facilitate or hinder the recovery process, and that is no small matter. This social network connects victim and family to communities because of the social systems they belong to or interact with in daily life.

CONCLUSION

A cultural perspective brings us closer to understanding the inner thinking of victims, families, and their neighborhoods. Understanding how culture shapes behavior post injury helps us develop expectations of what factors facilitate or hinder the recovery process, including increasing the chances of avoiding another gun injury. Creating this awareness is always challenging if we are to avoid stereotypes because culture is dynamic and subject to social forces.

6

Economic Perspective

Gun violence imposes heavy social, psychological, and financial burdens on both individuals and society at large. Some of these burdens are known—we learn of the emotional cost from news stories and moving personal accounts, and we have previously calculated the health care costs of treating gunshot injuries. But we know comparatively little about the relationship between gun violence and local economic health.

—*The Urban Institute (2016)*

INTRODUCTION

To comprehend gun violence consequences, we cannot ignore economics. An ability to assess economic costs assists policymakers and organizations in developing cost-effective strategies and more insightful evaluations of intervention strategies. Health problems take a personal physical toll but also an economic one. Gun injuries are no exception.

Society's views on the economic consequences of violence are in desperate need of updating. Injuries, and the legal consequences they bring, are grounded within a legal-medico context, calling attention to how to punish gun violence perpetrators, particularly when focused on an incapacity to work (González & Beauthier, 2020): "Classifying personal injury crimes and their punishments represents a historical challenge that has been subject to wide-ranging solutions. It is clear that there is a weak relationship between the duration of incapacity for work in the victim, be it paid work or everyday activities, and the criminal intent of the offender." The economic consequences of this incapacity are obvious.

Helping professions require grounding in social, cultural, and political dimensions of community violence beyond guns to be effective in carrying out their missions. An economic perspective may appear as heartless. How can we put a dollar figure on an injury or even death? The resources consumed divert attention away from other projects benefiting entire communities, bringing a zero-sum mentality to this dilemma. Injuries bring a human toll and alter a victim's daily life. Part of this toll falls into economic spheres (Chamberlin, 2020).

The Silent Epidemic of Gun Injuries. Melvin Delgado, Oxford University Press. © Oxford University Press 2022.
DOI: 10.1093/oso/9780197609767.003.0006

How can we put a price tag on quality of life when altered by a gun injury? We have no choice. Gun injuries require understanding injury patterns, helping predict long-term outcomes and costs on a variety of dimensions (Radiology Society of North America, 2019).

An economics perspective adds to other injury narratives. Gun injury cost burdens have increased while medical interventions have increased in saving lives (Dobaria et al., 2020), which comes with enormous economic expenses invariably paid by government. Again, there is a call for more research and for the federal government to assume an active role in collecting injury economic data to aid local governments (Chamberlin, 2020; Chaudhary et al., 2018).

CHALLENGES IN ARRIVING AT ECONOMIC COSTS

As with any major social problem, estimating gun injury economic costs requires a model that takes into account direct and indirect costs, as well as the shortening of a lifetime. Readers can appreciate the complexity of such a model. This model must embrace assumptions taking into account injury impact on victims' loved ones because they, too, have economic costs. A violent gun encounter between perpetrator and victim extracts a heavy economic price on both (Sekeris & van Ypersele, 2020), which needs to be taken into account in arriving at economic costs.

Separating economic costs from social consequences is artificial at best, with the following illustration of the killing of Chicago's Pamela Bosley's son broadening our understanding of how murder brings overlooked traumatic injuries and economic costs (Follman, Lurie, & West, 2015):

> On April 4, 2006, in a neighborhood on Chicago's South Side, Pamela Bosley's 18-year-old son, Terrell, was unloading a drum set from a van in preparation for a church choir rehearsal when a man walked up and opened fire on him and his bandmates. Terrell was rushed by ambulance to a nearby hospital, where he died a few hours later. He was one of Chicago's 384 gun homicide victims that year. Bosley describes Terrell, then a college freshman and the oldest of three siblings, as "my outgoing son." He was a starter on his high school football team, had performed as the lion in a school production of *The Wiz*, and played bass in jazz and gospel bands. He'd planned to major in music and tour the world.
>
> The few hours when Terrell clung to life in the hospital cost about $10,000, which was mostly covered by the family's health insurance. Later, Bosley spent thousands of dollars out of pocket on therapy and antidepressants for herself and another family member, who was hospitalized at one point for depression. Bosley also lost several thousand dollars in earnings during a six-month leave of absence from her operations job at a bank. She twice attempted suicide. "I could be okay one hour, then the next minute I could look at something and be broken down," she says. She regained some balance, but when she returned to her job, her coworkers' chatter about their kids, and her memories of Terrell visiting her at work, were too much to bear. She took a job at another bank.

In 2007, Bosley and her husband started a service to help gun violence survivors join support groups. Her second son is planning to become a medical engineer, and Terrell's youngest brother—who was just eight at the time of Terrell's death and used to pray nightly that no one else in his family would get shot—is now a thriving high school junior. But justice has been elusive: In 2008, a man was charged in connection with Terrell's murder, but prosecutors lacked sufficient evidence, and no one has been convicted.

Pamela's family shows the ever-increasing circle of damages caused by guns. Her son was a fatality with the family suffering injuries, too.

Various economic injury estimates embrace multiple factors, including outcomes. There is no escaping that these estimates are educated guesses, and they not have changed very much over the past three decades (Max & Rice, 1993):

> Estimating the cost of firearm injuries given available data really is "shooting in the dark." It is likely that the estimates presented . . . grossly underestimate the economic impact of firearm injuries in the United States. We simply do not know the number of people injured by firearms, and cost data specific to this population are inadequate. Yet these estimates are critical elements in any rational debate of firearm policy. (p. 181)

Although economic cost estimates are flawed, they still provide a window into an arena that increasingly takes on saliency on this issue.

This challenging magnitude has drawn congressional attention. In early 2020, Senator Warren and Representative Maloney requested that the General Accountability Office examine annual medical costs of gun violence (Maloney, 2020):

> Studies of the immediate medical costs of gun violence have found that gun-related injuries cost $2.8 billion in emergency department and inpatient hospital costs each year and that the largest share of these costs—nearly 35 percent—is borne by Medicaid. However, existing studies do not capture the longer-term medical costs of gun injuries—such as readmissions, rehabilitation, long-term care, physical therapy, behavioral health services, personal care, and disability—that fall on American taxpayers.

Paucity of comprehensive economic cost studies for treating gun injuries has compromised understanding of the burden being borne by taxpayers through federal health care programs such as Medicaid and Medicare.

ECONOMIC COST CATEGORIES

It is natural to think of gun injury economic costs involving health care, police, and criminal justice systems. These costs are associated with concrete acts and, as a result, receive considerable publicity. However, other systems are often involved, such as fire departments sending emergency medical technicians (EMTs); this results in local economic costs translating into higher taxes. The cleaning of ambulances after transporting a gun victim to an emergency room, too, has economic costs.

The National Institute for Criminal Justice Reform (2020b) has costs along six categories: (1) Crime Scene Response; (2) Hospital and Rehabilitation; (3) Criminal Justice; (4) Incarceration; (5) Victim Support; and (6) Lost Tax Revenue (sales, state, and local). These costs, based on a Stockton, California, study, translate into each gun fatality costing $2.5 million and each injury costing $962,000 over a lifetime. The impact on local taxes is salient. Hiring armed school guards, for example, rarely gets calculated in gun violence costs. Some costs rarely, if ever, get studied at all, such as the trauma and fears that stunt neighborhood development.

One estimate several years old has the annual cost at $2.8 billion for Emergency Department and inpatient care; when taking readmissions, rehabilitation, and lost wages into account, the price tag went up to $46 billion (Gani, 2017). This figure, when placed alongside treatment of patients with chronic obstructive pulmonary disease, which is the third leading cause of death in the country, illustrates its magnitude. One perspective that may be helpful in understanding the economic scope is the estimate that gun violence translates into every resident of the United States paying $700 per year (Skaggs, 2019).

Another viewpoint often overlooked is how gun violence affects local economic development. Gun violence impacts on local economies can be extensive, as in Minneapolis, Minnesota; Oakland, California; and Washington, DC. One study found an increase in gun deaths in neighborhoods translates into lowering the number of retail and service business establishments, reducing new job creation and local businesses sales (Irwin-Erickson et al., 2016).

These losses are rarely part of discussions on gun violence's economic ripple (Joint Economic Committee Democrats, 2017):

> While more difficult to measure, the indirect costs associated with gun violence are much higher. Lost wages and economic contributions from both victims and imprisoned perpetrators amount to an annual cost of $49 billion, and losses in quality of life are estimated to cost $169 billion. Though a lack of data makes it difficult to fully calculate, researchers have estimated that after accounting for direct and indirect costs, gun violence in America has an annual price tag of $229 billion, over two times the size of New Mexico's economy.

One study of gun violence economic costs over a 10-year period (2003–2013) identified 30,617 hospital injury admissions, translating to an annual rate of 10.1 admissions per 100,000, and average admission costs of $622 million, or $32,237 per admission, with the highest total costs being for unspecific gun type ($373 million) and assaults ($373) (Brunson, 2019). Not surprisingly, 25% of admissions were uninsured, with an annual cost of $155 million. Hospitals rely upon government insurance programs to fund gun injury treatment (Hansen,

2019; Peek-Asa, Butcher, & Cavanaugh, 2017). In cases of uninsured victims, states increasingly absorb gun injury health care costs, as in Maryland (Thurman, 2018).

Another estimate has the average annual cost of treating initial gun injury hospitalizations (2006–2014) at $6.6 billion over a 9-year time period (Kalesan, 2017). A national study of readmissions of children with gunshot injuries found hospitalization costs of over $382 million, with $5.4 million from readmission to a different hospital (Quiroz et al., 2020). This figure does not cover service costs by other institutions, including educational accommodations necessary to continue schooling.

Age- and intent-related differences in the burden and costs of gun injuries treated in emergency departments are not well documented (Kalesan et al., 2021). One study of hospital expenses estimated it costs $3 billion annually (Van Brocklin, 2017). Once a victim moves from an emergency room ($5,254) to in-patient status, costs increase by 1,800% or $95,887 on average. Estimates have gunshot patient hospital follow-up care costing $86 million per year (UPI, 2019). These costs do not factor in those victims dying en route to the hospital, those who refused to go to a hospital after being shot, or costs of subsequent outpatient physical therapy, trauma therapy, or in-home care.

Injury severity brings unique cost levels, as in the case of musculoskeletal injuries. Young children (under 10 years of age) with musculoskeletal injuries (muscles, ligaments, tendons, and nerves) are at increased risk of suffering severe outcomes with increased likelihood of growth disturbances, amputations, and impairments, requiring treatment related to femur injury, peripheral nerve injury, and further surgeries (Boschert et al., 2021). One study of financial burdens of musculoskeletal firearm injuries in children with, and without, concomitant intracavitary injuries found that 42% necessitated long-term follow-up with excessive financial costs (Evans et al., 2020).

The gun injury economic impact in a Louisiana urban trauma center (2007–2013), for example, found 3,617 patient encounters that met study criteria. The financial costs totaled $141,995,682; the hospital experienced a loss of $42,649,938 in nonreimbursable expense. Of the 3,617 patient encounters, 59% required orthopedic consultation, with 25% requiring inpatient surgical intervention, calling for specialized training and orthopedic surgical education (Russo et al., 2016).

A state-level economic viewpoint helps us appreciate the price gun violence extracts. One effort at arriving at costs of gun violence at a state (Minnesota) level found 922 annual shootings translate into $765 million per year in annual direct costs—$32 million in health care, $31 million in law enforcement and criminal justice, $4.5 million in employer costs, and $696 million in lost income (Minnesota Coalition for Common Sense, 2016), equal to 11% of that state's yearly operating budget.

Where can we attain more funds to support the injured, particularly for those who are financially insecure? The call for defunding police departments captures the need to reprioritize local budgets away from policing and raises needs best met through alternatives to policing, including helping those who are contending with gun injuries and the organizations that serve them. Camden, New Jersey, is an example of how novel policing techniques translated into decreases in gun vi-olence and health care system costs (Frisby et al., 2019). Nevertheless, gun injury financial costs, particularly beyond a narrow focus on hospitalizations, necessitate all levels of governmental assistance.

The National Spinal Cord Injury Statistical Center estimates that caring for someone who is quadriplegic (paralysis of arms and legs) is over $1 million in the initial year, with $185,000 for each additional year of life (Casey, 2018; National Spinal Cord Injury Statistical Center, 2017), which can be considerable as lifespans extend through innovations in caregiving. Arriving at accurate costs is challenging, however. Our knowledge of gun spinal cord injuries is unknown and so are the factors increasing the chances of getting this injury (demographics, bullet type and caliber, magazine size, gun type); without knowing this infor-mation we can't effectively treat these injuries (Casey, 2018) or arrive at accurate estimates of economic ramifications.

Focusing on immediate economic costs of gun injury health and social services provides but a brief snapshot of costs because long-term costs can be staggering. The lack of long-term data on initial and postinjury health care visits for gun injuries and associated costs severely limits our understanding of the economics of gun injuries (Ranney et al., 2020).

VICTIM SOCIOECONOMIC STATUS

It is important to have an economic perspective on the primary victims of gun vi-olence, particularly emphasizing economic position in society. Gun victims in the nation's poorest neighborhoods have a 6.9% greater chance of injury compared to their wealthiest neighborhoods, bearing a disproportionate burden, and are least resourced in dealing with the consequences of this violence (The Educational Fund to Stop Gun Violence, 2019).

There is agreement that social and economic inequities are at the root of gun violence for communities of color, calling for investments on improving health, promoting employment and training opportunities, quality housing, and pro-social programs, such as after-school, recreation, and community centers (The Educational Fund to Stop Gun Violence, 2020). Three quarters of gun victims were in neighborhoods with median incomes below $49,250 (Avraham, Frangos, & DiMaggio, 2018). We naturally associate economic means with insecurity indicators.

The relationship between urban food insecurity and gun violence is strong and is an independent risk factor for this violence, requiring further exploration (Smith et al., 2020). Low-resourced communities have few resources for gun in-jury recovery, which can mean losing employment, losing pursuit of formal edu-cation, paying medical expenses not covered by insurance programs, modifying homes where needed, and finding transportation for medical appointments. Transportation may seem less challenging when placed alongside the other factors, but it is not (Richardson et al., 2021).

Family income wields great influence on the chances of a member being shot, showing how money shapes life options. Those with lower household incomes have a higher chance of a gun injury, and those living in zip codes with a median household income at the bottom fourth of the scale ($40,333 or less) account for 52.9% of nonfatal gun injuries (Everytown USA, 2019a). These families have limited mobility out of their neighborhood even when it is dangerous. Residents are stuck in place (Sharkey, 2013), limiting their options in obtaining care for injuries. To understand why this is so requires a broad inquiry into society and its major structural forces.

COST OF HEALTH CARE

US gun violence costs $229 billion annually, or 1.4% of GDP (Joint Economic Committee of the U.S. Congress, 2019). This figure translates into $700 per individual in the country—more than we spend on the consequences of obesity, and almost as much as we spend on Medicaid (Follman et al., 2015). Gun injuries fall disproportionately on those least able to economically deal with the consequences. Households in the bottom fourth of the income scale are at a more heightened level (7 times more likely) of experiencing a gun injury than those in the top quarter (Everytown USA, 2019a). Gun injury rates increase with age, peaking among 20-to-24-year-olds (76.7 injuries per 100,000 people), and decreasing thereafter.

Health care costs influence care provision, which is essential on the road to recovery. There are other perspectives on health costs that get media attention and are covered in this section. However, peeling back the layers, we find there are other costs most of us never consider. Financial costs get reported using different time periods, making comparisons arduous. Nevertheless, an overarching picture sets the stage for why economics is integral to any discussion of gun injuries.

The US financial burden of initial gun injury hospitalizations (2006–2014) totaled $6.61 billion, or $734.6 million per year, with Medicaid ($2.3 billion) and Medicare ($0.40 billion) covering 40.8% of these costs and self-pay accounting for $1.56 billion, or 23.6% (Spitzer et al., 2017). These figures are prodigious, yet they substantially underestimate the actual health care costs according to experts in the field.

Gun violence costs health care systems $170 billion annually, with $16 billion for operations alone based on 262,098 victims requiring at least one major operation and not all hospital admittances (Science Daily, 2020): "The costs for hospitalizations, measured as median costs adjusted for gross domestic product, increased more than 27 percent over the 12-year study period, from $15,100 to $19,200. . . . The amount of time these patients spent in the hospital—a major cost driver—also increased from an average of 7.1 days to 12.6 days." Gun injury costs increased from improvements in transports to trauma centers having victims survive, with one cost estimate at $170 billion annually (Reinberg, 2020).

Race's impact on gun injuries is well established. African American/Blacks hospitalized for gun trauma had lower adjusted odds of in-hospital mortality when compared to White, non-Latinxs. Although having similar hospital lengths of stay and in-hospital morbidity, they had higher hospitalization costs and charges (Peluso, Cull, & Abougergi, 2020b). The difference may be due to severity of injury, existing pre-existing health conditions, and/or a history of not seeking health care prior to hospitalization for gun wounds. Social context prior to hospitalization shapes hospital experience.

One estimate has costs exceeding the $229 billion per year estimate cited earlier, with direct costs of $8.6 billion and over 50% spent on prisons, with the remaining portion devoted to improving victim quality of life; 87% of the costs paid by taxpayers average $700 per American per year (Follman et al., 2015). Even when a gunshot results in death, the expenses are considerable: (1) investigation: $2,200; (2) EMT transport: $450; (3) hospital stay, $10,700; (4) mental health for victim's family: $11,600; (5) trial for perpetrator: $2,300; and (6) perpetrator incarceration: $414,000 (Follman et al., 2015). This figure is considerably higher if the victim survives.

Social scientists often argue that funding violence prevention programs has been limited because these programs lack clearly identifiable stakeholders with a financial stake in interventions. This represents a novel gun violence perspective, particularly at the state level through Medicaid (Coupet Jr. et al., 2018). Gun injury costs are shared between states and the federal government, which translates into localities using tax funds for their share, with taxpayers ultimately paying for injury costs.

Increased survival rates result from a multitude of factors, with improved trauma transport, prehospital resuscitation, medical techniques, patient management, and therapies once admitted (Dobaria et al., 2020). Care fragmentation captures when patients are readmitted to a different hospital from where they initially received care; this is associated with increased likelihood of mortality (Passman et al., 2020). This finding stresses the importance of transporting victims to hospitals with care in the transport vehicle, with level 1 trauma centers standing out in importance once admitted. A literature review of access influences on trauma mortality found multiple factors influencing outcomes, with rapid treatment and transport to an appropriate trauma level setting being the best predictor (Jarrett & Devers, 2020). There is a recommendation to increase standardization of databases on time of transport, which is critical in saving lives and sets a baseline for measuring improvements.

Not all life-saving procedures cost the same, with patients with head-neck, vascular, and gastrointestinal operations falling into the high-cost category. As addressed in Chapter 3, these injuries affect many delicate internal structures when compared with injuries to lower extremities, requiring longer hospitalization and greater number of procedures, translating to increased economic costs and countless follow-up visits for victims, affecting them and their social network.

Saving victim lives increases with new advances, but it does so with financial costs, with increased funding demands being met by other sources. Readers can appreciate the consequences of a complete or partial paralysis. An economic perspective broadens the consequences, however. In 2016–2017, these costs were

$41.3 million, with Medicaid the primary payer for $32.4 million (or over 75%). An average hospital stay costs $180,481. Further, over half of these victims had at least one rehospitalization.

Pressure ulcers with spinal cord injuries (SCIs) and paraplegia resulting from gunshots (23%), for example, introduce a new dimension to treatment requiring specialized attention. Estimates have pressure ulcers among those who are paraplegics, with or without SCIs, costing $13 billion annually, with one pressure ulcer costing up to $40,000 (Chopra, Kaye, & Sobel, 2017).

PARALYSIS AS SUBGROUP

Gun victims who have been paralyzed are a subgroup uplifted for attention, and they are worthy of further attention from an economics standpoint. The economic cost of paralysis, depending on type, can run $1 million during the first year of treatment and $181,000 a year thereafter (Steffen & Harlow, 2014). When talking with individuals in high-gun-violence environments, one can easily get the impression that they have one of two options: It is either death or prison. This is a stark reality of life from a bifurcated viewpoint, but life is too complicated and nuanced to categorize into two camps. Phoenix's Jennifer Longdon's financial costs from a gun injury show the escalating nature of these economic challenges (Follman, Lurie, & West, 2015):

> On the evening of November 15, 2004, Jennifer Longdon and her fiance pulled into a strip mall parking lot in Phoenix to get some dinner when a truck sideswiped theirs and a man got out and began shooting. Her fiance took a bullet through the brain that left him profoundly impaired. Longdon was shot in the spine and left paralyzed from the chest down.
>
> The physical devastation was followed by financial ruin. Longdon, who lost her health insurance shortly after the shooting, has been hospitalized at least 20 times over the past decade. One especially bad fall from her wheelchair in 2011 broke major bones in both legs; she came close to having them amputated and had to have titanium rods inserted.
>
> There was also the $40,000 in modifications to her home, just so she could wheel through the front door, make dinner, or take a shower. And the $35,000 for a custom lift-equipped van (and the steep insurance rates that came with it). "I don't think people understand the way nickels and dimes add up to hundreds of thousands of dollars—millions of dollars—over the lifetime of an injury," she says. There were also the costs that might never be measured: "The loss of innocence for my then 12-year-old child. What it's cost him in terms of not being able to study for exams because Mom might be dying today."

It is fitting to end this illustration with how children's lives are changed when their parent suffers a gun injury. How Jennifer's son's life is altered will in all likelihood never be captured and learned from without research following his development. Lastly, no one was ever called to justice for this crime.

Living with paralysis is rarely mentioned and is considered a life prison sentence because of a loss of freedom. One former gang member paralyzed by a gunshot stated this point eloquently (Ralph, 2012): "They say when you gang bang . . . when you drug deal, the outcomes are either death or jail. You never hear about the wheelchair. I ain't know this was an option. And if you think about it, it's a little bit of both worlds cause half of my body's dead. Literally. From the waist down, I can't feel it. I can't move it. I can't do nothing with it. The rest of it's confined to this wheelchair. This is my prison for the choices I've made."

CONCLUSION

Social scientists and helping professions have examined the human toll guns take on our society and the disproportional effect this violence has on people of color. Examining the economic costs of injuries does not take away from humanistic perspectives. Economics introduces trade-offs between funding the consequences of gun violence and using funds for community programs that address the root causes of inequities.

Section III highlights community-centered approaches and case illustrations concretizing interventions on different injured subgroups. The stories in these chapters are inspirational, concretizing the potential of community and gun victims and highlighting why an embrace of resiliency and community assets must be central to any community-centered strategies. These chapters are humanizing and uplifting. It is so easy to give up on urban gun violence, concluding that there is nothing we can do to prevent them, and that is not the case.

Community-Centered Approaches and Urban Case Illustrations

Concretizing conceptual and empirical material in the previous chapters into two case examples brings immense rewards. This is not to say that what was previously covered is not important because of how it shapes conceptual, research, and practice responses. This section starts with a foundation on community-centered approaches before providing case illustrations, integrating key concepts and research findings.

Unlike case studies, which depend upon extensive documentation, these illustrations ("little gems") are snapshots of innovative projects with appeal to professionals interested in urban gun injuries. Illustrations highlight regional differences and goals, with particular relevance to collaborative practice. The first case examines a program in Chicago to prevent bleeding out. The second focuses on victims paralyzed by gunshots in New York City.

7

Community Interventions and Gun Injuries

> *I am not a victim, but a survivor.*
>
> —*Martha Childress (Shorr, 2017, p. 54)*

INTRODUCTION

Community centrality is manifested in gun injuries, shaping worldviews of who must be at the table in finding solutions to this public health problem. "Community involvement" is a catchall phrase, with many different meanings depending upon the values of those who embrace it and the institutions sponsoring these interventions. Values are like DNAs that interact within a sociocultural context.

According to Paddock et al. (2017), gun violence research calls for community interventions using four approaches to achieve a significant level of success: "Reduce easy access to firearms for people at high risk of engaging in violence. Improve trust between police and communities of color. Increase investment in families and communities at greatest risk of violence. Incorporate community engagement into prevention efforts." A fifth approach is needed. Survivors often struggle finding resources that consider their living situation in helping with trauma, and when available, they may hesitate to utilize services.

Benns et al. (2020) use a historical and place-based lens to further understanding of how redlining and racism, in this case in Louisville, Kentucky, has influenced present-day gun violence. Place-based violence interventions introduce environmental and social contextual changes, a key aspect enjoying wide acceptance by social scientists and helping professionals (Hohl et al., 2019). These interventions increase potential for highly innovative community-centered initiatives and flexibility in how they are conceptualized and implemented.

These interventions include major community stakeholders at the table where decisions are made. For example, in Milwaukee, a vacant lot was converted to a community garden, causing a reduction in crime (violent and nuisance) rates, with violent crime having the greatest reduction (Beam et al., 2021). Another study of remediating (greening or minimal mowing and trash cleanup) Philadelphia's

The Silent Epidemic of Gun Injuries. Melvin Delgado, Oxford University Press. © Oxford University Press 2022.
DOI: 10.1093/oso/9780197609767.003.0007

vacant urban land (541 lots) found a significant drop in gun violence (Moyer et al., 2019).

Chicago's initiative to increase local vacant land ownership resulted in a decrease in crime rates by 6.8%, giving credence to the role of vacant or distressed properties and the role local residents can play in reclaiming their communities (Stern & Lester, 2021). Vacant lots can serve as places for hiding guns (Everytown for Gun Surgery, 2016). Detroit (2014–2017) demolished over 10,000 distressed buildings, costing approximately $130 million and decreasing gun violence (Gardner, 2020).

Consideration of neighborhood greenness (Ceccato, Canabarro, & Vazquez, 2020; Pizarro et al., 2020) has shown promise in determining the extent of gun violent crime in Miami-Dade County, Florida (Moise, 2020). It bears noting that Han and Helm (2020), in a study (2012–2016) of 559 abandoned buildings demolished in Kansas City, Missouri, found no significant decline in crime. Vacant lots are opportunity points in addressing collective trauma and reducing gun violence. We must broaden our understanding of vacant lots beyond urban gun violence to examine how they influence health, for example. A literature review found 78% of the articles showed positive outcomes through the greening of vacant lots (Sivak, Pearson, & Hurlburt, 2021). Dissemination of findings on greening and converting vacant lots is critical in gun violence efforts (Hohl, Stegal, & Webster, 2021).

Finally, McCarthy (2019) introduces urban architecture and vacant lots in addressing personal trauma and community healing:

> Elements that are proven to help people heal will inform the architectural logic of interventions. Spaces for social cohesion will be centralized, anchoring both the building and community. Peripheral quiet spaces will foster reflection. Cloistered rooms for therapeutic services address the effects of trauma directly. Small additions will extrude beyond the existing envelope, providing opportunities for discovery and placing members in worlds outside the bounds of their neighborhood. (p. 1)

Community decision-making in creating these structures is essential.

COMMUNITY ENGAGEMENT

Participatory democracy is often associated with voting and representation, and it has a prominent place in gun injury solutions. This is also a central theme in this book, with this practice and philosophy finding fruition in countless ways. Residents are the experts of their own lives (Akosua, 2019): "And it takes the hood to heal the hood. We know what we need, we are the experts on this situation. This is our life and we need people who resemble us to say, 'Look bro, you don't have to be doing this.'"

Listening and engaging residents in high-gun-violence neighborhoods is not magical, but requires hard and systematic thought and work, creating collaborative initiatives as in Newark, New Jersey (Grossman & Clear, 2021). There are national organizations such as the Coalition to Stop Gun Violence (https://www.csgv.org/issues/impacted-communities/), among others, advocating for community engagement in solving gun violence.

The concept of community engagement creates a common language and space for decision-making on gun violence (Andrade et al., 2021). However, if we ask 20 community engagement experts to define it, we will get 20 different definitions. Readers wishing to have a definitive description of community engagement will be disappointed. Nevertheless, there are broad parameters that we can construct in the gun violence field. Everytown USA (2020b) provides six ways engagement can occur: (1) street outreach; (2) group violence intervention; (3) community-driven crime prevention through environmental design; (4) hospital-based violence intervention programs; (5) safe passage (providing safe routes from school); and (6) cognitive-behavioral therapy. These paths are not mutually exclusive, existing in various combinations and permutations, allowing local circumstances to dictate the most prudent approach.

Pizarro, Zgoba, and Pelletier (2020) found that gun use premeditation and motivation (rational choice theory) is a tool to gain compliance and facilitate the undertaking of a crime. Understanding gun-carrying premeditation and motivation to curtail violence may necessitate a combination of two strategies, with one focused on restricting gun access and the other on interrupting the cycle of violence. Street gang membership increases gun access and risk for victimization (Roberto, Braga, & Papachristos, 2018).

Thompson (2020) captures the typical activities of what community engagement means: "Community engagement involves conversations, sharing of insights and diverse perspectives that allow consensual problem identification among defined constituencies." There are many lessons the past several decades have taught us about community violence, with community engagement at the crux of these endeavors.

Intervention engagement is a value with the ultimate beneficiaries having a vital stake in the outcome, and involving them from the beginning ensures that we plan with, rather than for, communities. The former enhances empowerment and the latter paternalism. The choice is clear between these approaches. Collins (2019) raises caution in rushing to use an engagement strategy without a consensus on what this means for vested parties, particularly from a community standpoint:

Recently, a discussion around the need for more engaged scholarship has emerged within injury and violence prevention as a way to achieve several long-desired goals, notably better translation, uptake, and usefulness for injury research findings and their potential to improve health, safety and quality of life. However, there is a need for a more in-depth examination of the intent, methods and underlying assumptions of engaged research because approaches to engagement are differentially applied and understood across the field. As it is now, conceptions of engagement are headed towards potentially reifying hierarchies of knowledge that place scientific research in a more lofty and respected position than lay public, patient or otherwise "non-expert" ways of knowing. I assert that even while various publics are "engaged" in the conduct of research for injury prevention, these relationships retain the imbalance prevalent in expert/non-expert interactions. To better understand how scientific knowledge is overvalued in engaged relationships, we need to identify and reveal the underlying logics of these interactions as a step towards a more genuinely equal exchange.

A consensus on what engagement means requires that all interested parties have a place at the table, with community representatives at the head. There are many books on participatory research for good reason: There are many definitions of this concept.

Individual-level engagement tells us much about involving communities. Successful youth violence prevention workers, for example, have reduced violence by embracing multifaceted roles working with other prevention workers in collaborating and networking within the community. "Being a presence," emphasizing building relationships, being able to address conflicts and crises, and providing resources and advocacy to both clients and their families, is central to this approach (Free, 2020).

These themes are inherent in any effective community practitioner, highlighting the importance of an affirming and trusting relationship which is enhanced when workers share the background and experiences of the communities they serve, regardless of the age of the consumer of services. Hired staff must emphasize these qualities and bring "street credit," making them empathetic and responsive to the lived experiences of those they serve, facilitating engagement with communities. This quality is not obtainable in formal education.

TRAUMA-INFORMED CARE

What are trauma's long-term consequences? The answers assume prominence in how public health addresses gun violence, with survivors in central roles shaping research and interventions (Dicker & Punch, 2020). Experiencing trauma early in life increases the likelihood of entering into a criminally involved life, although there are protective factors that lend themselves to prevent this trajectory. Victims do not have to suffer physically to be victims (White, 2020). Understanding gun trauma in their lives is just as important as understanding their counterparts who experienced non-gun assaults.

Trauma-informed care rests on four key treatment pillars: (1) knowledge of the effects of trauma; (2) awareness of the signs and symptoms of trauma; (3) importance of avoiding retraumatization; and (4) and creation of appropriate corresponding policies and procedures (Fischer et al., 2019). These pillars need a contextual sociocultural grounding to make them relevant to urban centers and people of color. Trauma treatment engenders hope and desire to achieve a sense of safety, which, in turn, facilitates engaging with a community. Social workers and other professionals are keenly aware that trauma has consequences for victims, their families, and communities (Buchanan, 2014; Joseph et al., 2019; Ohmer et al., 2016).

Drawing on experiences of Philadelphia's African American/Black men recovering from traumatic injury helps identify their vulnerabilities and protective forces, with the latter frequently overlooked (Jacoby et al., 2020b). When we overlook protective factors, we fail to marshal them in the recovery process. Treating trauma among African American/Black male adolescents poses challenges in developing a "complex interweaving of circumstances in understanding the development of this vulnerable group" (Singletary, 2020b). Trauma-informed care requires contextualization.

Strategies diminishing psychological symptoms among African American males, for example, are predicated on research informing interventions that mitigate their distrust of service institutions (Jacoby et al., 2020c), and these "pre-existing" sentiments take on added significance when they face a gun injury and follow-up services. Neighborhood physical conditions, too, compound serving injured urban African American/Black men (Palumbo et al., 2019). Help-seeking pathways among African American/Black male trauma survivors suffering a violent incident will uncover multiple potential approaches influenced by posttraumatic stress disorder (PTSD) and depression, financial concerns, and worries about encountering discrimination (Rich et al., 2020). The barriers are far more consequential than facilitating services pathways, often requiring help from informal sources in postinjury life.

Trauma-informed care ultimately rests within a sociocultural context, and care providers reach out into communities to ensure they meet the needs of patients. A gun injury focuses attention on its consequences, but it is compounded further when a victim has prior untreated trauma. It is important to emphasize the trauma-informed care due to poor postinjury mental health outcomes, as with many injured African American/Black males (Rhee, 2019; Richardson et al., 2016).

It is never too early to introduce trauma-informed care, which is not limited to a few settings since those with traumatic experiences seek services throughout society, including informal or nontraditional settings. Trauma-informed care needs conceptualization along a lifespan. For instance, unborn children with injured mothers have extensive lifetime traumatic exposures because the mothers' injuries influence care seeking during this period and post delivery (Gokhale et al., 2020). Further, successful trauma care systems have conceptualized violence along a continuum, integrating knowledge of social determinants of health (Hameed, Knebel, & Rogers, 2020).

Critiques of trauma-informed care emphasize the need for corresponding social justice–informed responses (Befus et al., 2019; Fleurant, 2019). Further, there is a prodigious need for federal and state governments to create special programs to meet gunshot rehabilitation demands (Amnesty International, 2019), bringing a different set of needs compared to other injuries. Understanding these needs, including the stigma they bring when victims are innocent, shapes responses.

Trauma-informed care is patient centered. Nevertheless, we must avoid conceptualizing trauma from an individualistic perspective, broadening it beyond one victim to families and community. Community trauma bonds create interconnectedness and shared experiences that can be transformative if recognized and tapped, shaping intervention success (Duffey, Haberstroh, & Del Vecchio-Scully, 2020). Community trauma has ascended in importance within urban communities of color. Themes such as gun violence exposure, feelings of hopelessness, and perceived dismal futures call for enhancing resilience and mobilizing community assets (Lowenstein & Dharmawardene, 2020; Opara et al., 2020).

We need the expansion of trauma-informed care to communities; community-trauma-informed care has saliency because every victim has a network surrounding them that is also suffering from trauma. Such an encompassing concept necessitates development of a community-trauma research framework to inform gun violence interventions. Risk factors do not automatically translate into negative behaviors due to the presence of protective factors and community assets (Juzang, 2020). Trauma-informed care requires grounding within this context. Unfortunately, it is too easy to focus on risk factors without paying attention to mediating factors, and this presents a biased view.

Enhancing staff capacity requires investment in organizational training and supervision. Training trauma-center staff by long-time former youth patients focused on the Five Points of trauma informed care (TIC) (McNamara et al., 2021): (1) safety, (2) screening, (3) understanding context, (4) avoiding retraumatization, and (5) discharge planning. It increased trainee competencies in identifying and referring patients with trauma histories that normally would be overlooked for violence intervention services. Youth trainers, in turn, received training on presentation and storytelling skills, which are translatable to other community issues.

Trauma-informed care requires trauma-informed supervision, which must be culturally competent to maximize relevance and outcomes (Knight & Borders, 2018). Supervisors, in turn, need support from their administrators. All organizational levels need to support this goal, including institutions preparing future professionals.

Pathologizing (Johnston, 2020) gun injury victims and racialization of trauma (Armstrong & Carlson, 2019), when taking a victim purity stance, makes it arduous for victims to seek services because their negative experiences will circulate within their communities, influencing others from seeking services. Gun-injured children and youth have narratives to share with their peers, adults, neighborhood, and the nation. Tapping these narratives ascends in importance (Diaz & Shepard, 2019).

These narratives legitimize the pain and suffering they have encountered, giving voice to others experiencing similar fates but without their voices tapped in an organized fashion. Children and youth with gun injuries have narratives to share with their peers, adults, neighborhood, and the nation. Tapping these narratives ascends in importance when children and youth are victims of violence (Diaz & Shepard, 2019). These narratives legitimize pain and suffering, giving voice to others with similar fates but without their voices tapped in an organized fashion. Narrative therapy or story sharing, including poetry, song/rap, drama, dance, and writing, destigmatizes engagement for youth of color rather than conventional talk therapy (Cook, 2015; Delgado, 2018; Garo & Lawson, 2019; Green, 2020).

Royster (2017), for instance, describes the "Doll Project," a community/arts-based participatory research project encompassing art built on community strengths while addressing trauma through building leadership of those most affected by violence, helping to uncover trauma, and providing a mechanism for change without stigmatization. Art projects introduce innovation to community engagement and antiviolence campaigns.

NONTRADITIONAL SETTINGS

There is an increasing number of approaches for identifying and building on community assets on gun injuries. Outreach and education initiatives targeting nontraditional settings, such as barbershops and beauty salons, identify and address gun violence trauma. "The Confess Project," self-labeled as "America's first mental health barbareshop movement" (https://www.theconfessproject.com/coalitions), shows how nontraditional settings, in this case urban barbershops of color, are well positioned to provide comfort and help gun victims and their families.

Barbershops and other establishments are bedrocks within African American and other communities of color from social, cultural, and economic standpoints (Sanchez-Jankowski, 2008; Willett, 2000). These establishments provide a range of supports outside of the home and place of employment (Finlay et al., 2019). Readers may also encounter "natural helping networks," a social network term not necessarily cultural in origins (Ingram & Drew-Branch, 2017).

Nontraditional settings are community places where residents purchase a product, service, or congregate and in the process receive advice or concrete assistance on social and mental health issues in their lives, such as gun violence (Delgado, 1999). Nontraditional settings have deep cultural histories and meanings, bringing culture to aid in their appeal, and breaking down psychological and cultural barriers to service utilization. These settings are often invisible to outsiders, necessitating embracing a premise that all communities have assets. The question being, "Where are they located?" rather than, "Are there any?"

Those staffing these establishments render aid because the community trusts them and they often live in the community, further increasing their legitimacy and potential effectiveness. These establishments, for example, have been recruited to identify and refer victims of domestic violence (Weisberg, 2017) and to help in addressing COVID-19 (Shoptaw, Goodman-Meza, & Landovitz, 2020) and other health issues, such as detecting scalp and neck melanoma (Black et al., 2018).

What makes nontraditional settings attractive in community coalitions addressing gun violence? These establishments address four key pillars of accessibility: (1) geographical/physical location; (2) psychological (nonstigmatizing); (3) cultural (staffed by individuals who share the same economic, social, and cultural backgrounds of consumers); and (4) operational (accessible during periods when the community needs assistance). All four are essential to maximize access.

Chicago is home to major gun violence and innovative community approaches such as the Barber Cease Fire movement (Avila, 2019) that provides free haircuts for children and makes space available for community violence meetings: "In Chicago, professional barbers are exercising their basic rights, and contributing to their communities by creating a space for human interaction outside of their shops. The new era of American barbers preach longevity to the next generation, and all it takes is one haircut." A focus on male youth takes on added importance because gun violence disproportionately affects them. Barbershops, however, are not restricted to a particular community issue or concern (Thompson, 2017).

Barbers, as with Mr. Dow, may be gun victims themselves, too. They are often tied to gun violence in their communities and are expected to be turned to for providing the final haircut to victims of gun violence (Giles, 2020):

> Mr. Dow was 26 when he performed his first haircut for a dead client. In that case, it was an older man who had died of natural causes; circumstances that Mr. Dow said were much easier to manage than those of a shooting victim. He has continued to take on the difficult task of providing haircuts for clients who have been killed, for what he considers a straightforward reason —"because I cut their hair while they were alive." As his business has expanded, Mr. Dow has hired other barbers who have learned the trade of post-mortem hair cutting . . . Mr. Dow does his best to maintain that sense of safety. He tries to comfort his clients by talking to each of them while cutting their hair, even the young man who had been shot in the head. "I was talking to him while I was cutting his hair, like I do a lot of my deceased clients," Mr. Dow said. "I just said, you know, 'I hope you rest well.' "

These examples form a basis for expanding membership of community–provider partnerships and coalitions on gun violence, with barbershops and beauty parlors integral to a community's social fabric.

COMMUNITY ASSETS

A community assets approach, sometimes called "ground up," eschews overreliance on law enforcement because of how it stigmatizes residents and invites further affliction by the police (Byrdsong, Devan, & Yamatani, 2016). It stresses residents as instrumental parts of the solution. We can fixate on intergenerational and community trauma; however, we can also highlight intergenerational and community wisdom, strengths, and assets.

Demonizing gun perpetrators as violence prone should be avoided because of heavy racial overtones (Austin et al., 2019). Labeling these individuals as such moralizes a public health problem, making arriving at the best solutions arduous without emphasizing punitive approaches. Although readers might take issue with the fact that gun violence perpetrators often cause death, it is important to remember that they, too, are victims (Barnes, 2020).

We must focus on gun violence without losing sight of assets (McCoy, 2020):

> Urban communities are not simply bastions of violence. They are communities rich with neighbors who care about and support each other, and grassroots organizers that create and implement programs to interrupt and prevent violence. However, their efforts cannot be the only or primary response. We must remember that gun violence is never normal and it affects each of us. We must also acknowledge survivors and respond to their needs. We can no longer afford to ignore such a pervasive problem.

Developing localized gun violence incident maps for crafting interventions does not preclude identifying assets in the process to draw on indigenous resources (Castro et al., 2018).

Gun victims see a role for outside entities addressing this problem, but communities must lead these initiatives rather than be "junior" partners in these endeavors (Ndikum, 2018). Community stakeholder perceptions of gun violence causes must shape interventions (Garrett, 2018). Embracing community assets, however, does not translate into having the community solving the problem on their own.

Although the term "stakeholder" is popular in the literature, including the gun violence field, much research is needed to better understand this influence on gun violence outcomes (Scott, 2020):

> There is not a broad literature on the relationship of stakeholders to gun crime and gun violence. Stakeholders can encompass any individuals or groups with a specific interest in this topic. This can include political figures, community members, health professionals, and a variety of others. Despite the fact that these groups are often concerned with the state of gun crime and gun violence, we know very little about their influence on the outcomes. The research for which stakeholder involvement has been most obvious is in intervention attempts. (p. 3)

Moving from personal spaces of pain to collective organized spaces of healing is an important journey goal all communities must embrace (McCarthy, 2019). Unfortunately, there is agreement that access to gun trauma services is limited in availability (Stolbach & Reese, 2020). Although there is an extensive network of therapeutic services, no comprehensive system exists on survivor needs (Green, 2017), particularly in high-gun-violence communities.

A study of Philadelphia African American/Black male gun injury survivors uncovered five major themes (O'Neill et al. 2020): "(1) Isolation, physical and social restriction due to fear of surroundings; (2) Protection, feeling unsafe leading to the desire to carry a gun; (3) Aggression, willingness to use a firearm in an altercation; (4) Normalization, lack of reaction driven by the ubiquity of gun violence in the community; and (5) Distrust of health care providers, a barrier to mental health treatment." The theme of distrust is problematic for reaching survivors of color with histories of distrusting institutions (Saint-Hilaire et al., 2020).

Miami-Dade's "Walking One Stop" brings together helping professions within a high-gun-violence community (Green, 2016):

> Rawlins [coordinator] said that's why he invites community agencies that serve this population to meet the residents where they live. "Many providers get stuck in their cubicles thinking that someone will actually come to their cubicle for services and it's not happening," he said. "When they come out here and they see face-to-face residents that are traumatized by the violence, it reinvigorates them."

Aiding agencies to navigate entrance is an important function that can encourage these entities to become part of coalitions.

High-gun-violence communities require an extensive network to treat survivors and their trauma (Giordano, 2019). These networks have differences and similarities based upon local circumstances. Developing caring cultures is a step in developing community-centered systems of care (National Academies of Sciences, Engineering, and Medicine, 2019; Wojtowicz et al., 2019).

Gun injury interventions require focusing on subgroups (Bernardin, Moen, & Schnadower, 2021). We must not forget those who witness violence. Boston Medical Center's (2020) Child Witness to Violence Project provides therapy, advocacy, and outreach to children witnessing violence within their homes and community through a multidisciplinary team representative of and cognizant of how racial factors impede quality services. These projects need replication in nonhospital settings, with requisite modifications. Pinpointing active violence areas and time periods is a step in marshalling resources to address gun violence (Bowen et al., 2018).

This book emphasizes increased investment in and participation of victims, families, and their communities in constructing locally based approaches. Calhoun (2016) aptly captures how gun victim narratives must find a constructive outlet that validates them and their experiences:

> There is a story behind each bullet that impacts these youths, each with the potential to alter life as they knew it. As healthcare professionals, we have a responsibility to take a stance—to ensure these young people and their caretakers are equipped to return to the community with the best possible outlook on moving forward with a fulfilling, quality of life. (p. xi)

Courses can reflect narratives and circumstances that increase the relevance of the instruction, such as communities that do not have English as their primary language.

Gun violence defies narrow conceptualization between individuals but rather as a form of community violence (Gavine, MacGillivray, & Williams, 2017). Gun violence has a collective cause and consequence. Thinking of it from a broader community standpoint translates into killings, physical injuries, and exposure to violence, reducing the quality of life for victims, their families, and neighborhoods.

Embracing a community capacity-enhancement stance counters loss of a basic sense of physical and psychological safety, leading to increases in family and community functionality (Range et al., 2018).

Helping professionals often bring decades of experience in community work, opening the door for substantive contributions to this field of study. Urban communities are tired of having an ambulance arriving at a shooting scene and having this vehicle transformed into a hearse, or waiting an excessive period of time, as mentioned by one victim (Amnesty International, 2019): "I got shot three times. Three different times I mean. The second time I got shot, I drove myself to the hospital. My people called the ambulance and 20 minutes passed . . . I took my car and drove myself to the hospital. I knew I couldn't wait" (p. 24).

Ambulance crews play a significant role in transporting the injured from the front lines of the urban gun violence scene (Seim, 2020). A study of wounds causing fatalities found minimal differences in wounding patterns between urban and public mass shooting events when examining organs damaged, calling for rapid point-of-care transport to trauma centers and provision of care at the time of injury. Speed is essential for minimizing the consequences of a bullet. Van Brocklin (2018a) states: "Surviving a shooting, or losing a loved one to a bullet, can result in a lifetime of struggle. Especially when new horrors play on a loop." Health care delayed is health care denied.

Treating gun victims at the scene has taken on new urgency, creating innovative community-centered approaches. Intervention at a gun injury scene can mean the difference between life and death, representing the most propitious intervention point, other than actual prevention. We, as in this book, highlight the consequences of violence exposure within communities. However, ambulance staff, too, face consequences of direct violence and exposure (Cenk, 2019). There is a call for more research on law enforcement provision of care prior to EMS arrival and how best to support this life-saving endeavor (Klassen et al., 2018).

Transport speed from a shooting scene has ascended in importance. ShotSpotter technology, for example, can speed police and EMS transport of gunshot victims to hospitals, advancing a law-and-order agenda of apprehending perpetrators (Goldenberg et al., 2019; Hai, McKenney, & Elkbuli, 2020). This technology is in over 100 cities, but its success is mixed. In Minneapolis, this technology has existed since 2006, with some council members advocating its expansion into areas previously uncovered. In St. Louis, researchers concluded it has been successful. ShotSpotter has been particularly effective during evening hours and during the winter months when streets are not congested (Nelson, Sawyer, & McKinney, 2019).

Gun violence emotional trauma has ascended in importance (Wang, 2020). Helping professionals are acquainted with grief support for families following the death of a relative (Cooper, Stock, & Wilson, 2020; Delgado, 2021). Youth of color in school, for example, often confront an institution that is unable to address grief, but rather supports learning through order and structure. School, a prominent community institution in their lives, is incapable of addressing experiences and emotions on gun violence (Silva, 2019). They cannot afford to have other prominent institutions ignore grief. We must extend this work to injury survivors to help them and their social network deal with profound feelings of sadness accompanying this violence. Family support groups for those losing a loved one to guns are more commonplace; similar support groups for surviving victims, however, are not.

Community groups with outside support can undertake strategic initiatives. Involving residents as first responders, as discussed later in this book, for example, offers great promise for engaging helping professions in community-centered and participatory projects. Gun injury interventions must extend beyond starting at a hospital (Reed, 2019) and not be limited to health care professionals. Survivors are uniquely positioned to share their experiences and help gather information on how this experience has shaped their daily lives and dreams. There is no substitute for the "lived experience" in conceptualizing service delivery.

Taking a resiliency and community asset perspective on gun injuries opens up immense intervention possibilities, including affirming and participatory messages and actions. Martha Childress took a tragedy and converted it into a strength (Shorr, 2017): "I think what makes me strong and what makes me an inspiration is that I am a survivor. Being a survivor of gun violence doesn't make me weak or fearful, it makes me feel powerful. I got shot and survived a bullet. Most people don't live to say that, but I did" (p. 51).

Resiliency and asset approaches bring added dimensions to these efforts— namely, an uplifting stance that helps balance sad and tragic situations with a positive and affirming vision. Steven Smith, paralyzed when he was 16 years old, assumed a role as an anti–gun violence advocate, establishing a website (https://www.blessed4greatness.com/) to reach others facing similar challenges:

I am paralyzed from the shoulders down, also on a ventilator. The road has been long but I'm still standing tall even while sitting! In life we have ups and downs, but the way you move on is with perservence, strength, faith, and love. That in all aspects helps you deal. I believe I was given a second chance and I do not want to waste it. You should never want to waste the 1st chance, but if you get a 2nd one definitely do something positive with it. Everyone is put on this earth for a reason and we never know what the reason is but whatever you do, do it to the best of your ability and I think that will be good enough for God in my estimation.

Steven's second-chance attitude incorporates many of this book's themes, including a strong spiritual element. Further, it is significant that he had a transformative experience: he turned a horrific gun injury into an opportunity to become a community asset.

Ralph's (2014) book *Renegade Dreams: Living Through Injury in Gangland Chicago*, although focused on gangs and violence, introduces innovative community-asset interventions: "I saw how injury could be crippling, but could also become a potential, an engine, and a generative force that propelled new trajectories" (p. 17). Voisin's book (2019) *America the Beautiful and Violent: Black*

Youth and Neighborhood Trauma in Chicago presents the challenges communities of color face with violence and inequities, and the importance of communities in shaping solutions. Finally, Juette and Berger's (2008) *Wheelchair Warrior: Gangs, Disability, and Basketball* tells the story of Melvin (coauthor) and his use of athletics after being paralyzed by a gunshot, illustrating how this incident was a life-changing motivator.

Stories of the gun injured are rarely newsworthy, and this viewpoint has been lost to the public. Additionally, it has been lost to the scholarly field, too, including involving them in conceptualizing and helping to implement initiatives. This has skewed our view of injuries, as with Corie Davis (Turner & Wise, 2019):

> Corie Davis has been in a wheelchair since Aug. 30, 1999, when a physical altercation ended with a gunshot. "When the bullet hit me in my neck, I just went to the ground," Davis said. "My eyes closed and I went to the hospital, woke up [and] I was paralyzed." In the 20 years since, Davis has become a de facto mentor to many of the young men, including Gay, at the NRH's Urban Re-Entry Group, a support group for people with disabilities who are survivors of violence. "You just gotta take it one day at a time, and do what you gotta do to make yourself happy," Davis said. "I mean, you're going to have your ups and downs, but it is what it is."

Cory Davis has spent more than half of his life in a wheelchair, and his story has profound life lessons for utilizing a strength perspective. He has shifted roles from a victim to a helper, tapping the power of resiliency (Wise, 2019).

Postinjury life for survivors ascends in importance in this book and can be conceptualized as posttraumatic growth. The concept of posttraumatic growth has experienced widespread appeal because it taps a strength-resiliency set of values and theory that offers hope for survivors of violence. This concept has roots in personal tragedies, with gun violence taking on prominence for victims and communities, and finding a home in many of the affirming and empowering concepts covered in this book.

Recognizing the potential for survivors (posttraumatic growth) to join in the fight against gun violence offers great promise for community projects (Sanchez et al., 2020). Injury victims who survive near-death experiences have an opportunity to help themselves and potentially help other victims, too (Haider, 2017). Providing support in fulfilling this role requires an embrace of values, concepts, and a corresponding vocabulary that affirms and facilitates empowerment and participation in this journey.

The call for engagement in gun violence prevention initiatives is not limited to the usual suspects of helping professions, with the American College of Surgeons (2019) also advocating this position, illustrating the increased recognition this public health problem has caused:

> "*Engage* firearm owners and communities at risk as stakeholders to develop firearm injury programs." Many in the field (Bieler et al., 2016, p. viii) share

this position: "Improving community engagement in violence prevention is an immediate reform opportunity. Social service agencies, religious institutions, law enforcement, and other stakeholders in all three cities reported being constrained by inadequate communication and collaboration. Some reported that existing coordination and data collection is not often sustained, and that decision makers sometimes develop and implement plans in ways that are disconnected from the larger community."

The American College of Physicians' call for a public health approach with physicians assuming public advocacy roles to prevent gun violence (Cook, 2018) stresses this voice in shaping public discourse.

Case illustrations humanize statistics, but they often do not capture the potential of survivors to contribute to society. The case of Sherman Spears exemplifies this potential for taking a tragic experience and converting it to an opportunity for helping others while hospitalized. His efforts transformed the period of hospitalization for communities and helping professions everywhere (Van Brocklin, 2019a):

Sherman Spears was shot at a friend's apartment complex in East Oakland in 1989. Someone from the neighborhood had been seeking revenge on his friend, but found Spears first. He was struck by three bullets and hit his head as he fell to the ground, leaving him paralyzed. As the 18-year-old lay in a hospital bed recuperating, he felt disoriented and scared about his future. His parents were distraught, his friends wanted revenge, and he didn't feel like he could relate to the doctors and nurses. "I really didn't have time to process what I was going through," Spears recalled. "There was nobody I could really talk with about what was happening to me." While Spears adjusted to his new life, the experience of feeling lost in those first days remained vivid. A doctor connected him with Youth Alive, then a fledgling violence prevention organization in Oakland. With encouragement from the organization's director, he began going to the hospital and meeting with other young victims of violence. He talked to them about what they were going through, physically and emotionally, and let them know what to expect. Before leaving, he'd give them his phone number and tell them to call any time. The idea that Spears pioneered, of mentoring and supporting gunshot and stab-wound victims immediately after they are injured, is today known as hospital-based violence intervention.

Spears's embrace of a mentoring role tapped his experiential expertise in service to community. This empowered him and introduced an asset stance on an important public health issue.

Youth mentoring programs introduce older youth and adults into victims' lives in a nonstigmatizing manner (Wiley, 2020). Tapping youth lived experiences with violence is a patient-centered approach for intervention (Wical, Richardson, & Bullock, 2020):

Survivors recognized the need for male role models who understand their circumstances, as a dearth of suitable candidates may result in higher rates of violent injury. The shared lived experiences of successful role models are invaluable in providing survivors with the ability to discuss their mental health symptoms and desire to change. One survivor stated, "When you find a role model that been through what you been through, and was in the streets how you was, it be like, I can take some knowledge off of him. I can process it through like, yeah, he right." (p. 2)

It is impossible to advance this field without systematically obtaining knowledge and insights from survivors, regardless of their demographics and circumstances.

Comprehensive understanding of the causes of increased death after gun injuries leads to decreases in deaths and increases in survivors with profound injuries (Kent et al., 2017). We must broaden perspectives for developing interventions to save victim lives, with reducing distance and time to trauma centers in transporting victims increasing survival chances (Friedman et al., 2021). California, for instance, experienced a decline in gun fatalities during the 2005–2015 period, but injuries increased by 38.1% (Spitzer et al., 2020). Research shows that when police transport the violently injured to hospitals, survival rates are enhanced, calling for expanding this emergency service (Jacoby, Reeping, & Branas, 2020a).

Urban trauma care is not limited to health professionals. Police, too, are part of this team, and if acting in a respectful and caring manner at the scene and throughout the transport to a health setting, are well positioned to seek justice for victims (Jacoby et al., 2018). Conceptualizing police in this manner expands the team of gun injury providers, giving new meaning to "protect and serve."

Police perform gatekeeping roles in victim support services, calling attention to how victim–first responder interactions influence help-seeking behavior. Roman (2020) found police–victim interaction as biased by limiting access to victim services; this raises the importance for police implicit bias training, rights policies, and practices to assist gun victims in accessing social, legal, and support services. Roman et al. (2020), in another study, had far more positive findings. This study of police engagement of street crime and care provision found that prior arrests did not influence victim service utilization. However, police involvement did increase accessing victim services, which translated into receiving mental and health services. This finding is important because it provides a path to accessing services and publicizing services in multiple languages.

Access to police records when involving police shootings causing death or injury takes on added importance when families wish to gather detailed information on these incidents. Helping families obtain access is an important task that advocates can assist with (Powell, 2020), and it is considered an instrumental part of the grieving process. This task may seem rather minor. However, obtaining clarity on this type of police incident aids families in this emotionally arduous process.

Milliff (2019) studied emotional (information and cognition) responses to gun fatalities in Chicago with potential for application to gun injuries. This research focused on finding an association among victimization, anger, and retribution, concluding that the responses taken were influenced by the clarity of information victims have about the identity and motive perpetrators had for the violent act, and an understanding of their circumstances, mitigating a perpetrator's responsibility. Uplifting the importance of context must be purposeful on the part of providers.

Bringing loved ones into these meetings aids the recovery process and reduces the likelihood of seeking retribution and perpetrating the cycle of violence. Surviving a shooting is a crisis intervention moment, as noted by Kenyatta Hazelwood, Howard University Hospital Trauma Program Director (Spearman, 2020): "Sometimes it's a wake up call, and that wake up call is the impetus to them getting the help that they need that we have here at the hospital or whether it's through our violence intervention program or through their realization that they could have died from their injuries."

Chances of dying on arrival at an emergency room due to a gun wound are greater within urban communities of color when compared to White, non-Latinx communities (Sakran et al., 2018). Travel time to trauma care is positively associated with gunshot fatality, introducing spatial and temporal dimensions (Hatten & Wolff, 2020). Providing timely intervention at the site of a shooting necessitates timely transport to designated trauma centers to reduce mortality rates. We also need to pay closer attention to travel time from a variety of perspectives, including seeking follow-up services for treatment of injuries.

Increased gun injury prevalence requires better regional coordination among hospitals, trauma care systems, and assessments enhancing prehospital triage systems (Lale, Krajewski, & Friedman, 2017). Interfacility transfer of victims to level 1 trauma centers causes lower mortality in undertriaged victims, spotlighting the importance of patient volume and experience with gun violence (Renson et al., 2019). However, a Utah study of hospital type and patient transfer with vertebral fracture and spinal column injury found them with lower mortality rates but poorer health outcomes, with corresponding lengthier stays and costs. These results introduce a need for nuanced perspectives on what "success" signifies and the need to improve coordinated care of transferred patients and provision of surgical treatment and rehabilitation (Sherrod et al., 2019).

We cannot disentangle race from geography in urban trauma care, with a study of Chicago, Los Angeles, and New York, finding that African American/Black majority census tracts make them the only racial/ethnic group presenting consistent inequities in geographic access to trauma centers (Tung et al., 2019). An imbalance in trauma center distribution in Chicago, for instance, creates inequities in accessing trauma care within that city's trauma systems, with the location of an adult trauma center on that city's southeast side playing a significant role (Wandling et al., 2016). Detroit has one of this nation's highest per-capita homicide and violent crime rates (Clery et al., 2020), and there is an urgent need for faster victim transport procedures (Circo, 2019). Trauma centers are called on to address violence more broadly, introducing a prevention role (Scarlet & Rogers Jr, 2018).

Trauma centers have ascended in importance in treating gun injuries, with injury recidivism rates of 55% requiring interventions to stop or significantly reduce reinjury (see Chapter 7). These centers also draw on advances in reducing the incidence of cardiovascular disease and cancer, allowing identification of risk and protective factors that are public health hallmarks (Dicker et al., 2017). Trauma centers, too, are offering intensive case management to prevent retraumatization or possible death from guns, such as Washington, DC's Medstar Washington Hospital Center (Goldman, 2020):

> To do that, Wiggleton and his coworkers effectively befriend those clients, becoming their de facto support system and helping them take care of their most immediate needs from getting an insurance card, a driver's license, or their GED to making a follow-up appointment at the hospital. Their model is intensive, hands-on, individualized case management. The result, if done right, is an improvement in a host of risk factors that should reduce the patient's risk of reinjury. (p. 556)

Trauma center ascendency in importance cannot ignore community-based initiatives, which further the cause of saving gun victim lives.

Lack of connection and trust with hospital staff, as with violence prevention specialists, is a barrier in reducing violence. Having male role models, due to the high percentage of male victims, is an additional dimension to this team that increases the chances of success in reducing violence (Wical, Richardson, & Bullock, 2020). Recruiting staff from local communities is also a step to increase effectiveness. Ideally, these staff will bring personal gun injury experiences to enhance their credibility with the victims.

Engaging youth victims in specialized corps in hospital emergency rooms (screening and intervention) has been advanced in UK hospitals to reduce repeat attendances, aiding attending physicians in conveying "teachable moments" after admission (Jacob, Travers, & Hann, 2021; Wortley & Hagell, 2020). These individuals bridge hospitals and communities, facilitating transition for victims by aiding them to adjust and meet the challenges of living with such an injury. A variation on this model has support teams following patients back to their communities, rather than remaining institution bound. Their effectiveness is enhanced when reflecting a victim's racial and ethnic background, bringing knowledge of their communities in addition to negotiating institutional resources for the recovery process.

STOP THE BLEED

Blood is symbolic of life, and any campaign to save human life embracing blood as a key element brings saliency and symbolism to gun violence campaigns. We have urban communities across the country where it is not uncommon to have gun victims bleed out prior to help arriving, necessitating a coordinated response to prevent this from occurring in the first place and an enlistment of a wide engagement of people and professions in stopping it. Further, this approach lends itself to involving the citizenry and professions not normally associated with life-saving missions.

Uncontrollable bleeding is the primary cause of preventable death among trauma patients (El-Menyar et al., 2019). A 37-year review of gun violence and youth found their death rates steadily increased during this period (Manley et al., 2020), and the same can be said for nonfatal injuries. Estimates have over half of gun fatalities die at the scene of the shooting, calling attention to provision of aid at the scene of the violence (Carmichael et al., 2019). Untreated gunshot trauma can cause dying within minutes (Carman, 2019), increasing the importance of a rapid response at the shooting scene. It takes 5 to 8 minutes of bleeding out before a gun injury causes death, highlighting how every second and minute counts (Wetsman, 2018). All age groups benefit from compression and tourniquets (Friedman et al., 2019).

The American College of Surgeons Committee on Trauma (ACS COT) recognized the importance of bleeding control at the scene of injury and advocates for interventions stressing this goal (Stewart et al., 2017). Calling out physicians to be proficient as community first responders (Lewis, Carmona, & Roberts, 2020) is applicable to other professions.

Controlling bleeding at the scene has resulted in use of tourniquets, which are effective for treating extremity hemorrhages and result in few health complications (Beaucreux et al., 2018). Expanding the public arena for rapid point-of-injury hemorrhage control is highly advised based on battlefield experiences and the Stop the Bleed campaign (Goolsby et al., 2019). Although this recommendation is due to mass shooting incidents, similar efforts are applicable in high-violence urban areas.

Trauma care's primary objective is to prevent death from hemorrhages. The use of tourniquets, or combat application tourniquets (CATs), is a life-saving technique used at a shooting scene to prevent bleeding out (exsanguination) for children and adults (Harcke et al., 2019). Ironically, tourniquets use was de-emphasized after the Civil War because it was associated with gangrene and amputation. Many communities where this method has applicability are similar to war zones, requiring war-related responses to save lives, which is how urban neighborhoods often gets portrayed by media in reporting violent crime (Baranauskas, 2020).

The origin of this national campaign is traceable to the Hartford Consensus (April 2, 2013), a gathering of leaders from multiple health disciplines held by the American College of Surgeons. This gathering resulted in a published document that stimulated discussion, leading to strategies increasing victim odds survival (Jacobs et al., 2013). The Hartford Consensus promoted an integrated response (THREAT): (1) Threat suppression; (2) Hemorrhage control; (3) Rapid Extraction to safety; (4) Assessment by medical providers; and (5) Transport to definitive care.

Active violence incidents have increased throughout the country, necessitating innovative prehospital trauma care (Bobko et al., 2020). In 2015, a National Awareness Campaign established by The White House called "Stop the Bleed" set a goal of starting a grassroots movement encouraging bystanders to obtain training, become empowered, and equipped to provide aid before paramedics arrive at the scene. High mortality rates of penetrating injuries, such as those caused by a bullet, make it imperative to have victims diagnosed quickly and treatment initiated immediately. That urgency translates to providing assistance the moment a gun injury occurs.

One out of every five teens (15 to 17 living) in Chicago's South and West Sides has witnessed a fatal shooting, according to the University of Chicago Center for Youth Violence Prevention. From September 2011 until 2018, nearly 1,700 children under the age of 17 were shot in Chicago, and 174 were killed. Bystanders are uniquely positioned to provide life-saving aid until emergency personnel arrive at the scene, resulting in a sixfold mortality reduction in victims with peripheral vascular injuries (Teixeira et al., 2018). A youth-led anti–gun violence organization, as in Chicago, implicitly states that solutions are found among youth rather than adults (Delgado & Staples, 2008). This does not mean that adults cannot be involved.

A Baltimore study found gun victims willing to participate in Stop the Bleed programs, illustrating the potential of this campaign to reach into high-violence communities (Van Winkle et al., 2021). Wiard's (2018) literature review on bystander interventions on gun violence to reduce mortality concluded that these interventions are successful in decreasing morbidity and mortality.

This campaign's primary goal, called by various names, is to equip major public spaces for rapid point-of-injury hemorrhage control (Fredenburg & Warner, 2020; Goolsby et al., 2019). In addition, it trains a cadre of citizen responders throughout the county on basic life support (BLS), cardiopulmonary resuscitation (CPR), and basic first aid, including other life-saving training courses (Zarilli, 2019). Tourniquet efficacy interventions (e.g., CATs) are effective for adult victims. However, the success of arterial occlusion on preschool-aged children remains to be tested. Kelly et al. (2020) found this intervention effective for those with severe hemorrhage, expanding this life-saving approach. We must remember that infants can be victims of gun violence, too.

Empowering bystanders to aid gunshot victims, as in Chicago's Trauma Responders Unify to Empower (TRUE) Communities Course, supported by Cure Violence Chicago, the Chicago Committee on Trauma, and the Chicago Metropolitan Trauma Society, utilized a health belief model, with particular relevance in urban high-violence communities. This program had more than half of the trainees witnessing a shooting, and over a third witnessing a fatality.

These interventions increase self-efficacy, with implications beyond enhancing skills and bringing emotional help (Tatebe et al., 2019):

Within at-risk communities of Chicago, high school students are not spared exposure to the pervasive violence. We observed they are less likely to receive first-aid training before exposure to violence. Our course successfully improved self-efficacy and knowledge of trauma first-aid among Chicago's high school students, preparing them to be Immediate Responders.

Other cities have instituted similar programs.

Although Stop the Bleed focuses on bystanders, it can reach health professions, protection officers (Fagel & Benson, 2020), pharmacists (Moton et al., 2020), and school personnel (Ciraulo et al., 2020), among others, because of the strategic positions these professionals occupy within urban centers and their potential to save lives:

> Turning bystanders into first responders is a concept based upon a recognition that injuries can be serious but survivable, making rapid treatment essential. (Williams, 2019). There are specific areas within neighborhoods that experience heightened levels of gun violence, calling for greater research and tailoring of interventions (Magee, 2020b). There can be various models of stop the bleeding interventions. One such model is *The Trauma Program* at Beaumont Hospital, Royal Oak, Florida, which offers two classes designed for the public to assist in a bleeding emergency (Beaumont Hospital, 2018).

Stop the Bleed lends itself to a variety of settings, including schools in high-gun-violence districts (O'Malley, 2019).

A literature review on "bystander effect" found it might be severely limited in success when participants encounter active violent or dangerous emergencies (Levine, Philpot, & Kovalenko, 2020). However, the effectiveness of Stop the Bleed has shown success in a variety of settings. One large study (N = 1974) found the program successful in increasing confidence and skill proficiency (lay rescuers and medical rescuers) in using Stop the Bleed techniques (Schroll et al., 2020).

Another study found Stop the Bleed effective in all participants reaching mastery level for all four hemorrhage control skills within four tries, a vast majority (87%) determining a definitive sign of life-threatening bleeding, and almost three quarters (76%) feeling comfortable in using a tourniquet in a real-life emergency (Muret-Wagstaff et al., 2020). Stop the Bleed improves the confidence of medical professionals and community members, regardless of medical training, illustrating the range of potential participants (Andrade, Hayes, & Punch, 2020).

Research on Stop the Bleed high school participants found them successful in learning hemorrhage control via multiple methods, learning to identify wounds necessitating use of tourniquets, and displaying a willingness to provide aid from all modalities (Goolsby et al., 2018). Although it is understandable to evaluate this intervention from a trainee standpoint, we need to introduce impact objectives because trainees are in positions to teach their family and friends, furthering this intervention's reach. High school personnel, too, have increased their competencies through this program. A study of high school personnel's perceptions of self-efficacy and school preparedness also resulted in an increase in self-efficacy and school preparedness to control bleeding (Nanassy et al., 2020).

Stop the Bleed has been successful in bringing a potential for saving lives (Zwislewski et al., 2020), but there is room for improvement based on research findings. There is a call to make the training more effective using cost-effective mannequins that more realistically demonstrate pulsatile blood flow and cessation of hemorrhage (Villegas et al., 2020). Lay public bleeding control must embrace a singular design and training, with future initiatives not predicated on actual training being essential (McCarty et al., 2020).

Educating the public on preventing victim bleeding out has broad appeal, but this campaign requires easier and wider availability of Stop the Bleed kits (Tsui et al., 2020). A follow-up evaluation (6 months) of tourniquet application skills retention in one program, however, found 39% of participants unable to successfully apply a tourniquet and 26% unable to control life-threatening bleeding, raising the importance of refresher training within 6 months of the initial training (Weinman, 2020). Since outcome objectives are arduous to measure now, a series of process objectives are offered to measure success of Stop the Bleed interventions, facilitating improving facets of this intervention (Kirsch et al., 2020). It calls attention to the importance of longitudinal studies (Humar et al., 2020) for a more nuanced understanding of changes over time.

The Stop the Bleed campaign created highly innovative projects across the country that hold great appeal to helping professionals to reduce gun fatalities through early intervention (Kang & Swaroop, 2019), leading to localized efforts that take into account sociodemographic factors. These initiatives engage communities to train bystanders to provide hemorrhage control and scene safety. Chicago's Good Kids Mad City is such an example (Tatebe et al., 2019) and is discussed in Chapter 8. St. Louis, for example, initiated its own program called Acute Bleeding Control (ABC) targeting African American/Black males seeking to control bleeding until ambulance arrival.

The following bleeding control training effort shows how it unfolds and why it holds promise in an array of community interventions to prevent gunshot scene deaths (Andrade, Hayes, & Punch, 2019):

> The Acute Bleeding Control (ABC) program includes the entire BC course but has a strategic focus on groups most at risk of experiencing GV [gun violence]. We seek to bring appropriate BC training to communities affected by GV and to equip our participants with trauma first aid (TFA) kits. Participants assemble TFA kits from provided materials, including a tourniquet, hemostatic gauze, regular gauze, adhesive compression tape, trauma shears, a permanent marker, and gloves. Kits are personalized by offering different colored materials and properly fitted gloves. By allowing participants to create their own TFA kits, we ensure that they have interacted with the materials in their kit and know where to find them if the need arises. Individually sourcing the materials and assembling the kits onsite has allowed us to significantly reduce the cost of TFA kits.

Some may argue that this training is best suited for war zones where acute injuries are prosaic. However, I would argue that many urban neighborhoods resemble war zones, and such interventions would be welcomed if sponsored by the "right" organization enjoying the trust of residents.

Enlisting a volunteer civilian corps opens the door for other efforts to address injuries and, in the process, provides communities with opportunities to achieve empowerment (Swaroop, 2018). These campaigns must overcome barriers, such as fears of these responders becoming victims while helping, of making an injury worse, of causing the loss of a limb, of catching a blood disease, or even of causing a death. Nevertheless, this campaign brings the seeds to influence creation of other campaigns that build upon community assets, with far-reaching implications beyond gun violence.

Critics have argued that for all of its fanfare, Stop the Bleed may have limited applicability, with one study noting that only 4.6% of gun injuries would have benefitted from use of a tourniquet (Hsu et al., 2020; Lu & Spain, 2020). This is a weak argument and one has only to talk to the survivors benefitting from this program and their families to see its importance. Stop the Bleed both facilitates empowerment and saves lives.

Stop the Bleed is conceptualized along narrow grounds. Bleeding control initiatives have been spurred by gun violence, with the application of these techniques to other excessive bleeding situations (Sakran (2020), broadening this strategy's impact beyond neighborhoods. This makes this initiative more encompassing and destigmatizes participation.

INJURED GROUPS AS VEHICLES FOR SERVICES AND EMPOWERMENT: PARALYZED VICTIMS

Although we can extol the virtues of bleeding control to prevent deaths at the scene, we must not lose sight of the ultimate goal of preventing injuries in the first place (Masiakos & Warshaw, 2017):

> Caring for the critically injured, as our colleagues did in the early morning of June 12 is an essential element of a surgeon's core skill set. Much of what we know about treating these types of gun-shot wounds, increasingly seen in cities and towns across the nation, has been learned from battlefield surgeons who cared for similar wounds in Korea, Viet Nam, and the Middle East. We are trained to care for the worst traumatic injuries under some of the worst circumstances to save lives, and we are doing this well. Many of our fellow surgeons within the ACS [American College of Surgeons], however, believe that we are not only in the business of preventing death after injury; we also share a special obligation to prevent injury. We fear that the current debate about gun control has not adequately represented the concerns of the broader medical community about the public's health. (p. 37)

The imagery of battle injuries is appropriate. Prevention must always be our goal, but this does not devalue early intervention strategies.

Gunshot survivors have special challenges in regaining their lives after this tragedy. However, few groups elicit a greater emotional response than those paralyzed because of a bullet. A project in Chicago illustrates the potential of self-help on a gun-injured subset that is challenged in meeting their needs. The Crippled Footprint Collective consists of paralyzed ex-gang members injured in violent encounters, many gun related, showing how they embraced a mission to provide and receive support (Ralph, 2014):

Darius continues, "As you can see, all of us here have wheelchairs. And the reason we have wheelchairs is because we were out in the streets gang banging, selling drugs. We got shot, and ultimately we got paralyzed." Today, Darius says, the students will learn what happens to the body when the spinal cord is injured. By educating the students about the grim realities of being wheelchair bound, the Crippled Footprint Collective speakers hope to get current gang members thinking about their lives outside the gang—specifically, if they become paralyzed and have to care for themselves. Eventually, the gang deserts gunshot victims. If this message resonates with the students, Darius and company are well on their way to achieving their primary goal: Reversing the foundational belief that the perpetuation of violence unifies the gang.

The Crippled Footprint Collective also embraces a purpose to better the lives of participants facing physical and emotional challenges (Moser, 2014). COVID-19 introduced a need for safety precautions in addressing the immediate life-and-death moment, highlighting further challenges in reaching this group.

Collective action looks, feels, and acts very differently from action individually focused, and nowhere is this as salient as in addressing victims paralyzed by gun violence. Philadelphia, featured prominently in this book because of its gun violence tragedies, is a city not usually associated with Chicago and New York in discussions of the magnitude of gun violence (Beard & Sims, 2017). However, we can turn to it for generation of models aiding gun violence survivors and, in this case, those paralyzed.

In Philadelphia, a group of male victims gather to connect, share experiences, and describe actions that helped them to adjust to life post injury (Ubiñas, 2019):

Paralyzed gunshot survivors get together for the first time to talk about creating a support group at Temple University Hospital after one recent survivor, Jalil Frazier, came up with the idea. After they began to share their stories, they all wheeled closer together to continue their discussions. . . . As powerful as it is for the survivors to meet and bond, they need more—a dream team, if you will, of people who can help them navigate their emotional and financial struggles. It is, after all, one of the main reasons the group was started.

Sharing personal stories is a mechanism for establishing group bonding activities.

Washington, DC's Urban Re-entry is a group of men, primarily African American, who have been paralyzed by gunshot wounds. In this group setting, they share advice and support because of their unique needs (Wise & Turner, 2019):

Some patients defy their initial prognoses and regain significant use of their lower bodies. Others simply move away. Several, when faced with the prospect of living the rest of their lives in a chair, have taken their own lives or intentionally neglected their own health to die. "Getting a spinal cord injury or a serious brain injury puts you in a sort of catch-22, where you're both still alive, but not fully alive. Somewhat in prison. You're not in a typical prison, but you're kind of imprisoned by your disability."

This sober assessment of a victim with a paralyzed condition illustrates a state of mind that must be addressed before assuming a prominent role in life and within the community.

Projects are possible in partnership with communities, such as support groups, murals/art therapy, drama productions, and peace/healing gardens (Corazon, 2019; Delgado, 2021; Wolfe, 2017). Art and gun violence prevention, for example, can tap youth engagement (Samuels, 2020). It is unusual to find community art projects developed to commemorate those injured by guns; these types of projects are usually the focus of those killed. However, these projects can also focus on gun injuries, with victims playing a role in having their narratives shape how projects unfold. Art exhibitions by gun victims, for example, introduce a medium for expression that can help survivors share their challenges with audiences (Dell'Aria, 2020). Poetry, too, can be quite effective for addressing traumatic events (Giffords & McCann, 2017). These projects expand ways to tap community interests and talents.

Local efforts drawing upon gun victim experiences, with survivors willing to share their stories of recovery, bring an added dimension to community-centered approaches. Gun violence survivors benefit from helping others with the consequences of such an act, and they are also healed in this mutually beneficial process. Establishing a speakers' bureau of gun violence survivors brings experiential expertise to share (Fields et al., 2020).

Gun injuries empower victims by transforming tragedies into opportunities to assume new identities and positive community roles, as with victims paralyzed, by helping them seek alternatives to violence within their communities (Ralph, 2012): "Examining these forums ethnographically—and investigating the argument made by disabled, ex-gang members that their wounds *enable* them to save lives." The transformative process of these forums requires active verbal and emotional involvement of speakers and audiences, bringing highly interactive elements to this experience, tapping the collective power of audience and troupe to shape experiences.

Gender-specific interventions, too, are essential. Women gunshot victims bring unique challenges and are more likely overlooked because men represent the vast majority of injuries (Delgado, 2021; Izadi et al., 2020; Nagengast et al., 2017). LGBTQ groups, too, are called for because they are not exempt from gun violence.

Social workers and others with expertise in treating gun trauma can establish collaborations with local self-help organizations to conduct trainings and consultation with staff as needed (Delgado, 2021). The significance of photovoice gets increased when victims, their relatives, and/or friends use the power of images to convey narratives to others. Photovoice, for instance, is an excellent tool for engendering gun injury research and intervention projects (Delgado, 2015; Teixeira et al., 2020; Yang, Liller, & Coulter, 2018). They facilitate getting a picture, literately and figuratively, of a victim's daily activities, challenges, and rewards.

Photojournalists play a role in capturing injuries of survivors, balancing ethical considerations while informing the public (Miller & Dahmen, 2020). Photographs bring a visual dimension to gun violence not captured by statistics. The community violence legacy across generations can be captured through photographs and oral histories, as in Chicago, allowing capturing both the sorrow and resiliency (Lowenstein & Dharmawardene, 2020). Shorr's (2017) book, addressed earlier, titled *SHOT: 101 Survivors of Gun Violence in America*, is an inspirational endeavor uplifting the consequences of gun injuries on the daily life of survivors. The following description, although lengthy, is noteworthy (Winfield et al., 2019):

The ubiquity of gun violence has become the norm in the USA. We hear repeated horrors so much that it now produces a numbing effect. It is a helplessness that allows us to hear the news and say, "here we go again" and put it out of our mind. Gun violence is now something we expect to happen (but not to us).

I have been asked why I chose to photograph survivors. The project came about, as with any creative idea, because of a few different conditions. Years before when I was living in Greenwich Village with my husband and toddler, two men dressed as postmen came to the door of our home with a package. Suddenly, the "postmen" pushed in and when I recovered I had a gun pointing at me as the man holding it proclaimed "I'm not kidding, get inside." There is a terror that comes when it dawns on you that you are completely at the mercy of a person with a gun. If he wants to kill you he will, if he wants to hurt you he can and if he wants to be merciful to you, maybe he will. Adding to my fear was the fact that it was not just me but that he was controlling the fates of three of us. Luckily on that day the intruders were more interested in robbing us and our cooperation with them kept us physically safe.

The second thought to do the project came from my experiences as a New York City Teaching Fellow in the public schools in crisis. At school, many students wore laminated memorial cards around their necks with the photos of relatives and friends, usually young men, who were killed by gun violence. I started to think that all of those killed had achieved a "folk hero" kind of status in their communities. But I never heard anyone speak about those who were shot and survived. I thought that these survivors would be a very interesting group of people to speak with and to unite in their "survivorhood."

Lastly, I thought that our country was extremely polarized (now even more so) and I thought that if I could look at gun violence as a human problem with real survivors from all walks of life that we might be able to talk minus the politics about this public health crisis.

Photographs were taken, whenever possible, where the shootings occurred, introducing a contextual dimension because the scenes are everyday places, yet hold profound significance for victims, allowing a fuller description of their narrative.

Sports offer potential for victims paralyzed by gunfire. Life post injury takes on great significance and, even more so, when the activity is recreational but brings physical and emotional benefits. A vehicle combining multiple emotional and concrete benefits for those with gun injuries is one that interventionists perennially seek to maximize resources. Why include a sport as a case illustration, such as one focused on basketball? We often think of interventions as transpiring in various spheres of a victim's life. Sports, however, generally do not come to mind. Melvin, a former Chicago gang member paralyzed at 16 by a gunshot, answers this question very well (Juette & Berger, 2008):

> To those who "don't get it" about basketball, it may seem rather odd that a game—"a game that starts out as messing around, trying to accomplish something vaguely challenging and fun, throw a ball through a hoop, a fun, silly kind of trick till you decide you want to do better"—could carry so much meaning for some people. To me, it seems wonderful, however, that something as simple as a game could be so replete with possibility, that it could be a disabled man's salvation, a way for him and others like him to transform, even transcend disability, I think James Naismith would have been pleased. (p. 160)

A game, as noted by Melvin, is more than a game in a life of someone faced with numerous obstacles in adjusting to paralysis. A team-oriented game brings added benefits of involving a new social network.

A vibrant and responsive approach to victims paralyzed by guns offers an extensive array of options for engaging them. Sports may provide an opportunity for engagement that may open the door to provision of other services. A disengaged victim cannot benefit from rehabilitation services. Providing a reason for living after a tragic incident is at the core of meaningful recovery, and when this reason encompasses service to community, it takes on added significance.

We usually associate sports with an athletic gift and finding an organized outlet for this talent. We certainly do not associate an active sports agenda with those paralyzed by an accident or a gun injury. Yet wheelchair basketball is alive and well, and is found throughout the United States. It provides an avenue for paralyzed gun victims to tap their resiliency while resisting pervasive stereotypes about the disabled.

Basketball is more than a sport. It requires participants to have a schedule, a plan, and an ability to execute. This entails engaging a vast array of individuals who make up the team, as well as their supporters, expanding their social network beyond the immediate family. Sports, particularly team versus individual, bring excitement and comradery that take on even greater importance in the lives of individuals who are isolated. Sports provide an avenue to enhance self-esteem and possibly even reconstruct an identity that goes from "disabled" to a "competitor" or even to an "athlete."

Wheelchair basketball draws on a scientific base to enhance participant-training techniques to maximize performance at all levels from club- to elite-level competition (de Witte et al., 2016; Iturricastillo et al., 2016). As with any other sport, participant age, too, needs to be taken into account in programming (Bates et al., 2019; Carroll, Witten, & Duff, 2020). Opportunities to practice, compete, and congregate after a competition provide possibilities for developing friendships, sharing experiences, and equally important, information and strategies for navigating life.

HELPING PROFESSIONS AND GUN VIOLENCE

Helping professionals play an instrumental role in reaching gun victims, their social networks, and communities (Butkus, Doherty, & Bornstein, 2018; Kahn, 2018). Understanding social network influence on gun violence behavior has ascended in importance in both prevention and early intervention efforts (Delgado, 2021). This is not a call unique to health and human service providers. Nevertheless, there is a national call to social work, for example, to assume active and leadership roles in addressing gun violence (Sperlich et al., 2019):

> Some professional organizations, such as the American Pediatric Association and the American Psychological Association, have responded to this crisis by issuing comprehensive policy statements and practice guidelines for frontline workers; however, there has been relatively little guidance for social workers either from the research literature or from prominent social work national organizations. (p. 217)

Other professions such as nursing, community educators, recreational workers, and rehabilitation specialists, too, are facing this call to action, opening the door for innovative interdisciplinary collaboration.

Creating opportunities for multidisciplinary teams in gun injury research and prevention has resulted in needed research, highlighting a return on investment that can occur when there is a commitment to this field (Tobias, Miller, & Bermudez, 2020). This research translates into the possibility of the same professionals collaborating on interventions, setting the stage for long-term relationships that can create other forms of community interventions. Viewing time and effort in the establishment of coalitions or task forces as an investment allows relationship development in a sustainable manner.

Social work gun violence scholarship recognizes this act as a cross-cutting public health issue (Aspholm, St Vil, & Carter, 2019; Brown & Barthelemy, 2019; Delgado, 2018, 2020a; Hong & Espelage, 2020; Johnson & Barsky, 2020; Lyons et al., 2020a; Phalen et al., 2020; Richardson, 2019; Ward-Lasher et al., 2020), including collaboration with other professionals such as those in public health (Hardiman, Jones, & Cestone, 2019). This recognition brings an increased need for social workers to assume a more prominent role in the field.

This upsurge in social work scholarship has helped position it to make substantive contributions to this field and speaks to the urgency that gun violence has engendered within that profession. Our role in this major social issue can be multifaceted, covering the entire range of services, research, and scholarship. This call finds this profession, as well as others, assuming a vital part of multidisciplinary efforts.

Social service organizations and other institutions may argue that it is too expensive, time intensive, and unfeasible to make staff competent on gun violence, particularly if their mission is not health focused. One strategy is to deploy at least one staff member to become the resident gun violence expert, as advocated for in medical institutions (Webster, 2019): "As well as being local leaders at their medical sites, such a group of knowledgeable and motivated individuals could be part of the national advocacy cohort promoting local community actions to reduce gun violence" (p. 277). Helping professionals other than medical personnel, too, can assume similar roles in their organizations and become part of a local network of cross-disciplinary experts, enhancing community capabilities.

Local provider coalitions enhance their political influence, serving as a valuable support system for this demanding work. Training can reach a large cross-section, including efforts enlisting grassroots leaders to broaden this network. These coalitions open the door for local innovations that best meet their needs rather than models developed at the national level that may not be applicable locally without substantial modifications. Public health agencies and community organization coalitions, for example, must initiate cultural self-identity and educational intervention strategies as part of gun intervention strategies, reversing historical patterns that systematically undermine positive cultural self-identity (Mason et al., 2019).

THREE COMMUNITY APPROACHES TOWARD GUN VIOLENCE

Characterizing community approaches to violence is a tall task. There are, however, a number of strategies on urban community gun violence, with three standing out: (1) hospital-based violence intervention programs (HVIPs); (2) Cure Violence; and (3) Focused Deterrence. Each approach is covered elsewhere in this book but worthy of specific attention here to illustrate the breadth of approaches across the nation. Readers interested in any of these approaches have a vast literature to select from in furthering their pursuits. Each approach emphasizes key institutions, collaborative approaches, and extent of community-centeredness.

Hospital-Based Violence Intervention

Hospitals are often viewed as megastructures within a community-centered approach to gun injuries, wielding immense influence on how gun violence is conceptualized. Hospitals (with a history of trust and community relations) can assume pivotal roles in gun violence initiatives (Wojtowicz et al., 2019). Should we expect hospitals to be the sole or major provider of gun prevention initiatives? No.

They can still play a pivotal role in generating funding because of their political prominence and ability to develop an extensive network of community service providers better equipped and situated to reach gun injury victims (Richardson & Bullock, 2021). Successful hospitals have cooperative relationships with the communities they serve, having much in common with schools and similar community support.

Hospitals can be anchors in community violence initiatives by encouraging health providers to address broader social needs that reduce unnecessary hospital admissions (Stephens et al., 2017). The daily experiences these institutions have place them in a unique position to inform research and programs. Trauma centers with extensive experiences with gun injuries develop competencies better preparing them to meet victim needs, placing them at a distinct advantage in the gun violence field to increase survival rates (Fu et al., 2019).

Hospitals have assumed a prominent role in creating gun violence interventions (Dicker & Juillard, 2020; National Academies of Sciences, Engineering, and Medicine, 2019). Although I see them playing an important role in these campaigns because of their strategic position within communities and their resources (Snyder, 2018), they are only part of a bigger puzzle that requires other institutions and community organizations, formal and nontraditional, assuming prominent coalition positions.

As described earlier, inequities are shaped by social identity (e.g., race, gender, and sexual orientation), and the social determinants of health are the "terrain" on which the effects play out, which can be quite broad. Richardson Jr et al. (2020) examined the intersection of gun injury, incarceration, and trauma among young (18–30 years) Black men participating in HVIPs, calling for recognition of trauma symptoms and postinjury affective changes; this recognition must guide practice within these facilities to increase a successful recovery.

Gun victim distrust of law enforcement has consequences beyond the police. It even touches the institutions entrusted to treat their injuries, such as hospitals. This distrust provides evidence for how police brutality has assumed a prominent role as a social determinant of health in urban communities of color (Alang, McAlpine, & Hardeman, 2020). Police distrust makes it arduous for law enforcement to assume a helper role in gun injuries.

The American College of Physicians has advocated for the elimination of social determinants of health, including those on gun violence (Butkus et al., 2020), with hospitals assuming prominent roles in these endeavors (Stephens et al., 2017). It is not unusual to associate HVIPs with secondary and tertiary interventions, but we are now finding these settings addressing social determinants of violence. This is occurring along with advances in adapting best models findings and the important work done by national partnerships and organizations in helping to advance this public health model (Bonne & Dicker, 2020; Miller, 2020).

Gun injuries logically have hospitals/trauma centers assuming prominent positions within the violence prevention field, and that no doubt comes across in early portions of this book and bears emphasis in this section. These settings enjoy prestige and immense resources; although overstressed, they are positioned strategically to make significant differences within urban high-violence communities. The following is illustrative of what is possible (Clayton, 2018):

> The Wraparound Project and Caught in the Crossfire, the Highland hospital project started by Youth Alive! in the 1990s, are now two of dozens of HVIPs in the US that, in the moments after a shooting, dispatch social workers to help families and patients, and send local violence interrupters to communities and hospital bedsides to stop retaliation.

Hospital-based multidisciplinary teams, including use of interns as in South Carolina to address violence, are the latest trend (Hooker, 2021).

Solving gun injury cases is a low priority for police, in part because of lack of witness cooperation with investigations. Gun injuries also do not generate the same level of public outrage as killings. Police departments need to incentivize solving gun injuries to prevent future escalation that can result in killings. Neighborhoods fear gun violence regardless if it results in an injury or murder, but there may be more of a motivation for police to cooperate if it is a murder rather than an injury.

Is there a public demand for investing in HVIPs? A media campaign generating support for such interventions is needed to pressure lawmakers and policymakers to invest in these initiatives. A public campaign highlighting the importance of solving gun injuries is necessary and must involve trusting partnerships between police and well-respected community institutions.

Genuine cooperation, if not actual partnerships, strengthens police–community partnerships (Braga, 2016):

> The police alone cannot implement many of the responses discussed in the guides [problem-oriented guide]. They must frequently implement them in partnership with other responsible private and public bodies including other government agencies, nongovernmental organizations, private businesses, public utilities, community groups, and individual citizens. (p. ii)

Each agreement is dependent upon prior cooperation and a level of trust.

Quick staff deployment is key to preventing the escalation of retaliatory gun violence. According to Everytown USA (2020c), an estimated 25% of victims survive. Those aged 24 or younger will experience another shooting incident within 10 years of the initial injury. Effective problem-solvers must forge genuine partnerships and invest effort in making these partnerships work. Each guide identifies individuals or groups with whom police might work to improve overall response. Current debate on defunding police departments or reallocating funding to social services and community development does not diminish the importance of this cooperation.

There are no national statistics on retaliatory violent disputes (Klofas, Altheimer, & Petitti, 2020), even though gang violence plays a prominent role within high-violence communities. Rochester, New York, for example, during the period of

2010–2012, can trace 60% of the shootings to retaliatory incidents, and in 2015, one retaliatory dispute resulted in 7% of gun assaults. These retaliations are often the result of gangs and their drug business.

Gun injury data require understanding differences or similarities on intent-specific firearm injury groups in order to develop a more integrated portrait of gun injuries. If we are going to be able to develop interventions with high probabilities of achieving success, this requires introducing data from medical, criminal, and vital records (Mills, 2017). Local interventions are dependent upon local data, and the more comprehensive, the higher the relevance.

Hospitals have made significant strides with substance use/misuse through their high-violence intervention programs, illustrating the importance of viewing these two problems as closely related. Affinati et al. (2016), in a literature review on adults presenting with intentional violence injuries, concluded that the state of science has not kept up with demand on hospitals due to data limitations and inadequate data access.

Not every hospital has a trauma center. Not every trauma center has an HVIP. These programs are critical to the violence prevention field because they rest at the critical intersection of health care and community-based violence interventions (Monopoli et al., 2021). Taking an ethical stance on the duty of trauma systems to confront gun violence uplifts this public health issue by making violence prevention and intervention integral with trauma care in this country (Scarlet & Rogers Jr, 2018):

> Most trauma centers do not possess the resources, workforce, or systematic approach necessary to address the social underpinnings of violence. In many cases, a decision to either screen for risk of violence or offer social support or referral to resources that could provide support is not standardized and is left to the discretion of treating clinicians. . . . HBVI programs have emerged as a promising method of breaking the cycle of violence. These programs are structured to address what we now understand to be proximate causes of violence, with an emphasis on the social determinants of health. HBVI programs leverage access to trauma care at the time of injury, which, for many, might represent their only access point to the health care system. (pp. 483–484)

The prominence of trauma centers brings the potential to leverage their propitious position on access and translates this advantage into a national call for HBVI programs in every US trauma center. Provision of case management and advocacy services enhances their potential contribution, positioning institutions to embrace prevention.

HBVI programs, however, are still in need of cost-effectiveness studies to advance these approaches (Chong et al., 2015). Further, HVIPs have not benefitted from systematized research on outcomes regarding proximal goals and activities of the programs themselves (Grossman & Choucair, 2019). Fortunately, there is empirical evidence supporting HVIPs for advancing this field. An HVIP literature review on effectiveness and costs found they have achieved success in reducing recidivism rates and contained costs. Creating a greater understanding of consumer use of HVIPs is a worthy undertaking, as in Boston, which uncovered a profile (African American/Black, stable home situation, younger age, mental health diagnosis, gunshot wound, and more severe injuries) predicting service use (Pino et al., 2021).

According to the NYC Department of Health and Mental Hygiene, their Cure Violence campaign resulted in a 60% reduction in gun injuries (Gunter, 2018). Nevertheless, there is a need to understand the ancillary benefits that these programs provide to individuals with intentional injuries, providing best practices regarding case management, hospital leadership buy-in, and hospital-community-based organization collaborations (Nusbaum et al., 2020).

Nevertheless, there is a lack of systematized research on outcomes and a need for broadening how healing and recovery are defined (Monopoli et al, 2021). Further, relatively few of these programs have undergone thorough evaluation. An evaluation of Philadelphia's Turning Point program, an inpatient violence intervention program, found success in changing the attitudes of gun victims toward violence (Loveland-Jones et al., 2016). A long-term follow-up of this intervention, however, is required to advance this dimension of the field of gun violence forward.

The experiential legitimacy hospital case managers bring when they have suffered gun injury(s) enhances their effectiveness (Decker et al., 2020): "(1) understand and relate to their sociocultural contexts, 2) navigate the initial in-hospital meeting to successfully create connection, 3) exhibit true compassion and care, 4) serve as role models, 5) act as portals of opportunity, and 6) engender mutual respect and pride." These themes should not be surprising in the field, and they are not restricted to hospital settings with equal applicability to other community-based efforts. These themes call for initiatives seeking the participation and employment of those injured by guns to be part of programs and teams addressing gun injuries.

Pediatric trauma centers were establised approximately 60 years (1962) when the first one was established at Brooklyn's Kings County Hospital Center; 14 years later (1976) witnessed the publication of the *Resources for Optimal Care of the Injured Patient* by the American College of Surgeons, setting criteria for a pediatric trauma designation (Daley, 2021). Pediatric hospitals have assumed prominence in gun violence along four domains (clinical care, advocacy, education, and research) because almost 7,000 children are hospitalized by gun injuries on an annual basis (Silver et al., 2020). These settings have assumed great prominence on gun injuries within high-violence urban communities of color, as covered in Chapter 3.

Cure Violence

The title "Cure Violence" brings a heavy medical connation, and it is a strategy that has enjoyed nationwide popularity. Cure Violence represents a public health approach by how it attempts to change attitudes toward gun violence (Butts et al., 2015). Cure Violence, formerly known as "The Chicago Project for Violence

Prevention" and "CeaseFire," is a public health campaign with Chicago origins in the 1990s (Eisenman & Flavahan, 2017). This campaign's vision brings the nation's public health and medical systems together (Cure Violence Global, undated):

> Violence behaves like a contagious problem. It is transmitted through exposure, acquired through contagious brain mechanisms and social processes, and can be effectively treated and prevented using health methods. To date, the health sector and health professionals have been highly underutilized for the prevention, treatment, and control of violence. Now is the time to mobilize our nation's healthcare and public health systems and methods to work with communities and other sectors to stop this epidemic.

Communities assume a prominent place within this campaign. A small group is largely responsible for this violence. Focusing resources on this group translates into a higher probability of maximizing reduction in gun violence.

A Cure Violence strategy treats gun violence as a contagious disease, which lends itself to a public health model (Slutkin, 2017):

> For all other contagious health problems—such as HIV/AIDS, influenza or Ebola—when outbreaks spread, they are correctly identified as epidemics and public health methods to control and eradicate epidemics are rapidly deployed. With recent Ebola outbreaks, the world rallied to collect resources and develop a sophisticated public health response. However, a similar public health response and resource deployment is not occurring with epidemic violence because the problem is still fundamentally misdiagnosed. We have been negligently slow to recognize violence as a contagious health issue. (p. 1)

Gun carrying motivation takes center stage in coordinated efforts to prevent gun violence (Robertson, 2018). Carrying a gun for safety taps a different motivation than carrying one when angry or to protect a drug enterprise (Evans, Garthwaite, & Moore, 2018), serving as motivation for influencing construction of interventions. Both motivations are required to reduce casualties. Cure Violence is one of the most well-known gun violence campaigns that use violence interrupters to minimize violence from retributions (Marx, 2020). Relationship-based street outreach strategies form the heart of the approach (Giffords Center, 2020a), and it is the glue in many successful gun violence prevention strategies. Strategic outreach is only possible when those doing it are familiar and have earned the trust of their community (Bower, 2019).

Cure Violence targets groups with a high probability of engaging in violence, such as those released from prison, a marginalized group with a history of engaging in violent acts, including use of guns (Willoughby et al., 2020). Perpetrators of gun violence, too, often have been direct or indirect victims themselves. Violence interrupters share the backgrounds of those they seek to alter the behavior of, and they are best represented as a secondary-level intervention. Hiring staff with "street culture" competence entails investing in and helping them acquire professional competencies to carry out their tasks (Brisson, Pekelny, & Ungar, 2020).

This intervention makes no pretense of being a prevention program. Rather, it seeks to minimize the escalation of gun violence when a gun incident occurs, preventing even greater injuries and fatalities. Outreach workers place emphasis on engendering trust with residents considered at high risk for engaging in gun violence as seen in Cure Violence campaigns (Crifasi, McCourt, & Webster, 2019):

> Cure Violence and similar programs employ street outreach workers in impacted communities to build trust with those at highest risk, mediate disputes, promote nonviolent alternatives to conflict, and facilitate connections to social services and job opportunities. Such programs appear to promote norms that avoid the use of guns to settle disputes, and often lead to reductions in gun violence. Other programs such as Rapid Employment and Development Initiative work with similarly high-risk individuals and offer employment opportunities and cognitive behavioral interventions to change thinking patterns that leave such individuals vulnerable to using guns in response to conflicts. (p. 9)

Quick staff deployment is a trademark of Cure Violence campaigns, connecting residents with services and opportunities to shift their thinking to a worthwhile future. Cure Violence has shown immediate and short-term success (Giffords Center, 2020b; Tomberg & Butts, 2016); long-term success, however, has proven elusive.

Riemann (2019), in critiquing Cure Violence (discussed in Chapter 7), argues that Cure Violence medicalizes violence by focusing on individual pathology (at-risk identities), ignoring structural (political) forces in having them achieve a healthy state and identity. Embracing a community assets viewpoint counters this tendency to pathogize communities and, therefore, victims. Another critique is that by recruiting former members of gangs, for example, to contribute to their communities, we glorify these individuals and, therefore, violence.

Focused Deterrence

Focused Deterrence Violence Prevention is a multiagency initiative finding saliency in addressing community violence, with gun assault prevention a prominent part of this strategy. Marshalling resources across sectors to concentrate on violence warrants further attraction of resources on this goal normally not targeting guns (Braga & Weisburd, 2015).

Braga and Kennedy (2021) advance a three-part framework from which to classify violence-focused deterrence strategies: (1) group violence intervention programs; (2) drug market intervention programs; and (3) individual offender programs. It is important to note that perpetrators of gun violence can fall into various permutations and combinations of these categories. Significant publicity is generated by this approach, with success dependent on the "word getting out" and having the intervention embraced. Focused deterrence concentrates resources, and this strategy rests on having legitimacy across sectors, including the community (Kennedy, 2019).

This initiative has shown promise in stemming gun violence by "focusing" on individuals and groups considered at highest risk for gun violence (McGarrell, 2019), although it is labor intensive because of its emphasis on developing and sustaining collaboration between entities not accustomed to working together. This focus was never intended to be solely punitive by also stressing that a community cares about them (Nieto & Mclively, 2020). Further, some basic assumptions underpinning this strategy require in-depth examination (Haberman & Link, 2020). For example, an assessment of Philadelphia's Focused Deterrence on gang activity strategy found this approach achieving significant reduction in gun violence in targeted communities, with findings at the gang level mixed (Roman et al., 2019).

Successful collaboration rests on open communication and a shared agreement on the problem to arrive at a shared vision of a solution (Gebo & Bond, 2020). Collaboration can reduce intervention costs (Cerdá, Tracy, & Keyes, 2018). The preferred route for creating gun violence programs is by coordinating efforts and combining resources, with the following organizations and agencies critical in these collaborations (Yang, 2020):

(1) Academic community collaboration to enhance violence prevention by translating evidence-based research into effective practice; 2) Clinical-public health partnership to improve and evaluate the quality of medical support for youth victims; 3) Local violence prevention agencies to help deliver health care service to families and communities; and 4) Funding agencies, which will make important contributions by prioritizing their funding opportunities to youth violence prevention. (p. 3)

This broad coalition illustrates the reach of these initiatives and the potential for changes at the local level. Sustaining them, however, is labor intensive and a long-term ordeal.

Focused Deterrence and other approaches in this section capture basic premises and distinct approaches but also provide valuable insights into how the etiology of this violence shapes language and approaches. More approaches on preventing gun violence may draw upon scarce resources, but the tentacles of this problem require multifaceted strategies. Gun violence deterrence is viable, but we must also embrace alternatives to instill hope (Abt, 2019).

CONCLUSION

Community-centered approaches to gun injuries encompass a broad arena and can be exciting and frustrating at the same time. This chapter reflects my bias in how community-centeredness is operationalized to address this public health problem. Obviously, no two communities are identical, requiring flexibility in constructing initiatives that take local conditions into account. This chapter set the foundation for the next two chapters by providing examples of gun injury programs that capture the excitement and challenges this field presents.

8

Case 1

Chicago's Good Kids Mad City and Stop the Bleed

> *There must be a rhythmic alteration between attacking the causes and healing the effects.*
>
> —*Dr. Martin Luther King, Jr. (1958)*

INTRODUCTION

Community involvement in gun violence prevention and treatment is one strategy that many of us wholeheartedly embrace. How we achieve this goal opens the field to a wide range of options, which is exciting but frustrating because there is no one universal model. There certainly is room for innovation in the gun injury field.

Coverage of Stop the Bleed (STB) in the previous chapter makes selecting a case illustration on this topic a natural next step. STB brings a concrete example of a community-centered gun violence approach, empowering residents to assume health provider roles. Chicago is a site for viewing this campaign because of its prominence with gun violence and its efforts to stem this tide. The organization selected for attention in this chapter is youth-led, bringing this dimension to life. Youth activism after the Parkland, Florida, shooting received national attention (Braun, 2019). However, we can find youth-led activist organizations throughout the country, as evidenced in Chicago.

BRIEF HISTORICAL OVERVIEW

Community-led efforts have resulted in numerous community organizations confronting gun violence. Chicago's Good Kids Mad City (GKMC) is such an example, and more specifically, it is a youth-led organization (Delgado & Staples, 2008). This type of youth-led organization makes logical sense because gun violence has disproportionately influenced youth of color. Youth gun violence prevention organizing can take various forms (King, M., 2021).

The Silent Epidemic of Gun Injuries. Melvin Delgado, Oxford University Press. © Oxford University Press 2022.
DOI: 10.1093/oso/9780197609767.003.0008

However, invitations to join this organization are based on decisions of the youth in charge with specific tasks in mind. In the case of GKMC, they have an adult mentor (Kofi Ademola) with a background in Chicago's Black Lives Matter movement. Readers may have difficulty reconciling youth in command when adults are supposed to possess expertize and wisdom on gun violence. These adults often do not live in their communities or socially navigate daily life under hostile and life-threatening circumstances. GKMC youth assume the role of first responders by learning to apply tourniquets, using a jacket or shoelace, and providing CPR, to prevent victims from bleeding out (Ali, 2019): "Darrion, a soft-spoken rising ninth-grader who lives in the South Side neighborhood of Englewood, was there to learn life-saving skills in case he comes across a shooting victim."

MISSION

It is important to pause and consider the title of the organization because of how it stresses the need to differentiate between youth of color who are engaging in positive pursuits from counterparts taking a different and less healthy path (Bellware, 2019):

> The name Good Kids, Mad City, Norwood notes, is supposed to reinforce the fact that the group is comprised largely of young people who are hurt by vio-lence. "We definitely ripped off Kendrick Lamar," she said laughing. "But we want to shift the narrative that we're bad kids, bad people. We're good kids, but in a mad city—a city that doesn't support us. We're showing people what it is to be good kids.

We must never lose sight that there is no such thing as a "bad community." We must seek out those who have made a commitment to creating a better and safer community, highlighting their efforts and success for the outside world.

Understanding CGKM's STB campaign requires an appreciation of how this initiative ties in with its overall mission. Although this chapter emphasizes the STB campaign, this organization has a broad approach on violence, highlighting social change strategies (Boyle, 2020): "GoodKids MadCity, a youth-led organi-zation, was formed to empower young people to confront the violence they face in their communities. The group focuses on restorative justice, creates healing spaces for young people impacted by trauma and works to fill resource gaps in neighborhoods." Goodkids Madcity Englewood has established a number of partnerships within Chicago, one of which is with the Evening Star Missionary Baptist Church, where they volunteer and assist with packaging and distributing food to the community.

The organization also identifies relationships with the police as a key element in addressing violence within the community (Boyle, 2020):

> Engaging leaders around the past and present community climate of policing and allowing them to analyze and organize around ways to de-escalate tension

and raise communication/relations with police, while also working on or pushing for legislation that advances the safety of pedestrians on a daily basis from police bias and brutality.

Police–community relationships are key in addressing gun violence.

Their mission is achievable through a variety of ways. One involves social protests, such as a campaign to reallocate 2% of Chicago's police budget to mental health services to reduce gun violence (Boyle, 2020). Its multifaceted approach on gun violence helps it achieve a high-profile status in Chicago. GKMC offers a range of victim supportive services, such as meetings for youth paralyzed by gun violence. The STB program is not an outlier to this organization's mission, increasing its effectiveness in reaching potential participants and obtaining community support. A multidimensional organizational mission allows expansion and contraction of activities to meet presenting needs.

ONE STORY

In 2018, 19-year-old Delmonte Johnson was shot and killed as he walked to a store (Sacks, 2018). Delmonte was an active churchgoer and a member of youth organizations, including Chicago Gay Men's Chorus (CGMC), further accentuating the point that gun victims do not have to have histories of violent behavior. Darrion, his brother, a ninth grader and member of CGMC, describes the role he is prepared to carry out the moment he encounters a shooting victim (Ali, 2019a):

> Darrion Johnson knows exactly what to do if he sees someone get shot. On a Friday night this year on the South Side of Chicago, Darrion, 14, watched while another teenager practiced putting pressure on a fake bullet wound, checked for a pulse and began two-handed chest compressions on a teen pretending to be a gunshot victim. Along with a dozen other teens, Darrion was attending a first responder training course held by a youth-led anti-violence organization, Good Kids Mad City. "Just don't panic," Darrion firmly repeated aloud to the instructor, a member of Good Kids Mad City, as she explained that it can take just minutes to die from blood loss after being shot, while an ambulance can take up to 30 minutes to arrive.

Darrion's motivation to join a life-saving program was due to his brother's death after he suffered a gunshot wound; Delmonte died just 45 minutes after being shot. Darrion wondered whether he could have saved his brother since the shooting took place a short distance from him at the time. This doubt remains with him for a lifetime. Unfortunately, there are numerous stories that can be shared, such as those of GKMC members who have seen their share of gun violence.

DISCUSSION

Participant engagement in the STB campaign is often painful. Youth participating in anti–gun violence activism face the possibility of retraumatizing themselves, as recounted by one Chicago activist member of Goodkids Mad City (Apantaku & Emmanuel, 2020):

> I know when I'm traumatizing myself, when I start to overthink about the situation and just constantly think about it, because then I'm reminding myself of the trauma that just happened. And I feel like as long as I think about it, it gives me a chance to not forget about it. I'm traumatizing myself every time I relive that moment.

The importance of this health issue demands a sacrifice for the good of a community.

A long-term perspective provides insights into how this intervention shaped the future and career paths of gun violence survivors. Further, taking an impact perspective of how their participation benefitted their relatives and friendship networks needs recognition by providers, researchers, and academics. These accomplishments are worthy of highlighting. Gun injuries are complex, bringing pain and joy in cases where these injuries transformed lives into community assets.

CONCLUSION

Stopping the flow of blood is a powerful symbol and captures the public's imagination. STB holds the promise for moving this field into new arenas and age groups. Other interventions are possible when communities enter into trusting partnership with organizations. This Chicago campaign transformed bystanders into community assets, in a city in serious need of gun violence success stories. The promise of youth-led efforts, including the organizations they run, can be supported by other organizations through provision of financial and technical support. Having youth-led organizations at the table ensures that all segments of a community are actively engaged in pursuit of worthwhile interventions.

9

Case 2

Open Doors (Paralyzed Gun Victims)

This wheelchair life is not a game ever since I got paralyzed my life hasn't been the same. But I will never be ashamed to let the whole world know my struggles and pain.

—*Micah Harris "El Beardo"*

INTRODUCTION

Readers will witness the potential of the arts to transform lives that have been dramatically altered by gun violence, with victims becoming survivors and part of a new cadre addressing this violence. This chapter illustrates what is possible when we are bold and creative in designing interventions.

As discussed in Chapter 3, the challenges victims face warrant a case illustration. Victims paralyzed by gun violence present some of the gravest challenges for the field, as well as costing society immense funds over a lifespan. Those engaging in gun violence think they understand the risks, with a bimodal view of life as death and prison. Another bimodal perspective is operative, with disability rarely talked about as a potential outcome of violence, reinforcing a binary view of life as either being able-bodied or dead (Hinton, 2017). Paralysis, and dealing with its complications, necessitates an extensive and highly sophisticated system of care.

The Open Doors program offers a unique focus on the arts with its multifaceted approach, including the prominent role played by participants in wheelchairs. Readers will be inspired by how the pain of a life-altering gun violence event has been channeled through use of poetry or the spoken word.

REACHING AND SERVING PARALYZED GUN VICTIMS

Why has the Open Doors program been successful? A key element is the comradery and common purpose that the group provides; this is an essential quality of its attractiveness to members (Hu, 2019).

The Silent Epidemic of Gun Injuries. Melvin Delgado, Oxford University Press. © Oxford University Press 2022. DOI: 10.1093/oso/9780197609767.003.0009

Members of the group often say that they previously lived risky lives that exposed them to violence. They believed they were destined for either prison or death; they did not expect a life confined to a wheelchair. But by supporting one another, the men have found a cathartic outlet. Lawrence and other members hope sharing their stories will inspire compassion and, most importantly, deter young New Yorkers from repeating their mistakes. Lawrence writes of his mission in one of his poetic works, "Then looking up in the sky asking God why did he save and didn't just let me die. At that moment, he spoke to me and said, 'You're special to me, there's more for you in store. Just wait and you'll see.'"

We now have a foundation from which to appreciate this case illustration.

BRIEF HISTORICAL OVERVIEW

Roosevelt Island's history is almost as old as that of New York City. It has been home to the City's unwanted groups, which have varied over this period. It started by housing various institutions, such as a penitentiary, a smallpox hospital, and a mental hospital (Van Brocklin, 2019c).

In 2014, Jennilie Brewster moved to Roosevelt Island and shortly after volunteered at Coler. Her artistic background (a writer and painter) and interest in contemplative care, a caregiving approach rooted in Buddhist practices of mindfulness and compassion, introduced a unique person to a unique set of circumstances (Van Brocklin, 2019c). Responding to the request by a resident to start a writing group eventually led to the creation of Open Doors in 2016. At first, performances were limited to the Island. However, eventually the players ventured out to other parts of New York City.

The Open Doors Program, with the help of Steven Willis, a young slam poet, led to an 8-week workshop for members. Poetry elicits a profound and visceral response and is a mechanism that captures and conveys the rules and rituals of living in "the hood." Poetry is usually not associated with a therapeutic or rehabilitation method among urban residents, but it can be appealing to people of color if led and structured in a manner that respects their lived experience. It must also be open to the language they are familiar with to share these experiences with others.

POETRY AND THEATER

New York City's Fortune Society sponsored a Realty Poets performance, and the title used to advertise this event captures what makes the Reality Poets worthy of attention in this book: "Open Doors Reality Poets: An Afternoon of Realness, Resilience & Healing." I draw attention to resilience and healing because these two concepts are rarely associated with gun injuries, no less than with those paralyzed by guns. Open Doors views itself as an arts and justice program based in the Coler Rehabilitation & Nursing Care Center (https://www.angelicaprogram.org/

main-st), a 500-bed facility. This program is a network of activists, poets, rappers, and disability rights advocates. Various combinations and permutations of these roles illustrate the dynamic aspects of participant lives. Reality Poets, according to Joel Francois, a Reality Poets coach (Nishimura & Robledo, 2019), summed up the poets' embrace of their message: "All of the Reality Poets have lived a life shaken by violence, bravely they offer up these stories. They do not lie in the telling of these stories, they do not soften the blows. Your comfort is never priority over truth."

Although Open Doors seeks to prevent gun violence, it is also a survivor mutual support group (Manhattan, 2019). It has the potential to allow expression of the psychic pain associated with life-altering injuries, thus providing an opportunity for healing both the writers and their audiences. This workshop was well received and led to the establishment of the Reality Poets. This group meets on a regular basis and performs throughout New York City, conveying a message of hope and helping to change the hearts and minds of those who may resort to gun violence. The performances also reached audiences that may have personally experienced gun violence or experienced it in their social network.

Those paralyzed by gunshots have an array of methods to achieve therapeutic benefits, as evident in the following New York City performance, if we expand our vision for the rehabilitation field (Van Brocklin & Fernandez, 2018):

> "*There be days that I wished I got shot in my head Instead I'm stuck in a f---ing hospital bed. Could have left me in a pool that was all red, Instead, I'm left with a body that's half-dead.*" LeVar Lawrence spoke those words last week into a microphone at an art gallery on New York's Roosevelt Island. Lawrence, who was shot in the neck in 2005, was there to perform poetry and spoken word alongside five other gunshot survivors. The performers are members of OPEN DOORS, an organization under the Angelica Patient Assistance Program, a New York–based nonprofit. The group began as a writing workshop three years ago and now provides men who've survived gun violence with mentorship and resources to create art, pursue education, and learn other skills. From their wheelchairs, wearing hoodies and baseball caps, the men shared stories of death, pain, regret, and resilience. One speaker, Micah Harris, described himself as a "world-class orator, wheelchair warrior." Another, Andres Molina, talked about finding his compassionate side. "We all have something inside from all that pain that we went through," said Molina. "That stays there."

Gunshot paralysis does not mean a victim's voice is silenced; survivors can find their voices trough different vehicles.

COVID-19 has had a tremendous impact on the nation's nursing homes, and New York City has not escaped, with places such as Colar's Open Doors also experiencing consequences (Schwartz, 2020). This pandemic has increased the importance of safety for gun victims and provision of opportunities to engage in productive activities while keeping them safe.

Open Doors provides gunshot survivors with instrumental and expressive support, mentorship, and resources to seek an avenue for healing through self-expression. The members give to others by offering a writing workshop that teaches poetry and the spoken word (Hu, 2019). Their mission is to support each other and, by sharing their stories, help others learn and be inspired by their lives, as noted by Lawrence in one of his poems.

Theater can be a constructive vehicle to channel survivor creativity, and in this case, the actors are in wheelchairs in a play titled *Fade*, which is sponsored by Coler's Open Doors program, which premiered to a full house on June 29 and 30, 2019. Open Doors is a city-operated nursing home and rehabilitation center (Van Brocklin, 2019b):

> The lights come on in a darkened community theater, illuminating five men in wheelchairs. Childhood friends who survived the crucible of gunfire in New York City, now they've come together in this abandoned building to decide what to do next. One of the characters, the youngest in the group, touches his back with a grimace. He's in pain, and his medical bills are stacking up. Another, wearing a black cap backwards, is still struggling to accept that he can no longer command the same type of respect on the streets. The play, titled "Fade," centers on a group of shooting survivors who decide to pool their resources to open a barber shop. Giving shape-ups and advice to a collection of locals becomes their focus, even as they're plagued by memories of what brought them here. Towards the play's end, the crew is crushed at the news of a teen slain in their neighborhood. The man in the black cap cries, for the boy but also for himself, and for the life he never got to live. "My coffin just got wheels. Everyone who loved me buried me when they saw me like this," he says. "Now I got my life back, but only half."

Traumatic life experiences can find affirming and positive life-altering outlets. Discovering victim talents broadens their possibility of engaging in activities that make them grow and contribute to others at the same time.

Open Doors published a poetry book titled *Wheeling and Healing*, consisting of a range of poems (Nishimura & Robledo, 2019):

> The work is an eye-opening observation of the many things people often take for granted, and the many ways trauma seeps into daily reality. The Poets create works that, with the occasional comic twist, highlight the struggles of people who find themselves suddenly not fully physically able. There are also poems that tackle darker topics about the difficulties of requiring a wheelchair for mobility, and on a spiritual note, question the higher powers who allowed them to outlive their incidents.

These poems are a mechanism for expressing hopes and fears. There are other projects at various stages of planning or completion—a video, a rap album, Spanish album, and their own record label, too. These projects illustrate the creative capacity of members; with support, the group would like to broaden their reach across New York City and the country, serving as an inspiration for other cities.

The poetry is a stark view of violence and life post injury as members remember it. These losses are multiple and not limited to their physical injuries. Nevertheless, the poems reflect strengths and the ability to bounce back from a tragedy that put them in a wheelchair. Poetry taps into urban street traditions that stress verbal dexterity and intellect, sometimes referred as "ranking" or "the dozens," that entail quick verbal insults, usually involving someone's mother. This language opens the door for engagement across age groups to share an urban experience.

Participants create strong bonds through engagement in these activities, and these bonds extend beyond participation in activities, creating social networks based on shared experiences, interests, mutual aid, and place of residence. These bonds are arduous to create, or re-create as the case may be, because of the consequences associated with paralysis. The strengths of these bonds are evident when members leave Roosevelt Island to live elsewhere in New York City and return to participate in activities. Creating the opportunity and climate conducive to engendering these bonds is difficult to achieve, thereby making Open Door unique and worthy of attention.

ONE STORY

We will never have a problem finding masterfully written and informative books on gun victim fatalities, such as Younge's (2016) *Another Day in the Death of America: A Chronicle of Ten Short Lives*. However, there is a desperate need for a book on victims that survived gun violence. Books with a specific focus on a range of injuries enhance our understandings in profound ways, helping the field of prevention and treatment. Academics must humanize victims while moving the field forward with scholarly contributions. One is not possible without the other.

The story of El is much too common and illustrative of how bullets do not care who they strike. Alhassan "El" Abdulfattaah's poem describes the incident that cost him physical mobility. El was born and raised in Fort Greene (Brooklyn), New York City. His poem makes reference that he still has much to see (https:// www.opendoorsnyc.org/realitypoets):

> Phone blew up in hand no one to call.
> Lady holding my bloody hand praying don't know her from a hole in the wall.
> MTA [Mass. Transit Authority] man making the 911 call saying to myself it's
>   not my time yet I still got more to see.
> Ambulance arrive putting me on the stretcher
> Seeing all eyes and stares didn't make it no better
> Went to the hospital friend came to visit
> Told him look at my life now and see how I'm living.

Capturing that split moment in time when a shooting transpires stays with a victim a lifetime. However, it does not mean that it destroys their dreams. El, for example, is a student at City Tech in Brooklyn pursuing study in Computer Science and a core member of the Open Doors' design team.

DISCUSSION

Readers can see how hope is identified, manifested, and channeled into activities that create community and a sense of purpose. Poetry and theater hold special appeal in society. However, they are rarely associated with low-resourced urban communities, let alone with performers in wheelchairs. Nevertheless, that is exactly what has transpired on Roosevelt Island. The arts have been underutilized for addressing gun-related violence, but that is changing.

A desire to help others learn from participants with debilitating illness or injury is at the center of rehabilitation. The lessons learned through this journey must to be shared with others. Human beings have an innate desire to be part of groups, and nowhere is this more apparent than when facing an existential crisis, such as a life-altering gun injury or another illness. In thinking of group activities for individuals who suffer from some form of paralysis, we must consider program models that are open to bringing together individuals with different causes of this condition. Jay, a member of the Open Doors Reality Poets (Disability Visibility Project, 2020), did not become disabled through a gun injury but because of a rare lung disease, yet he can become part of ventures that emphasize gun violence:

> I live in Coler Nursing Home and Rehabilitation Center on Roosevelt Island in NYC, where I met my brothers, the Reality Poets. We are all members of OPEN DOORS, a network of artists, activists and advocates motivated by community building, gun violence prevention and disability rights. Through OPEN DOORS, we formed the Reality Poets, a collective of truth-telling artists bonded through a shared mission to spread a message of realness, resilience and healing.

A shared vision and mission provided by a group effort as articulated by Jay can be made available to all those sharing a similar state of being and wishing to be part of such a group effort.

Not all gun victims who have paralysis wish to be part of group ventures, or those suffering illnesses that result in similar outcomes, so it is a mistake to assume that is the case for all. Having a menu to select from in engaging in community-centered activities allows each potential participate to select the best medium to meet their needs while serving a broader community.

CONCLUSION

The creativity found in the Open Doors program is rare. This program illustrates how tapping creativity provides a venue or door for gun victims to draw upon their experiences to reach a wider audience in the hopes of preventing future gun violence. Participants illustrate that the future for those confined to a wheelchair is still alive and they have much to contribute to society.

The Open Doors program also opened the door (no pun intended) to including individuals in wheelchairs who ended up in them through different means other than gun violence. Such efforts foster the building of bridges across groups, allowing the free exchange of ideas and support.

Recommendations for Advancing the Field

Gun injury survivors are the living embodiment of the hope and pain often commonplace in this nation's high-violence neighborhoods. We can endeavor to recognize the pain while celebrating the hope. This balance is never easy to achieve but essential in successfully treating injuries. This final section presents a distillation of the current knowledge on the topic of urban gun injuries and translates it into action steps. This section consists of three chapters covering professional education, research, and finally, practice. I treat each chapter as its own entity, but in reality, all three are highly interrelated, with considerable overlap among them.

Recommendations for Advancing the Field

10

Professional Education

Getting shot hurts . . . No matter how hard I tried to breathe it seemed I was getting less and less air. I focused on that tiled ceiling and prayed.

—*Reagan (2016, p. 157)*

INTRODUCTION

This book has a central goal of advancing the field of urban gun injuries. No single profession in this field shoulders the burden of this challenge, increasing the importance of education in creating a cadre of providers, researchers, and academics interested in addressing urban gun violence. Educating providers represents what is arguably the cornerstone of any strategy to moving the field forward in the next decade (Kwong et al., 2019) to remain relevant to urban research, policy, and practice. Professional education represents an ideal venue to create a cadre of providers at all levels of service provision who are equipped to address this public health issue.

Gun injuries and death need to ascend to greater national importance, with specialized attention in the training of physicians (Webster, 2020) and other professions (Szajna & Shaffer, 2020; Wheeler, 2019). Medical professionals, as highlighted throughout this book, are increasingly coming to the fore in this field.

The Stop the Bleed campaign, which was covered in Chapter 8, highlights the potential of this movement within high-violence urban communities. Professional education programs have started to integrate this curriculum into their trainings, such as in medical schools (Chernock et al., 2020) and nursing (Connerton, & St Clair, 2020), to highlight but two. Other professions, too, must do so, and it takes on greater significance for those providers that are community based because this campaign has the potential to create a cadre of providers that can come together to share and learn, as well as serve as a source of support.

IS THERE AN ETHICAL AND MORAL IMPERATIVE?

What is our ethical or moral duty to gun survivors and their communities? Is addressing gun violence a responsibility that comes with providing care or service,

The Silent Epidemic of Gun Injuries. Melvin Delgado, Oxford University Press. © Oxford University Press 2022. DOI: 10.1093/oso/9780197609767.003.0010

or does it need to go beyond these narrow, but nevertheless, important boundaries? Answering this question necessitates having the time to do so, the context that values a deliberative process, and willingness to act upon our convictions. Yes, as professionals, we subscribe to an ethical code of conduct. Using this code on a daily basis translates into actions unifying us across professions, facilitating bold, brave, and innovative actions.

Morality and gun violence invariably get framed from a perpetrator-victim perspective. Joseph and Reese (2020), however, discuss the thorny question of morality and gun injuries. They emphasize our responsibilities to the injuried and their collective body, raising our motivation for engaging in this service and activism:

> Given the extent of the problem, one may postulate that practitioners who care for victims of violence are morally obligated to be active in legislative advocacy and/or to build comprehensive programs at their institutions to address violent injury. Does placing that burden on a practitioner end up increasing moral distress in that practitioner? Perhaps revisiting the origins of our work can help answer these questions. (p. 139)

ADVOCACY EDUCATION AND TRAINING

As discussed in Chapter 10, there is a call for an increased advocacy role for advancing a gun injury prevention and treatment agenda. If this call is to succeed, it must assume a role in the education we receive. Lack of formal education or training on gun violence advocacy may serve as a major barrier to active involvement in advocating for a safer society (Webster, 2019). Effective advocacy does not just happen; it necessitates strategic thinking and understanding pressure points to be successful. Academic advocacy requires we familiarize ourselves with the boundaries dictating this role and potential hazards (Lynch, Bateman-House, & Rivera, 2020):

> Academics who wish to advocate for policy changes supported by their research outside the typical venues of academic journals, op-eds, and blog posts may consider reaching out to those in a position to effectuate policy change, including federal and state legislators, the president, governors, mayors, executive agencies and committees, the courts, and others. In many cases, however, academics may not be aware that these activities could have legal consequences, both for them individually and for their institutions. At the other extreme, they may be intimidated by ambiguous or restrictive institutional policies. (p. 44)

Advocating for laws and policies is a professional responsibility requiring a knowledge base and skill set to succeed, with these strategies and techniques part of our professional education.

Advocacy roles require having this education as part of training, as with pediatricians (Dodson et al., 2021). Advocacy education must not be limited to degree programs but also a part of internships/residencies and continuing education. This stance bridges residents and providers. "Homicide activism" is a major social force in addressing gun violence (Leeolou & Takooshian, 2020). "Family activism," too, is emerging to help families cope with their pain (Cook, 2021). "Injury activism," unfortunately, has not generated similar levels of outrage, calling for a cadre of activists and advocates on this issue (Zoller & Casteel, 2021). Helping professions are in a position to advance this agenda.

EDUCATING HELPING PROFESSIONALS

Introducing curriculum on gun violence during the education of professionals is an important step in creating a cadre of providers equipped to save gun victim lives, and this is happening but needs to be expanded (Onufer et al., 2020). Surgeons in training now receive gun injury hands-on training. Washington University School of Medicine's course titled "Anatomy of Gun Violence (AGV)" is a multidisciplinary curriculum that trains surgical residents to treat gun victims, including a bleeding control unit. Students, in turn, can act as instructors to other health providers (Hamilton, 2020):

> The AGV curriculum also contained a section on STOP THE BLEED®, training residents with hands-on practice of direct manual pressure, tourniquet application, and wound packing to stop severe bleeding. Residents assembled their own bleeding control kits during the training and, in their second year, if they had previously completed the STOP THE BLEED® session, they had the opportunity to serve as an instructor for a session with hospital environmental services workers. Serving as an instructor gave residents the chance to teach lifesaving skills to members of their community.

Introducing gun injuries in surgical curriculum (technical and nontechnical skills) bodes well for future effectiveness.

POTENTIAL FOR POSTTRAUMATIC GROWTH

Can anything good come out of a gun injury? This question will generate a range of complex responses, and most of them will be negative. Yet this book is replete with examples of much good emanating from a tragedy. Concepts such as strengths, growth, grit, natural supports, assets, protective factors, and resiliency, and many others, can be found throughout this book. There is a fundamental difference between rebounding and thriving, with posttraumatic growth capturing the latter. The saliency of this possible growth opens up an arena for the field.

Resiliency, for example, is a construct that is associated with a post–adverse episode that does not result in a trauma (Nugent, Sumner, & Amstadter, 2014). Life is a highway of roadblocks and setbacks, and resilient individuals manage to socially navigate these barriers. Posttraumatic growth, in turn, reflects victim advances beyond a pretraumatic episode conceptualized as troubling or challenging.

Recovery and growth from gun violence clearly fall into a posttraumatic growth construct (Kaufman, 2020):

> To be sure, most people who experience posttraumatic growth would certainly prefer to have not had the trauma, and very few of these domains show more growth after trauma compared to encountering positive life experiences. Nevertheless, most people who experience posttraumatic growth are often surprised by the growth that does occur, which often comes unexpectedly, as the result of an attempt at making sense of an unfathomable event.

Posttraumatic growth can transpire in a variety of domains, highlighting the complexity of this construct. Kaufman (2020) notes that there are potentially seven areas of personal growth from traumatic experiences, with implications for research and practice: "(1) Greater appreciation of life; (2) Greater appreciation and strengthening of close relationships; (3) Increased compassion and altruism; (4) The identification of new possibilities or a purpose in life; (5) Greater awareness and utilization of personal strengths; (6) Enhanced spiritual development; and (7) Creative growth." Readers will see these potential outcomes throughout this book focused on urban context and the experiences of people of color. Community practice in the violence prevention field can enhance and help direct these outcomes for the benefit of survivors and their communities.

CONCLUSION

This chapter's educational recommendations will prove challenging to implement because they require serious rethinking of how best to meet the needs of gun victims, their families, and communities. We need a cadre of health and social science academics to move the gun violence field forward by expanding their contributions to empirical evidence and conceptual analyses (Lynch, Bateman-House, & Rivera, 2020). This necessitates investments in recruiting and preparing future academics. Professional accrediting societies must lead, and professions will follow. This education can occur in formal institutions of higher learning and through continuing education.

11

Research

When I was shot, it changed my life—and the lives of everyone in my family—completely. We'll live the consequences for the rest of our lives, as will the families of other survivors, a group that grows by hundreds every single day. As survivors, our stories are all different, but together, they show how urgently we need action to prevent gun violence.
—Karina Sartiaguin, the unintended target of a 2010 drive-by shooting outside of her school, Central High School, in Aurora, Colorado (Everytown USA, 2020c)

INTRODUCTION

A nation that ignores gun injuries is, in reality, a misinformed country on gun violence, and this makes it arduous to find viable solutions. Grasping gun violence's reach requires research capturing its visible and invisible consequences, and there is never a singular victim, but rather victims. There is no magical solution for gun violence; it necessitates quantitative and qualitative research that is highly participatory and community capacity enhancing, too.

The quest to better understand African American victims of violent injuries, for example, calls for advancing research methodologies predicated on important nuances resulting from culture, lifestyles, and environmental factors (Vil, Richardson, & Cooper, 2018). It is difficult to fathom how gun violence exposure cannot assume a more prominent place within these types of measures when viewed from an urban perspective, and more so in high-violence contexts. I realize the challenge to capture this dimension. Nevertheless, that fact that it is hard to do must not be an excuse.

There is no disputing the importance of accuracy in research, but it takes on added significance in gun violence. Sondik (2021) sums up the rationale for accurate gun violence data:

Perhaps the key characteristic that data on gun violence—or for that matter any data used to inform public policy decision-making—is public trust; trust

The Silent Epidemic of Gun Injuries. Melvin Delgado, Oxford University Press. © Oxford University Press 2022.
DOI: 10.1093/oso/9780197609767.003.0011

that the data are an accurate representation of what it purports to be. In the case of information on firearms, accurate data are essential to both the public health and public safety sectors in order to realize the full scale and scope of gun violence. This in turn informs prevention, interventions, policy and legislature that can improve the lives of all Americans by creating a safer and less violent country. (p. 20)

Improving overall injury data will bring reliability of gun-related data (Hemenway, 2020). There is no escaping the role that data-driven solutions have for reducing gun violence (Joseph, Bible, & Hanna, 2020). The longer gun victim advocates work in this arena, the greater the acknowledgment that research can play a prominent role in shaping advances in this field.

The Gifford Law Center (Nieto & Mclively, 2020) posits that federal funding for community violence "has been insufficient, unfocused, and sometimes harmful, contributing to the mass incarceration." It is not surprising that a similar conclusion is applicable to gun violence research. Expanding the evidence base by bridging research and practice goes a long way in addressing gun violence, including reaching out to the criminal justice system (Rodriguez, 2018), which will prove arduous from philosophical and practice standpoints.

High-violence communities should not be viewed as separate from those seeking to serve them, within and outside of the academy. Significant advances rest on a foundation of multidisciplinary and community-centered partnerships to advance research. This is not to say that gun victims have not been the focus of research attention. However, this attention was interrupted by a pullback of funds from the Centers for Disease Control and Prevention (CDC). Fortunately, that has changed.

Perpetrators of gun violence must weigh in concerning their reasons for engaging in this violence (Cook, Pollack, & White, 2018). This will be challenging when involving perpetrators in correctional custody, for example. A comprehensive understanding of gun violence cannot be complete without knowledge obtained from perpetrators.

If we construct a trauma-care bucket list, gun violence would occupy a prominent spot (Søreide, Weber, & Thorsen, 2020):

Defining a research agenda or even a priority among the topics viewed as "research worthy" may prove hard. Clearly, priority gets based on perspective, which again is all about location and situation of the beholder. So, for a US trauma researcher, among all of the important research themes that may come up on any given agenda, the prevention of deaths and injuries caused by gun violence cannot escape attention, as it is a major trauma-related research challenge from so many perspectives. (p. 2051)

Finally, Shultz, Ettman, and Galea (2018) argue that population health science has a role in informing gun violence research because it is the nexus of various perspectives that include social and medical sciences, economics, and demography.

A narrow research conceptualization informing gun policy and practice is ill advised. Gun injuries transpire within a broad social context, with many different forces at work shaping their manifestations and the ultimate goal of health equity (National Academies of Sciences, Engineering, and Medicine, 2017).

RESEARCH TRANSFORMING PRACTICE

Gun research must not limit itself to informing practice but transforming it (Carlson, 2020b). Being able to accurately assess violence measures is critical in evaluating the effectiveness of policies and interventions (Sondik, 2021). This takes on added significance at the local level. Effective policies on gun injuries are not possible without corresponding data to guide them (Stern & Zhang, 2017, p. 73):

> As a result, policy makers, law-enforcement officials, public-health experts, urban planners, and economists are all basing their work on information that is unproven or incomplete. Without more data—without identifying who commits shootings, where, how, and against whom; without plotting their rise and fall, to correlate with potential contributing factors; without analyzing those questions on a national, regional, local, neighborhood, and individual basis—it's impossible to tell which public policies and interventions could be most effective at reducing gun violence.

Repositories of gun injury data can play an important role in shaping locally driven solutions. There is a call for development of a regional hospital-based violence data repository to close a gap in knowledge on identification and outcomes of gun victims, as done in St. Louis, with particular attention to those with nonfatal injuries (Mueller et al., 2020). We can house these repositories in level 1 trauma centers, which, in turn, assume active roles in shaping localized responses.

The number of gun injuries is staggering, and research has not kept up with demand for knowledge (Powell & Sacks, 2020):

> Injury due to firearms is a serious health issue in the USA, leading to nearly 40,000 deaths annually and many more non-fatal injuries. Despite the significant impact on morbidity and mortality, relatively little research funding is dedicated to understanding the impact of firearm-related injury and to developing strategies to mitigate harm. It is certainly understandable why we need to elevate the public in any form of public safety. (p. 2182)

Helping professionals have been slow to embrace this stance.

Developing an awareness and competency for navigating data sets, such as those that are police centered, requires tailoring of requests as outlined by Ryley (2020). We must learn why the percentage of solved shootings is so low and start with requesting an agency's documents listing all the fields in the databases they seek to use. Further, preventing and intervening early in gun violence, without question, is not achievable without the public or community, with research embraced as a natural part of urban existence, and integrated into daily life whenever possible.

Any coordinated response to gun violence from federal and state departments requires sharing of facts and data. Transformation of gun research is a sharing process. Enhancing access to federal data for research purposes requires data sharing on gun traces, such as that possessed by the Bureau of Alcohol, Tobacco, Firearms, and Explosives (Tiahrt Amendment to the 2003 and 2004 federal budgets) (Roman & Van Ness, 2020). A change in presidential administration can result in Executive Orders to facilitate this access, and there is hope that the Biden presidency will accomplish this.

Gun injury research findings require dissemination in a manner that increases the likelihood of academics reading this material. A call for journal special issues on gun injuries offers great promise for moving this field forward (Sacks & Chow, 2018; Steinbrook, Stern, & Redberg, 2016) by providing academics with an opportunity to publish with other like-minded colleagues from across the country. But caution is needed that a call for increased research funding must not translate to garnering multiple forms of data from multiple sources, eschewing developing reliance on one source. Developing a composite picture requires data to reduce biases, requiring coordination and a systemic sharing of data, as addressed in the following section.

Gun violence is a national problem severely underfunded from a research standpoint. This may have to do with the disproportionate number of low-resourced communities and communities of color bearing the worst of this violence. Interventions must minimize or prevent use of guns (Oliphant et al., 2019). Community initiatives need guidance by scientific research, as presented in this testimony to Congress (Morral, 2019):

> 1. We know little about gun violence and its prevention compared to other safety and health threats, because the federal government has not had a comprehensive program of research in these areas for decades. 2. Additional, high-quality research is needed to craft policies that could contribute to reducing gun injuries, deaths, and violence. 3. There are many ways Congress could help build a robust and transformative gun policy research enterprise. (p. 2)

We can take the points raised by Morrel and apply them to secondary and tertiary interventions because of the role research plays in developing interventions of national significance. Further, the issue of racism enters into this deliberation (Bernstein, McMillan, & Charash, 2019), becoming part of broader health inequities deliberations, including gunshot outcomes (Cheon et al., 2020; Delgado, 2021).

For research to inform practice and policies, it must be meaningful in purpose and scientifically sound. Mind you, no research study is without limitations, and gun injury research is no exception. Few in the field would argue against more research on this subject (Hofmann et al., 2018). That call is not controversial. The research called for on gun injuries in marginalized communities of color will be dramatically different from that in resource-rich communities, with local circumstances shaping methodology and questions. Urban communities of color are disenfranchised in our society (Burrell et al., 2021). Research, therefore, cannot ignore this reality.

INTEGRATION WHENEVER POSSIBLE

The promise of evaluating potentially dangerous situations, such as those found in high gun-violence-prone neighborhoods, includes navigating model fidelity adherence, sample recruitment and retention, documentation, data collection, and other dimensions associated with quasi-experimental designs (Brisson, Pekelny, & Ungar, 2020). Research findings are only as good as efforts made to have them provide an accurate reflection of a phenomenon at the time it is undertaken and the degree to which respondents trust researchers (Van Brocklin, 2018b):

> Gun violence reporting [research] is not all about statistics and physical injury. It's equally important to learn about the day-to-day realities gunshot survivors face. A single bullet can cause chronic pain, life-threatening infections, and isolating fear. It can also transform someone's personal relationships, hamper their ability to get around their home or city, and place a huge financial burden on them or their family relationships, hamper their ability to get around their home or city, and place a huge financial burden on them or their family.

Postoperative infections, for example, are a serious concern, complicating the healing process and even resulting in death (O'Brien, Gupta, & Itani, 2020). The knowledge garnered through these interactions shape interventions with lasting meaning in helping victims recover, and possibly helping others with similar challenges.

Understanding the breakdown of services in helping those injured by guns receive services and rehabilitation requires understanding the continuum of service provision from the initial injury to reintegration into their communities. Woodside, Caldwell, and Calhoun (2020) provide a definition of a service provision breakdown, which was originally intended for the service industry but has equal applicability to health and human service provision and gun injuries: "A service breakdown is an event/action that prevents the completion of a service act/process in typical or reasonable time. Service breakdowns range in severity from a loss of the time or money to correct the breakdown to the loss of life of the customer/client/patient or service associate." To gain this knowledge, a wide range of research is needed, including program or service evaluation, as well as research that is qualitative and quantitatively maximizing input from gun victims and their social networks.

Urban gun injury researchers must aspire to achieve meaningful integration within the community and ideally engage gun social networks (Bourgois et al., 2019). This necessitates they be trusted, requiring a role description that has them attending courts, hospitals, houses of worship, and other institutions as part of the support system of victims. A spiritual or religious perspective on gun victims has not gotten due attention in examining the multifaceted ramifications of this experience, calling these institutions to play a systematic role in aiding survivors and their families (Rosell, 2020) from spiritual, advocacy, and service provision perspectives.

RESILIENCY

Resiliency must always be part of the conversation on gun injuries. A concerted effort at identifying resiliency aids researchers and practitioners in countering vicarious trauma. It is possible to pay attention to dire consequences of gun injuries without losing sight of resilience. Kotlowitz's (2019) *An American Summer: Love and Death in Chicago* spotlights that city's tragedies and resilience, often easy to overlook in a rush to document hurt and how victims and families respond, including how they experience trauma and guilt. How gun victim families adjust, forgive, and grow has much to teach us about bouncing back for victims, their families, and communities.

One cannot help but marvel on how some gun injury survivors transform their lives in response to violence that could have easily altered their life in a negative manner. Sometimes they have even assumed the position of leading spokespersons on the subject of gun violence, bringing experiential legitimacy to this subject. There is a dramatic difference between expertise earned from books and that earned from being a survivor. Ideally, having both only enhances this legitimacy.

Readers, I am sure, are versed on this construct in a wide range of adversities. Nevertheless, the field of gun injuries has not benefitted from this type of attention (Van Brocklin, 2018b). For example, mentoring has shown promise in tapping both victim and neighborhood resiliency (Wiley, 2020), also opening up the door for creation of a cadre of gun survivors wishing to go this route in contributing to their communities.

Although the likelihood of obtaining gun injury research funding is greatest when focused on the consequences of this traumatic act, it does not mean that valuable information is not obtainable on their resiliency and social network; this information is critically important for developing a balanced picture of victims and obtaining insights into systems-of-care failures. Gun injury victims, particularly, when overcoming barriers to adjusting to their new status and identities, have much to teach us on how to structure services (Van Brocklin, 2018). Resilient gun victims likely have social networks helping them in the recovery process (Wang, 2016). How these networks are developed and sustained influences treatment plans and outcomes.

MAXIMIZING PARTICIPATION

Making research relevant to urban gun injury victims and their families is dependent on obtaining their participation in the research, providing valuable insights into their lives post injury. Getting their participation in research requires addressing their perceptions of research goals and overcoming their distrust of researchers (Bruce et al., 2016). Seeking community input may find differing opinions that must be reconciled. Attempts at eliciting community stakeholder perspectives on youth violence will often elicit overlapping responses and divergent views (Ross et al., 2021).

Van Brocklin (2018b), a leading reporter on gun violence, issued recommendations for media reporters on gun victims that also apply to researchers:

> If you don't already belong to a community that experiences high rates of gun violence, it's helpful to identify the gatekeepers in communities that do, and who are likely to know survivors: activists, prevention workers, violence interrupters, and the heads of anti-violence nonprofits. . . . Say you're sorry for what happened to the person you're interviewing. Give them a heads-up before asking a tough question. Be prepared for a wide range of emotional responses––despair, anger, bitterness, confusion, numbness—and approach any response with kindness and understanding. The anniversary of a person's shooting can be an especially tough time, so be extra sensitive about reaching out or setting up an interview around then. If they aren't up for talking, give them space, but also check in later.

These recommendations are commonsense. When these interviews fail because of callousness or ignorance, it sets back the movement for understanding gun injury survivors. Monetarily incentivizing participation adds motivation when discussing injury victims from low-resourced communities. However, this funding is insufficient. We must endeavor to hire residents, and preferably those who have suffered gun injuries, to be part of research teams.

Enlisting residents in assuming reporting roles opens the door to innovative approaches such as the Philadelphia Center for Gun Violence Reporting. This organization stands out as a model calling for initiatives to improve reporting by enlisting credible reporters from communities most affected by gun violence to increase accurate reporting (Philadelphia Center for Gun Violence Reporting, 2021). These efforts not only serve as a vehicle for empowerment but can also build on community assets.

POTENTIAL RESEARCH AGENDAS

Research is a discipline best viewed from a variety of approaches. For example, Metzl (2019) advances the importance of integrating cultural meanings into shaping gun research:

> Addressing guns symbolically allows guns researchers to address questions that are counterintuitive in addition to ones that are self-evident. In what ways to guns shape identities?, might be one such question. Or, in what ways are guns sublime? Such an approach also provides means of better understanding, and

then pushing back on stigmatizing stereotypes that surround gun violence—
such as the stereotypes of race promoted by the NRA, or assumptions that
"mental illness" alone causes mass shootings. (p. 4)

Metzl, McKay, and Piemonte (2021) propose a research agenda on gun injuries
based on mass shootings and multiple-victim shootings, which takes into account
structural factors with particular significance for people of color.

Hospitals represent an arena for further attention from advocates. Hospital
prominence in treating gun injuries requires they receive increased attention
in obtaining research funding. Hospital-based violence prevention programs
originated in the 1990s, but despite their almost 30-year history, research on their
effectiveness is still sorely lacking because of insufficient funding on gun violence
research (Richardson & Bullock, 2021).

There is merit to research on the visible consequences of gun violence. Our
challenge is to broaden the field by better understanding the overlooked (invisible) aspects of gun injuries that have profound consequences for victims and
their social network. Gun violence erodes trust among survivors (Wu, 2020),
impeding their sense of social and psychological wellbeing, an injury that will go
unreported. This outcome also compromises community wellbeing. When gun
violence occurs during childhood, it translates into a lifetime of difficulties in
establishing trusting relationships, which is essential for seeking fulfillment and
stability in life.

Rosenberg (2021) identified four critical questions that should guide the structure of gun research sponsored by the CDC and National Institutes of Health that
readers may find meaningful:

First, what is the problem: How many people get shot, who are they, where
does it happen, what is the relationship between the shooter and the victim,
what other types of damage are incurred, and are the shootings increasing
or decreasing? Second, what are the causes: What is the role of alcohol and
drugs; what is the role of gangs, poverty, and systemic racism; what is the role
of mental illness, robbery, and domestic violence; what is the role of private
gun ownership (both positive and negative) and easy access to guns? What
are the factors that protect us, such as stable families and safe environments?
Third, what works: Which practices, interventions, policies, and laws work
best to prevent these deaths and injuries? And fourth, how do you do it: How
do you implement the findings and translate them into policies, legislation,
and practices that can be scaled up?

I do not think readers are interested in seeking answers to all four questions.
However, I am sure that at least one question will hold appeal. I find the third
set of questions particularly appealing, for example, including emphasizing what
increases resiliency and transforming an injury into an opportunity to give back
to their community.

In turning to specific research on the case illustrations used in Section III, Stop the Bleed campaigns offer tremendous promise for urban communities. Expanding this campaign requires a broad research agenda beyond the coverage it received in Chapters 7 and 8. A Delphi study on developing a research agenda for layperson prehospital hemorrhage control initiatives uncovered five thematic areas (Goralnick et al., 2020): (1) epidemiology and effectiveness, (2) materials, (3) education, (4) global health, and (5) health policy. Each cluster guides future research, particularly when viewed from an integrative stance. Unfortunately, this case illustration and comparable efforts in other settings have not benefitted from a national campaign to evaluate them.

Research specifically focused on urban youth must ascend in focus and importance. Waterman's (2020) taxonomy of potential developmental consequences of traumatic events, such as gun injuries, for example, has immense implications for urban adolescent groups: "(a) identity resilience, (b) identity affirmation, (c) identity delay, (d) identity threat, (e) identity loss, (f) identity alteration, (g) identity replacement, (h) trauma-shaped identity, and (i) trauma-centered identity." This taxonomy has implications for shaping research, practice, and educating a cadre of gun violence specialists interested in youth.

Finally, researchers of color have a pivotal role in advancing a research agenda. If gun violence research is a career killer for academics, what does it say about the career chances of researchers of color helping lead the way in addressing this national epidemic? There is no argument that we are in a unique position to bridge the divide between community and institutions entrusted to treat gun injuries. Brokering brings its challenges, leaving researchers of color to negotiate the politics of funding and implementing an agenda well understood by victims and misunderstood by institutions serving their needs.

A CALL FOR A NEW GENERATION OF RESEARCHERS

These recommendations are not surprising. Federal funding of gun violence research is essential in addressing health and safety because of the sheer amount of funds made available, and the nature of the peer review process helps increase scientific validity of findings (Rajan et al., 2018). Lack of research funding translates into a corresponding paucity of scholarly publications, raising ethical concerns, too, on the imperative of advancing this field (Cone et al., 2021; Siegler, 2020). A national health epidemic without substantial federal research support says a great deal about the significance of how this problem is viewed.

Limited governmental research support also translates into researcher hesitancy for fear that careers will be compromised (Taichman, Bornstein, & Laine, 2018). One public health official captured the risks quite well (Follman, Lurie, & West, 2015): "Do you want to do gun research? Because you're going to get attacked. No one is attacking us when we do heart disease." This observation is significant. Stark and Shah (2017) concluded based on a literature review of gun injuries and death over a 10-year period (2004 and 2014) that less than one twentieth as many scientific publications on gun injuries and deaths existed when compared to other causes of death in the United States with similar numbers of people killed.

Chien et al.'s (2020) gun violence literature review from 1981 to 2018 found a similar significant downward trend, but with a publication ratio increasing since the 2010s. Increased scholarship translates into informing current and future practitioners and researchers. Alcorn's (2017) study of US gun violence research publications (1960–2014) concluded that thousands of studies were published during the study period, with the annual number of publications not increasing between 1998 and 2012. However, it increased in 2013 and 2014, but there were few active career researchers. Paucity of federal funding for gun violence research does not preclude funding from other sources because of the immensity of the problem (Hills-Evans et al., 2018):

> Consistent financial and leadership support from academic and private sector institutions is currently lacking but desperately needed to overcome this lack of federal funding for research. Other disease-specific organizations, from breast cancer to suicide prevention, have been successful in raising public awareness and research funding . . . ; gun violence prevention foundations could learn from this model to raise the financial resources needed to attract research interest and proposals, to motivate communities to stand in solidarity to address this public health crisis, and to help initiate collaborative research teams in hospitals, clinics, and communities around the country. Without reliable funding, motivated investigators will continue to be unable to build careers dedicated to gun violence prevention research.

Availability of nongovernmental funding brings potential for supporting innovative projects, critical in advancing this field in communities of color. General absence of a public funding agenda relative to its importance as a public health problem speaks volumes on this issue as a "third rail" in politics.

Galea et al. (2018) refer to this period as the "lost generation of firearms research," issuing a challenge for developing a 21st-century research cadre in response to this lacuna. It bears noting that federal funding for FY 2020 was increased (Jaffe, 2020). Researchers need to play an increasingly important role in shaping a public health agenda, with other helping professions sharing similar values on gun injuries and deaths (Galea & Vaughan, 2019):

> In their response to gun violence, public health practice and academic public health have lagged substantially behind in their study of one of the defining epidemics of our time. Public health practice has been hampered in no small part by political forces. . . Challenges with firearm violence research include that we need more scholarship in the field and that we need to know enough to guide action. Public health action must almost always act in the absence of imperfect information, and in the area of guns we know plenty to move us past the current state of policy inaction. And yet, growing population health science provides tendrils of hope that we are at a crucial time in the field, which

was first seen in 2017 and has now been sustained throughout the past couple of years, portending a future when we have abundant science that can inform an engaged public health practice and when the US firearm epidemic starts to beat a retreat. (pp. 1490–1491)

This renewed hope for well-crafted policies and interventions on gun victims depends on innovative research that increases the number of stakeholders setting an agenda.

CONCLUSION

I understand that not every reader is interested in the field of research, although our field would benefit from practitioners playing an increased role in this endeavor. Research needs to help paint a picture of the problem and the solution. Reliable gun injury data must be obtained at all three governmental levels: local (neighborhood and block), state (including county), and national (Everytown USA, 2020c). More specifically, to understand urban violence, we need to focus our attention on city blocks within neighborhoods (Vargas, 2016). Business as usual, however, is ill advised if we are to paint a complex picture of this phenomenon rather than a simplistic rendition. Bringing together researchers and practitioners in urban community partnerships is a noteworthy goal as outlined in this chapter.

12

Community Practice

*I'm pretty sure there are multiple videos on the internet about anti-gun vio-
lence and stuff, but obviously it is not helping anything. So I feel like a new
approach like stepping out into the communities and gaining a personal relationship with them and when I say that I don't just come talk and then I never
see you again in my life. That is pointless and that is going to make me dislike
you. Make it a schedule. Make it something that is actually beneficial and not
something that is just going to happen one time.*
                    —*Unidentified youth (Beck, Zusevics, & Dorsey, 2019, p. 69)*

INTRODUCTION

There is an overlap between researching gun victims and community practice. Nina Vinik, director of the Joyce Foundation's Gun Violence Prevention &
Justice Reform Program (2021), arrived at two conclusions: " 'First, there's no
"silver bullet" that will end gun violence. We need a comprehensive approach that
includes stronger gun laws, community-based solutions, and more attention to
the root causes of violence', . . . 'And second, the federal government has a long way
to go to adopt research-informed solutions to this public health and safety crisis.' "

Readers, too, have seen how gun violence disproportionally impacts a significant segment of this nation and why this is so, with the promise of a public health
stance. Gun violence must not eschew engaging in dialogue and embracing a
change agenda on race and class. Failure to do so relegates public health to the
sidelines of community-centered solutions.

If guns are the source of so much destruction and pain, then what happens if
we eliminate them? Would gun violence occur in high-resourced communities?
What would happen if women perpetrated this violence toward men? Social
forces are not easily dismissible because of their strength and tenacity in resisting
change.

Community practice is dependent upon our vision of what community means.
If we place it at the center of these efforts, then institutions, such as hospitals
and others, are there to serve communities, rather than the other way around.

The Silent Epidemic of Gun Injuries. Melvin Delgado, Oxford University Press. © Oxford University Press 2022.
DOI: 10.1093/oso/9780197609767.003.0012

Community practice is labor intensive. The slogan "bottom-up" ascends in importance in community practice, translating into a goal of bringing together diverse resident groups in pursuit of a common goal (Howard & Rawsthorne, 2020). Community practice is predicated upon an ecological foundation (Wilkinson et al., 2018), with relationships (Westoby, 2017) forming the glue allowing what appear as disparate elements to pursue a common agenda.

Readers will never be at a loss finding a definition of community practice that best suits them and the communities where they practice. Community practice is an undertaking of purposefully structured activities premised on an explicit set of values that emphasize the worth of residents and their capacities to create positive change in their lives and communities.

A moral basis for shifting a vision for our roles from caring for victims to ultimately healing and transforming them better characterizes the journey we must ultimately embrace to make significant differences in their lives, network, and community (Joseph & Reese, 2020). Healing broadens our arena beyond an individual. Trauma surgeons, for example, must transcend their medical roles and assume advocacy and activism roles, joining ranks with other professions, because their role must encompass the goals of treatment and prevention, bringing clinical and public health ethics to bear (Peetz & Haider, 2018). This stance brings added legitimacy (educational expertise) to the gun violence cause.

Gun violence gravity requires creation of a cadre of experts to develop and staff these initiatives (Fischer et al., 2020). This cadre brings unique skill sets and focus. However, their marching orders get shaped by our knowledge base. This base draws on co-created knowledge, providing a construct allowing residents, victims, providers, and researchers/academics to come together to shape how best to prevent and treat those injured by guns.

It is possible to be creative and develop projects, such as those that are service centered and increase youth civic engagement (Wray-Lake & Abrams, 2020). Social justice–service learning projects initiated by schools to engage youth dealing with gun violence and its consequences offer great promise (Delgado, 2016a). These projects have injured students playing active or consultative roles. Participants benefit from enhanced knowledge and skills while educating others, with engagement, education, and service working together.

Creating opportunities to give back to communities provides former gang members injured by guns, for example, the opportunity to have a role in these initiatives. Providing a platform to former gang members to share their experiences with violence is a viable strategy for performing a public duty to deter gang involvement (Whitney-Snel, Valdez, & Totaan, 2020); this allows them to perform educational roles within their respective communities and gives them a legitimized voice in the anti–gun violence movement.

NON–LAW ENFORCEMENT APPROACHES

Violence prevention approaches must encompass a wide variety of strategies. Goodwin and Grayson (2020) call for investing in supporting communities of color to help them address gun violence, helping to shift from a law enforcement/ criminal justice stance to a public health stance. Law enforcement has assumed various levels of engagement in urban gun violence, as to be expected.

Although many models for violence reduction practice have law enforcement playing different roles at varying degrees, the potential for reducing violence without involving the police has been explored and seven key strategies were identified (John Jay Research and Evaluation Center, 2020): (1) improve the physical environment; (2) strengthen antiviolence social norms and peer relationships; (3) engage and support youth; (4) reduce substance abuse; (5) mitigate financial stress; (6) reduce the harmful effects of the justice process; and (7) confront the gun problem.

The first six recommendations seem self-explanatory and were covered throughout this book. The last recommendation—confront the gun problem— is the only one specifically identifying gun violence that needs clarification. We need increased controls in legal gun access for those with violent backgrounds, and we must limit access to guns (e.g., impose waiting periods, require training), to reduce access to guns by young people. This gun control recommendation was touched upon in this book, but not to the same extent as the other six recommendations.

Local circumstances dictate which of these seven recommendations has the greatest saliency. Clearly, law enforcement as the leading, and possibly only, approach has not worked in this nation's urban communities. The remaining practice recommendations in this chapter broaden the scope of potential interventions to prevent and treat gun injuries.

REDUCING BARRIERS TO SERVICES

Surviving a gun injury translates into a journey of receiving a multitude of services that often require visiting various service institutions that may well be outside of one's community. Urban gun violence has increased the recognition that emergency medical transportation, and other barriers, such as financial, personal safety, and program credibility, will play a critical role in increasing survival rates.

Lack of transportation and other barriers shape service access post injury and are deserving of attention (Richardson et al., 2021): "These barriers included reluctance to use alternative forms of transportation services (i.e., bus or subway) due to potential encounters with rivals, increased risk of repeat violent victimization, the need to carry a weapon for protection, stigmatization, and symptoms associated with traumatic stress" (p. 43).

These services are essential in helping survivors regain their lives, even under difficult personal circumstances. Services must surmount potential barriers that fall into four categories (Delgado, 1999): (1) geographical/physical (services located within neighborhood and accessible physically for those who have mobility impairments); (2) psychological (whether the setting is emotionally safe); (3) cultural (understanding values, language, and other factors); and (4) operational (hours, fees, methods of payment, documented status, and paperwork requirements). It is insufficient to have two or three out of the four, for example. You need all four to maximize meaningful positive outcomes.

Slim (29 years old) articulates the importance (practical and symbolic) of making transportation available for receiving treatment (Richardson et al., 2020):

> [Uber Health]'s big, they just got me right here. I just caught an Uber to come right here through [CG-VIP], know what I am saying? So, whoever is giving [funding] for this situation, give them some more, give them some more money for that. It ain't going to hurt you, y'all got it. Because that sh*t helps . . . if you tell somebody, I got paradise over here and you living in hell, but I got paradise over here, if you come over I got paradise for you, but you gotta buy your own ticket over here though to get there. While we living in hell holes, I am glad you let me know about paradise over here (program), or I'd have to still be over there while you are eating, drinking, Grey Poupon and sh*t. I'd still be over there. (p. 5)

Transportation transcends moving from one place to another, highlighting why it holds such meaning and why creativity in tearing down barriers aids the rehabilitation process.

Urban indigenous establishments introduce possibilities for integrating helpers within settings that are accessible to victims. Internships and service-learning projects (Delgado, 2016a), utilizing community settings such as barbershops and beauty parlors, expand the terrain for practice and research on gun injuries within communities of color.

One means of effectively reaching gun victims is to support self-help organizations addressing this violence (Delgado, 2021). The Center for American Progress (2020) makes a similar recommendation:

> Given the success of community-based programs in addressing the disproportionate impact of gun homicides within specific communities, federal, state, and local governments should further source and resource these programs so that the programs can expand their efforts to contribute to an overall reduction in nonfatal gun-related crimes.

These efforts or initiatives build upon community capacity, opening up possibilities for future collaborative undertakings.

GENERATION OF RESOURCES TO COMBAT GUN VIOLENCE

To say we live in turbulent times is an understatement, and nowhere is this more apparent than in the nation's struggles to think about law enforcement, with efforts to defund police departments taking center stage during the summer of 2020. Efforts to defund the police and divert or reallocate funds have captured the nation's attention. Shifting from policing to neighborhood safety by reconceptualizing public service, with corresponding funding toward new approaches, is occurring across the nation (Pearl, 2020). Other less publicized efforts are ongoing and deserving of attention. For example, Kevin Wicks speaks to the importance of experiential legitimacy in hiring staff in localized efforts to address gun violence (Hylton & Eggers, 2020):

> In recent years, gun violence has consumed most of Kevin Wicks' life in Chicago. He's been shot 10 times, and both his son and his stepson have been killed in shootings. Since April, Wicks has made preventing gun violence his job. That's when he was recruited by Chicago CRED, which hires people to prevent violence between cliques and gangs primarily in the Roseland neighborhood. At night, Wicks goes out in the neighborhood and talks to young men in the middle of active conflicts. . . . CRED, which stands for "create real economic density" and provides counseling and job training, was founded in 2016 by former Education Secretary Arne Duncan. But after the death of George Floyd and historic protests demanding police reform this year, the organization has positioned its work as an alternative to public safety. The organization's belief is that law enforcement and arrests are often not the best long-term solutions and that someone like Wicks, with all of his experiences with violence, might actually be better than the police at keeping his community safe. Wicks said he never sees Chicago police officers do violence de-escalation or prevention work in Roseland. . . .
>
> "We don't want them to come down here, because they bring tension," he said. "They pull up after the shooting is over with and stand there 30 deep and then tell you it's a crime scene. OK, why is all the police standing here if this is a crime scene? Go do some crime work!" Instead of pursuing the perpetrators, the officers just end up hassling witnesses, Wicks said.

Alternatives to law enforcement provide communities of color with interventions that best match their community's needs, reducing the consequences of gun violence.

New funding sources have increasingly supported programs to hire social workers, medics, and other service providers, away from police officers, signaling a shift from conventional law enforcement to a public service approach (Walters, 2020). This shift reduces the number of violent situations that can result in gun use. Sources of this funding are varied, but they generally consist of taxes and reallocation of funding from law enforcement budgets.

The following are examples of how current efforts are transforming the landscape in this area. Miami-Dade County issued a tax on restaurants to help house those who are homeless. New Orleans outsourced responses to minor traffic accidents to a private company. Eugene, Oregon, developed a team (medic and a crisis worker) for deployment to 20% of all 911 calls. Colorado uses taxes from selling marijuana to establish mental health teams to accompany police on mental health calls. Seattle, a city in the forefront of shifting funds away from the police to social services, instituted LEAD (Law Enforcement Assisted Diversion), an initiative that frees law enforcement from making arrests of sex workers and those with small-time drug possessions. These and other efforts emphasize serving the public rather than protecting the public.

ADVOCATING AND LOBBYING

There is a desperate need for professions to lobby and educate elected officials on gun violence (Andrews et al., 2019; Strong et al., 2018; Taichman et al., 2017). Other than those who are injured and their loved ones, no one fully grasps the far-reaching consequences of gun violence in this country. Gun violence practice models involve communities. Children advocating for an existence free from gun violence do not have the same sway as their adult counterparts, but they still must be enlisted in this endeavor (Naik-Mathuria & Gill, 2020); children are killed by gun violence in high numbers, and they have a stake in this public health epidemic.

AN ARMY OF COMMUNITY ADVOCATES

Gun violence survivors must be given a voice in this public health arena (Jones, 2020; Rothschild, 2018). The concept of "advocate" is one understood regardless of profession or academic discipline. I have used the term "army" purposefully because we are at war with this epidemic, and the term conjures up images of taking this issue seriously. It is easier to think of advocates as adults because they bring the wisdom of experience and are naturally taken more seriously in our society. However, enlisting the support of other groups, such as youth, as part of a concerted community effort is essential. Professional athletes who have survived gun injuries, too, are in a strong position to act as advocates and spokespersons (Aradillas, 2021).

Such a cadre has gained attention overall, with particular significance in gun violence (Wong & Raphael, 2020), representing a broadening of what it means to seek community engagement. Athletes, too, particularly when having local roots in the communities they seek to influence, can assume advocate roles (Bembry, 2016).

EXPAND VIOLENCE PREVENTION INITIATIVES

Why expand violence prevention initiatives? It is fitting to include gun survivor testimonials in this chapter. Calabrese (2017) reports on six gun violence survivors and how their injuries reshaped their lives in new and unimaginable ways:

> The moment they were shot, as frightening and painful as it might have been, marked a new beginning. Their lives were forced onto another path and each of them had to adjust to a body that no longer felt like theirs. For many of the survivors, the new beginning came in the form of a purpose: Johnson has turned her trauma into a source of comfort for other women affected by domestic violence. Jones supports gun control and founded a startup to help

people with limited mobility like himself book travel plans. Brough organizes blood drives — recognizing the crucial need after he had lost so much blood when he was shot. All six survivors came to their own realizations of what it means to be "surviving." For Cusimano, it's not as simple as waking up or walking out the door. She clings to the idea that it's about "making something" out of her life — and it's a life that she has learned to be proud of: "I get my sense of power from myself."

These six testimonials are part of thousands of gun violence survivor stories that remind us why prevention initiatives must not be lost in search of better ways to treat injuries.

This book emphasized several major gun violence initiatives. There are, however, gun violence initiatives that emerge throughout the course of time at the local level. Readers, for example, may be curious about gun buyback programs and the role they play in reducing gun violence. Unfortunately, these programs, although popular and generating considerable local publicity, have generally proven ineffective in reducing gun violence (Carpenter, Borrup, & Campbell, 2020).

The community violence field is sorely limited in specialized practitioners, but this represents an opportunity for expansion and innovation with an urban focus, within and outside of trauma centers. Gun violence youth programs, for example, have not benefited from rigorous research designs to determine their effectiveness (Ngo et al., 2019), and this charge must be met if this field is to have an impact on gun injuries. The concept of safety co-production captures the potential of ecological changes for reducing violence in certain public places, which can involve use of lighting, removing distressed buildings, and creating green spaces; it is accomplished through reliance on inclusive practices (Ceccato & Assiago, 2020). This concept opens the door for involving professions we typically do not think about in gun violence campaigns.

Pinpointing a moment when a victim can either embrace a new positive identity or proceed along a path that will in all likelihood result in greater impairments, prison, or even death is a critical juncture for helping them make an affirming decision (Green, 2019):

> Investment in prosocial masculine identities and the divestment from his criminal identity that helps sustain desistance. On the other hand, following VAI, some men are unable to overcome the stigma associated with their injury and seek to overcompensate for their masculinity through hypermasculine coping (e.g., gun carrying). Unfortunately, these coping mechanisms only help to reinforce the offender's criminal identity and lead to further entrenchment in criminal behavior. (p. 306)

Thinking of this moment as a crossroads is an understatement. Nevertheless, this point, if properly identified and addressed, can dramatically shape a life from being a community liability to a community asset (Delgado & Humm-Delgado, 2013).

Violence intervention programs exist in a small fraction of the nation's trauma centers, and this has hindered development of best practices, with 25 hospital-based violence intervention programs meeting the needs of 10% of the nation's trauma centers. Not surprisingly, there is a desperate need to expand these programs across the nation (Coupet, Huang, & Delgado, 2019). Trauma centers are a critical component in any gun violence care system. Hospitals currently bear the brunt of gun injury treatment, making them ideal places for initiating hospital-based violence prevention initiatives (Kaufman & Delgado, 2020).

Neighborhood violence interventions must not just focus on individuals but also seek environmental changes (Kondo et al., 2017), such as reducing alcohol availability, improving street connectivity, and providing green housing environments (Kondo et al., 2018). Urban abandoned lots are potential sites for crime and also convey to residents they are not worthy of physical surroundings that are affirming and enhancing of their wellbeing (Teixeira, 2016). These environmental conditions must incorporate the development of community-centered violence prevention programs (Jay et al., 2019). Place-based strategies alter the environment to reduce violence, enjoying wide appeal in the gun violence field (Abt, 2019).

Neighborhood physical conditions influence substance abuse and violence, which includes guns. The Neighborhood Inventory for Environmental Typology (NIfETy) is an environmental observational assessment tool that identifies and assesses these conditions and is used with Google Street View (GSV) (Nesoff et al., 2020).

PROFESSIONAL MEETINGS AND ASSOCIATIONS

The mere mention of professional meetings and associations with this movement may seem odd. Advocates for curtailing gun violence can take a variety of steps to move the field forward. Professional societies, for instance, can economically reward states with tough gun access laws by having their professional conferences held there (Leavitt, 2018).

Urban activist organizations and advocates bring a dimension to the field that has generally gone unrecognized (Rothschild, 2018). Community activism covers a range of issues, from street violence to those focused on social justice, as in Chicago (Doering, 2020). Fostering activism at the local, city, state, and national levels is another example because of pressure exerted in gaining political support and resources (research and interventions), including those youth centered and directed (Giles, D., 2020; Rahamim, 2018). Social action engagement can be therapeutic and achieve social change. It is important to mention that African American/Black women started the Black Lives Matter movement, and the role women played in anti–gun violence activism is widely acknowledged (Delgado, 2021).

Gatherings of gun violence experts are increasing, bringing greater attention to this subject. In 2019, a major meeting of 44 major medical and injury prevention organizations and the American Bar Association, hosted by the American College of Surgeons (ACS), addressed three objectives (Bulger et al., 2019):

1. Identify opportunities for the medical community to reach a consensus-based, non-partisan approach to firearm injury prevention; 2. Discuss the key components of a public health approach and define interventions this group will support; and 3. Develop consensus on actionable items for firearm injury prevention using the public health framework.

These meetings shape gun injury research, but they require inclusion of other professions and organizations. Further, community leader gatherings can transpire regionally and nationally to share and enhance their abilities to confront gun violence at the local level. Task forces, for example, can be a step toward inaction. There is no denying that systematic and lasting change is not possible without them.

In 2019, a call for action on gun violence was issued by the American College of Physicians and six other medical and public health organizations, with the following being a partial list of organizations joining this call (McLean, 2019): Alliance for Academic Internal Medicine; American Academy of Allergy, Asthma, and Immunology; American Academy of Neurology; American Academy of Ophthalmology; American Academy of Physical Medicine and Rehabilitation; American Association of Clinical Endocrinologists; American College of Cardiology; American College of Chest Physicians; American College of Obstetricians & Gynecologists; American College of Preventive Medicine; American Geriatrics Society; American Medical Group Association; American Medical Women's Association; American Psychological Association; American Society of Hematology; American Society of Nephrology; American Thoracic Society; Association of American Medical Colleges; and C. Everett Koop Institute at Dartmouth. This list will expand in the future. The creation of multidisciplinary gun violence coalitions is supported by many in the field (McLean et al., 2019).

He and Sakran (2019) argue that the elimination of the Centers for Disease Prevention and Control (CDC)'s gun research moratorium was not enough, and that this organization must create a multidisciplinary task force to advance gun violence knowledge:

We call on the CDC to create a firearm injury prevention task force that includes physicians, community leaders, and advocacy organizations who will set priorities in research and advocacy. Although more data are critical to developing solutions that can be specifically tailored toward this public health crisis, an opportunity for clinicians to act now exists. (p. 195)

This call for a coalition must also prominently involve the community.

Another example is the American Foundation for Firearm Injury Reduction Medicine (AFFIRM). AFFIRM (2020) is a nonpartisan coalition of over 40,000 health care professionals, public health experts, and researchers seeking to eliminate and treat gun injuries. This coalition captures the spirit of this recommendation by bringing together providers on the frontlines with researchers, advocates, and others interested in solving this health problem.

Community engagement is a science and art form that applies to gun violence initiatives. The American College of Surgeons Committee on Trauma's firearm strategy team (FAST) has operated through an embrace of three guiding principles that readers can relate to for community engagement (Talley et al., 2019):

ACS COT has worked to develop a consensus strategy on how best to reduce the firearm injury death and disability on three guiding principles: (1) Advocate and promote a public health approach to firearm injury prevention; (2) Implement evidence-based violence prevention programs through the network of ACS COT-verified trauma centers; and (3) Provide, foster, and promote a forum for civil dialogue within our own professional organization with the goal of moving toward a consensus on programs or interventions aimed at reducing firearm injuries and deaths. (pp. 198–199)

The first two principles found traction within the ACS and other professions. I focus on the third principle because it can create forums encouraging free exchange of ideas leading to innovative gun violence prevention and treatment strategies. Having gun violence on professional conference agendas, and particularly those that are interdisciplinary, has great potential for moving the field forward. Wojtowicz, French, Alper, and the National Academies of Sciences, Engineering, and Medicine (2019) advocate for forums, emphasizing engagement of affected communities in these gatherings to break down barriers between the scientific community and victims of gun violence.

It is fitting to end this section with public health playing a catalyst role (Sathya, 2020):

It is our duty to change the public discourse around guns and focus on this as a public health issue. Similar to the tobacco debate decades ago, when it was taboo for doctors to ask about smoking, we must shift the paradigm and view gun-violence prevention as part of the routine health care we deliver.

COMMUNITY SERVICE PROVISION INTAKES

It is customary for service seekers to undergo a structured intake. These encounters represent opportunity points to identify gun violence in their lives. Practitioners will often find a strong association between social support, violence, and social service needs, including health care, in communities of color, as in Baltimore (Chandran et al., 2020). This relationship sets the foundation for a coordinated response to gun violence, particularly when injuries require ongoing service provision from multiple sectors. Social service organizations in high-gun-violence communities must add questions on this form of violence when conducting intake interviews of those seeking help.

Intakes are an opportunity point for gathering gun violence information and the extent of gun violence that may go unreported to authorities. For instance, we should know the lifetime number of gun injuries for those presenting for services that are not gun violence centered. Cumulative trauma may be lifelong, with none occurring in the past year yet wielding great influence in a person's life, calling for an expanded period to gather data on. Gun violence permeates a community's life; thus, all service systems must play a role in addressing this epidemic. This information helps us develop a better understanding of gun violence trauma at the local level.

CELEBRATION OPPORTUNITIES

Recognizing gun violence consequences, while maintaining hope for the future, is a balancing act, including triumphs against incredible odds that survivors and their communities face. An embrace of community assets and capacity enhancement is a theme in this book. Finding opportunities to celebrate accomplishments must be uplifted in communities with histories of tragedies, including identifying posttraumatic growth when present (Delgado, 2016b). Too often, a community's collective memories are of tragedies. Celebrating urban community life opens the door to honoring survivors who may become community assets by serving as role models for others with similar life situations. These celebrations can be broad or small.

KATHY SHORE

In personal correspondence with Kathy Shorr on her work with gun injury survivors, she responded to my question on key message(s) she wished readers of this book would take away from addressing gun injuries:

> Anyone, anywhere can be shot or a victim of gun violence. No one is immune and gun violence has to be treated as a national health crisis/emergency. This was one of the principles of my book, the attempt to show all kinds of people, all ages, many ethnicities, different socio-economic classes, from high and low profile incidents, gun owners and non gun owners from across America—all survivors of gun violence. Random shootings, crimes, accidents, gang violence, domestic violence, uncontrollable anger, mass shootings etc.—if we don't look at this collectively I don't think we will solve the problem.
>
> The dignity and power of the survivors—all wanting to participate in the project so that no one else will have to experience this nightmare. Their unselfishness and courage was inspirational for me. My overwhelming feeling when I left the great majority of people photographed was the positive energy, determination and transformation that they chose to accept and incorporate into their lives, taking this terrible experience and turned it into post traumatic growth. One of the reviews of my book likened seeing their physical scars to the Japanese art of Kintsugi or the repairing of broken pottery with

metallic powder thereby making a stronger and more beautiful piece of pottery than the original. From my experience, this is exactly what I witnessed photographing the strength and beauty of the survivors.

Gun injuries affect people for life. Even if the injury is non life threatening, the emotional trauma remains and can be triggered by such everyday things as a loud noise. Three of the survivors photographed (that I know of) have passed away from the injuries they sustained three–four years later. If you have been left paralyzed, you remain so for the rest of your life. Loss of limbs, brain trauma etc. require all new skill sets to live and are never as comfortable or acceptable as your previous physical/mental state.

Readers may be familiar with the construct of posttraumatic growth, which has great applicability in addressing gun injuries in communities of color (Lee et al., 2020; Orejuela-Davila, 2020).

CLOSING COMMENTS

It is an intense and all-consuming experience to write a book on any subject. However, when a book devotes itself to hurt, pain, tragedy, and social justice, as the nation struggles with police violence, no author is prepared to appreciate this toll. This book both inspired and frightened me because of how gun violence dramatically changes lives. We too often think of violence in the extremes, but that middle ground covers so much territory. Further, we can simply be at the wrong place and time to be a gun victim.

I am not so naïve to think that doing away with guns will eliminate violence, because it will not. But it will limit the damage caused and the number of causalities needing treatment. Guns make violent situations more lethal, having an instrumentality effect (Braga et al., 2020b). Inequities destroy hopes and dreams, both essential in creating a future for a community and nation.

Paralysis is an outcome few people think about when discussing gun injuries. For many, this outcome is worse than death or prison, and this is why paralysis receives special attention in this book. I was planning on writing about it, but as the book unfolded, it ascended in saliency because of the challenges it poses to survivors and their loved one. The challenges of meeting their needs, too, struck me because it requires thinking out of the box, which is always difficult. It is hard to miss the pain in survivors' eyes as they share their stories. This agony may never go away, but by helping others it can be transformed into wisdom.

We have marching orders that respect different gun injury foci and roles because there is no single path. It is a journey with detours and temporary setbacks, but it is one worth taking nevertheless. For some of us it means taking a stance at an agency or community level. Others approach this journey from a research, policy, or scholarly standpoint. Further, these roles are not mutually exclusive. Regardless of the role we play, we need others to join us in this march on ending gun violence and corresponding injuries to individuals, families, communities, and the nation.

The US gun violence epidemic continues as this book goes to press (Bates, 2020; Tedesco et al., 2020), compounded by COVID-19. This final section marks the end of this book, concretizing a national vision on gun injuries without losing sight of fatalities. I sincerely hope that the reader's view of gun injuries has broadened and that this book has provided an impetus or renewed energy to address this issue.

We must endeavor to capture the joy of life after suffering a near-death experience; it is never too late to make a difference in other people's lives and their neighborhoods. These perspectives and stories must not get lost by a focus on pain because the lives of survivors are so much more than the moment in time a bullet entered their bodies or the sound of a bullet retraumatizing them. Having a life transformed from a statistic to an asset needs celebrating.

Entrapment in a painful memory frozen in time is a prison sentence. Those seeking first-hand testimonies of how gun violence dramatically alters lives can go to Everytown USA (https://momentsthatsurvive.org/stories/). Violence is never an act claiming one victim; it claims multiple people and with lifetime consequences. Breaking through to create new dreams that empower and aid others will leave survivors and their communities in a better place.

Gun violence healing must encompass community healing. We do not have to know the victim. The residents where the violence occurred, the neighborhood where the victim lived, the institutions and organizations where they attended—they, too, experience this pain and need healing. This quest for healing emphasizes our humanity, belongingness, and hope. I have not argued in this book that gun injuries demand more attention than fatalities. They should receive equal treatment by providers and academics. I realize that this goal is a challenge, but it is worthy nevertheless.

REFERENCES

Abaya, R. (2019). Firearm violence and the path to prevention: What we know, what we need. *Clinical Pediatric Emergency Medicine, 20*(1), 38–47.

Abaza, R., Lukens-Bull, K., Bayouth, L., Smotherman, C., Tepas, J., & Crandall, M. (2020). Gunshot wound incidence as a persistent, tragic symptom of area deprivation. *Surgery, 168*(4), 671–675.

Abdalla, S. M., Keyes, K. M., & Galea, S. (2021). A public health approach to tackling the role of culture in shaping the gun violence epidemic in the United States. *Public Health Reports, 136*(1), 6–9.

Abdallah, H., & Kaufman, E. (2020, December 30). Gun violence is a public health crisis just as important as the epidemic: Opinion. *The Philadelphia Inquirer.* https://www. inquirer.com/news/gun-violence-shootings-philadelphia-crisis- 20201230.html

Abdallah, H. O., Zhao, C., Kaufman, E., Hatchimonji, J., Swendiman, R. A., Kaplan, L. J., . . . Pascual, J. L. (2020). Increased firearm injury during the COVID-19 pandemic: A hidden urban burden. *Journal of the American College of Surgeons, 232*(2), 159–168.

Abt, T. (2019). *Bleeding out: The devastating consequences of urban violence—And a bold new plan for peace in the streets.* New York, NY: Basic Books.

Affinati, S., Patton, D., Hansen, L., Ranney, M., Christmas, A. B., Violano, P., . . . Crandall, M. (2016). Hospital-based violence intervention programs targeting adult populations: an Eastern Association for the Surgery of Trauma evidence-based review. *Trauma Surgery & Acute Care Open, 1*(1).

Ahiagbe, A. A. (2020). *Missing targets: The ethical necessity of firearm injury prevention education* (Doctoral dissertation, Temple University, Philadelphia, PA).

Ahlin, E. M., Antunes, M. J. L., & Watts, S. J. (2021). Editorial introduction: Effects of gun violence on communities and recent theoretical developments. *The Journal of Primary Prevention, 42*, 1–3.

Akosua, M. (2019, December 9). "It takes the hood to heal the hood": Tackling the trauma of gun violence. *The Guardian.* https://www.theguardian.com/us- news/2019/ dec/09/gun-violence-richmond-bay-area-healing

Alang, S., McAlpine, D. D., & Hardeman, R. (2020). Police brutality and mistrust in medical institutions. *Journal of Racial and Ethnic Health Disparities, 7*, 760–768.

Alcorn, T. (2017). Trends in research publications about gun violence in the United States, 1960 to 2014. *JAMA Internal Medicine, 177*(1), 124–126.

Ali, S. S. (2019a, July 17). They're like soldiers: Chicago's children are learning to save lives amid the gunfire. *NBC News.* https://www.nbcnews.com/news/us- news/ they-re-soldiers-chicago-s-children-are-learning-save-lives-n1018196

Ali, S. S. (2019b, August 7). For some in Chicago, gun violence is a daily reality, leaving the same trauma as mass shootings. *NBC News.* https://www.nbcnews.com/news/us-news/some-chicago-gun-violence-daily- reality-leaving-same-trauma-mass-n1040231

Alizadeh, A., Dyck, S. M., & Karimi-Abdolrezaee, S. (2019). Traumatic spinal cord injury: An overview of pathophysiology, models and acute injury mechanisms. *Frontiers in Neurology, 10,* 282.

Altholz, R. (2020). *Living with impunity: Unsolved murders in Oakland and the human rights impact on victims' family members.*

Alvis-Miranda, H. R., Rubiano, A. M., Agrawal, A., Rojas, A., Moscote-Salazar, L. R., Satyarthee, G. D., . . . Zabaleta-Churio, N. (2016). Craniocerebral gunshot injuries: A review of the current literature. *Bulletin of Emergency & Trauma, 4*(2), 65–74.

American College of Surgeons. (2019, August 7). American College of Surgeons comments on the continual occurrence of firearm deaths and injuries in the United States. *News from the American College of Surgeons.* https://www.facs.org/media/press-releases/2019/firearm080719

American Foundation for Firearm Injury Reduction Medicine. (2020). Our mission: A public health approach to solving firearm injury. https://affirmresearch.org/

American Medical Association Resident and Fellow Section. (2020). https://www.ama-assn.org/system/files/2020–10/nov2020-rfs-report-e.pdf

Amnesty International. (2019). Scares of survival: Gun violence and barriers to reparation in the USA. https://www.amnesty.org/download/Documents/AMR5105662019ENGLISH.PDF

Anderson, E. (2000). *Code of the Street: Violence and the Moral Life of the Inner City.* New York: W.W. Norton & Co.

Anderson, E., & Kryzanski, J. (2020). Prognosis and futility in neurosurgical emergencies: A review. *Clinical Neurology and Neurosurgery.*

Andrade, E. G., Hayes, J. M., & Punch, L. J. (2019). Enhancement of bleeding control 1.0 to reach communities at high risk for urban gun violence: Acute bleeding control. *JAMA Surgery, 154*(6), 549–550.

Andrade, E. G., Hayes, J. M., & Punch, L. J. (2020). Stop the bleed: The impact of trauma first aid kits on post-training confidence among community members and medical professionals. *The American Journal of Surgery, 220*(1), 245–248.

Andrade, E. G., Hayes, J. M., Wood, I., & Punch, L. J. (2021). Reducing the incidence and impact of gun violence through community engagement. In M. Crandall, S. Bonne, J. Bronson, & W. Kessel (Eds.), *Why we are losing the war on gun violence in the United States* (pp. 255–264). New York, NY: Springer.

Andrews, J., Jones, C., Tetrault, J., & Coontz, K. (2019). Advocacy training for residents: Insights from Tulane's internal medicine residency program. *Academic Medicine, 94*(2), 204–207.

Antonucci, M. U. (2019). Firearm injury prevention. *Annals of Internal Medicine, 171*(4), 304–305.

Apantaku, E., & Emmanuel, A. (2020, July 20). Liston: Chicago youth leaders Miracle Boyd and China Smith reflect on activism, trauma, and growth. *Injustice Watch.* https://www.injusticewatch.org/commentary/2020/miracle-boyd-china-smith-chicago-police-protests/

Apelt, N., Greenwell, C., Tweed, J., Notrica, D. M., Maxson, R. T., Garcia, N. M., . . . Schindel, D. (2020). Air guns: A contemporary review of injuries at six pediatric Level I trauma centers. *Journal of Surgical Research, 248,* 1–6.

Apte, A., Bradford, K., Dente, C., & Smith, R. N. (2019). Lead toxicity from retained bullet fragments: A systematic review and meta-analysis. *Journal of Trauma and Acute Care Surgery, 87*(3), 707–716.

Aradillas, E. (2021, January 28). Professsional athletes who survived gun violence share stories of pain—and purpose. *People.* https://people.com/crime/professional-athletes-survived-gun-violence-share-pain- purpose/

Arceo, S. R., Runner, R. P., Huynh, T. D., Gottschalk, M. B., Schenker, M. L., & Moore Jr, T. J. (2018). Disparities in follow-up care for ballistic and non-ballistic long bone lower extremity fractures. *Injury, 49*(12), 2193–2197.

Armstrong, M., & Carlson, J. (2019). Speaking of trauma: The race talk, the gun violence talk, and the racialization of gun trauma. *Palgrave Communications, 5*(1), 1–11.

Aronowitz, S. V., Mcdonald, C. C., Stevens, R. C., & Richmond, T. S. (2020). Mixed studies review of factors influencing receipt of pain treatment by injured black patients. *Journal of Advanced Nursing, 76*(1), 34–46.

Aspholm, R. (2020). *Views from the streets: The transformation of gangs and violence on Chicago's South Side.* New York, NY: Columbia University Press.

Aspholm, R. R., St Vil, C., & Carter, K. A. (2019). Interpersonal gun violence research in the social work literature. *Health & Social Work, 44*(4), 224–231.

Assari, S., Caldwell, C. H., Abelson, J. L., & Zimmerman, M. (2019). Violence victimization predicts body mass index one decade later among an urban sample of African American young adults: Sex as a moderator and dehydroepiandrosterone as a mediator. *Journal of Urban Health, 96*(4), 632–643.

Associated Press. (2020, August 3). 8 dead in weekend Chicago shootings. *The Boston Globe,* p. A2.

Athavale, A. M., Fu, C. Y., Bokhari, F., Bajani, F., & Hart, P. (2019). Incidence of, risk factors for, and mortality associated with severe acute kidney injury after gunshot wound. *JAMA Network Open, 2*(12), e1917254–e1917254.

Attridge, M. M., Holmstrom, S. E., & Sheehan, K. M. (2020). Injury prevention opportunities in the pediatric emergency department. *Clinical Pediatric Emergency Medicine, 21*(1).

Austin, J., Schiraldi, V. N., Western, B. P., & Dwivedi, A. (2019). *Reconsidering the "violent offender."* New York, NY: Columbia/Academic Commons. https://academiccommons.columbia.edu/doi/10.7916/d8–556r-jv97

Avila, A. (2019, November 24). Chicago deep cuts. *The Culture Crush.* https://www.theculturecrush.com/feature/chicago-deep-cuts

Avraham, J. B., Frangos, S. G., & DiMaggio, C. J. (2018). The epidemiology of firearm injuries managed in US emergency departments. *Injury Epidemiology, 5*(1), 1–6.

Bachier-Rodriguez, M., Freeman, J., & Feliz, A. (2017). Firearm injuries in a pediatric population: African-American adolescents continue to carry the heavy burden. *The American Journal of Surgery, 213*(4), 785–789.

Balcazar, F. E., Magaña, S., & Suarez-Balcazar, Y. (2020). Disability among the Latinx population: Epidemiology and empowerment interventions. In A. Martínez & S. Rhodes (Eds.), *New and emerging issues in Latinx health* (pp. 127–143). New York, NY: Springer.

Ballesteros, M. F., Sumner, S. A., Law, R., Wolkin, A., & Jones, C. (2020). Advancing injury and violence prevention through data science. *Journal of Safety Research, 73*(June), 189–193.

Baltimore Sun Editorial Board. (2021, March 4). Not even the children are safe from Baltimore gun violence: Commentary. https://www.baltimoresun.com/opinion/editorial/bs-ed-0305-por-children-baltimore-crime-20210304-a3qgda3ydrf65es72c3tgxwuzy-story.html

Baranauskas, A. J. (2020). War zones and depraved violence: Exploring the framing of urban neighborhoods in news reports of violent crime. *Criminal Justice Review*, *45*(4), 393–412.

Barao, L., Braga, A. A., Turchan, B., & Cook, P. J. (2021). Clearing gang-and drug-involved nonfatal shootings. *Policing: An International Journal* .

Barna, M. (2020, February/March) US hospitals stepping up to end violence among youth: Interventions addressing gun traumas. *The Nation's Health*, *50*(1), 1–20.

Barnes, P. (2020). *Living with killing: The lived experiences of young Black men in South Chicago* (Doctoral dissertation, Walden University, Chicago, IL).

Barret, J. P., & Barret-Joly, J. (2020). Acute management of facial burns, acute versus long-term, surgical versus non-surgical face transplant. In M. Jeschke, L. P. Kamolz, F. Sjöberg, & S. Wolf (Eds.), *Handbook of burns, Vol. 1* (pp. 459–464). New York, NY: Springer, Cham.

Barton, A., McLaney, S., & Stephens, D. (2020). Targeted interventions for violence among Latinx youth: A systematic review. *Aggression and Violent Behavior*, *53*.

Bates, J. (2020, December 30). 2020 will end as one of America's most violent years in decades. *Time*. https://time.com/5922082/2020-gun-violence-homicides-record- year/

Bates, L., Kearns, R., Witten, K., & Carroll, P. (2019). "A level playing field": Young people's experiences of wheelchair basketball as an enabling place. *Health & Place*, *60*.

Bauchner, H., Rivara, F. P., Bonow, R. O., Bressler, N. M., Disis, M. L. N., Heckers, S., . . . Rhee, J. S. (2017). Death by gun violence—a public health crisis. *JAMA Psychiatry*, *74*(12), 1195–1196.

Bayouth, L., Lukens-Bull, K., Gurien, L., Tepas III, J. J., & Crandall, M. (2019). Twenty years of pediatric gunshot wounds in our community: Have we made a difference? *Journal of Pediatric Surgery*, *54*(1), 160–164.

BBC. (2014, November 29). Chicago's violence provides training for military doctors. https://www.bbc.com/news/av/world-us-canada-30243321

Beam, D. R., Szabo, A., Olson, J., Hoffman, L., & Beyer, K. M. (2021). Vacant lot to community garden conversion and crime in Milwaukee: A difference-in-differences analysis. *Injury Prevention*, *27*(5), 403–408.

Beard, J. H., Resnick, S., Maher, Z., Seamon, M. J., Morrison, C. N., Sims, C. A., . . . Goldberg, A. J. (2019). Clustered arrivals of firearm-injured patients in an urban trauma system: A silent epidemic. *Journal of the American College of Surgeons*, *229*(3), 236–243.

Beard, J. H., & Sims, C. A. (2017). Structural causes of urban firearm violence: A trauma surgeon's view from Philadelphia. *JAMA Surgery*, *152*(6), 515–516.

Beardslee, J., Mulvey, E., Schubert, C., Allison, P., Infante, A., & Pardini, D. (2018). Gun- and non-gun–related violence exposure and risk for subsequent gun carrying among male juvenile offenders. *Journal of the American Academy of Child & Adolescent Psychiatry*, *57*(4), 274–279.

Beaucreux, C., Vivien, B., Miles, E., Ausset, S., & Pasquier, P. (2018). Application of tourniquet in civilian trauma: Systematic review of the literature. *Anaesthesia Critical Care & Pain Medicine*, *37*(6), 597–606.

Beaumont Hospital. (2018, March 8). *Stop the Bleed*. https://www.beaumont.org/health-wellness/press-releases/learn-to-stop-the- bleed-after-firearms-related-violence

Beck, B., Zusevics, K., & Dorsey, E. (2019). Why urban teens turn to guns: Urban teens' own words on gun violence. *Public Health, 177*, 66–70.

Befus, D. R., Kumodzi, T., Schminkey, D., & Ivany, A. S. (2019). Advancing health equity and social justice in forensic nursing research, education, practice, and policy: Introducing structural violence and trauma-and violence-informed care. *Journal of Forensic Nursing, 15*(4), 199–205.

Beharie, N., Scheidell, J. D., Quinn, K., McGorray, S., Vaddiparti, K., Kumar, P. C., . . . Khan, M. R. (2019). Associations of adolescent exposure to severe violence with substance use from adolescence into adulthood: Direct versus indirect exposures. *Substance Use & Misuse, 54*(2), 191–202.

Bellware, K. (2019, September 3). Meet the young activists fighting Chicago's gun violence with lobbying and group hugs. *Teen Vog*. https://www.teenvogue.com/story/good-kids-mad-city-chicago-gun-violence- activists

Bembry, J. (2016, August 25). Hard stories on violence and loss: The undefeated holds conversation in Chicago on athletics, guns and the path forward. *Black History Always*. https://theundefeated.com/features/the-undefeated-holds-conversation-on-athletes-guns-and-violence/

Benateau, H., Chatellier, A., Caillot, A., Labbe, D., & Veyssiere, A. (2016). Computer-assisted planning of distraction osteogenesis for lower face reconstruction in gunshot traumas. *Journal of Cranio-Maxillofacial Surgery, 44*(10), 1583–1591.

Benner, P., Halpern, J., Gordon, D. R., Popell, C. L., & Kelley, P. W. (2018). Beyond pathologizing harm: Understanding PTSD in the context of war experience. *Journal of Medical Humanities, 39*(1), 45–72.

Benns, M., Ruther, M., Nash, N., Bozeman, M., Harbrecht, B., & Miller, K. (2020). The impact of historical racism on modern gun violence: Redlining in the city of Louisville, KY. *Injury, 51*(10), 2192–2198.

Berardi, L. (2021). Neighborhood wisdom: An ethnographic study of localized street knowledge. *Qualitative Sociology, 44*, 103–124.

Berdychevsky, L., Stodolska, M., & Shinew, K. J. (2019). The roles of recreation in the prevention, intervention, and rehabilitation programs addressing youth gang involvement and violence. *Leisure Sciences*.

Berg, M. T., & Mulford, C. F. (2020). Reappraising and redirecting research on the victim–offender overlap. *Trauma, Violence, & Abuse, 21*(1), 16–30.

Bernardin, M. E., Moen, J., & Schnadower, D. (2021). Factors associated with pediatric firearm injury and enrollment in a violence intervention program. *Journal of Pediatric Surgery, 56*(4), 754–759.

Bernstein, D. S. (2017, December 19). Americans don't really understand gun violence. *The Atlantic*. https://www.theatlantic.com/politics/archive/2017/12/guns-nonfatal-shooting-newtown-las-vegas/548372/

Bernstein, M., McMillan, J., & Charash, E. (2019, December). Once in Parkland, a year in Hartford, a weekend in Chicago: Race and resistance in the gun violence prevention movement. *Sociological Forum, 34*, 1153–1173.

Beseler, C., Mitchell, K. J., Jones, L. M., Turner, H. A., Hamby, S., & Wade Jr, R. (2020). The Youth Firearm Risk and Safety Tool (Youth-FiRST): Psychometrics and validation of a gun attitudes and violence exposure assessment tool. *Violence and Victims, 35*(5), 635–655.

Betz, M. E., Bebarta, V. S., DeWispelaere, W., Barrett, W., Victoroff, M., Williamson, K., & Abbott, D. (2019). Emergency physicians and firearms: Effects of hands-on training. *Annals of Emergency Medicine, 73*(2), 210–211.

Bieler, S., Kijakazi, K., La Vigne, N., Vinik, N., & Overton, S. (2016). *Engaging communities in reducing gun violence.* Washington, DC: The Urban Institute.

Black, N. R., O'Reilly, G. A., Pun, S., Black, D. S., & Woodley, D. T. (2018). Improving hairdressers' knowledge and self-efficacy to detect scalp and neck melanoma by use of an educational video. *JAMA Dermatology, 154*(2), 214–216.

Blumberg, T. J., DeFrancesco, C. J., Miller, D. J., Pandya, N. K., Flynn, J. M., & Baldwin, K. D. (2018). Firearm-associated fractures in children and adolescents: Trends in the United States 2003–2012. *Journal of Pediatric Orthopaedics, 38*(7), e387–e392.

Bobko, J. P., Badin, D. J., Danishgar, L., Bayhan, K., Thompson, K. J., Harris, W. J., . . . Fortuna Jr, G. R. (2020). How to stop the bleed: First care provider model for developing public trauma response beyond basic hemorrhage control. *Western Journal of Emergency Medicine, 21*(2), 365–373.

Boeck, M. A., Strong, B., & Campbell, A. (2020). Disparities in firearm injury: Consequences of structural violence. *Current Trauma Reports, 6*(1), 10–22.

Bohan, S. (2018). *Twenty years of life: Why the poor die earlier and how to challenge inequity.* Washington, DC: Island Press.

Boine, C., Siegel, M., Ross, C., Fleegler, E. W., & Alcorn, T. (2020). What is gun culture? Cultural variations and trends across the United States. *Humanities and Social Sciences Communications, 7*(1), 1–12.

Bonne, S., & Dicker, R. A. (2020). Hospital-based biolence intervention programs to address social determinants of health and violence. *Current Trauma Reports, 6,* 23–28.

Bonne, S., Tufariello, A., Coles, Z., Hohl, B., Ostermann, M., Boxer, P., . . . Livingston, D. (2020). Identifying participants for inclusion in hospital-based violence intervention: An analysis of 18 years of urban firearm recidivism. *Journal of Trauma and Acute Care Surgery, 89*(1), 68–73.

Borg, B. A., Krouse, C. B., McLeod, J. S., Shanti, C. M., & Donoghue, L. (2020). Circumstances surrounding gun violence with youths in an urban setting. *Journal of Pediatric Surgery, 55*(7), 1234–1237.

Boschert, E. N., Stubblefield, C. E., Reid, K. J., & Schwend, R. M. (2021). Twenty-two years of pediatric musculoskeletal firearm injuries: Adverse outcomes for the very young. *Journal of Pediatric Orthopaedics, 41*(2), e153–e160.

Boston Medical. (2020). *Boston Medical's Child Witness to Violence Project.* (https://www.bmc.org/programs/child-witness-violence-project)

Bourgois, P., Hart, L. K., Karandinos, G., & Montero, F. (2019). Coming of age in the concrete killing fields of the US inner city. In J. MacClancy (Ed.), *Exotic no more: Anthropology for the contemporary world* (pp. 19–41). Chicago, IL: University of Chicago.

Bowen, D. A., Mercer Kollar, L. M., Wu, D. T., Fraser, D. A., Flood, C. E., Moore, J. C., . . . Sumner, S. A. (2018). Ability of crime, demographic and business data to forecast areas of increased violence. *International Journal of Injury Control and Safety Promotion, 25*(4), 443–448.

Bower, B. (2019, November 4). Can neighborhood outreach reduce inner-city gun violence? *Science News.* https://www.sciencenews.org/article/neighborhood- outreach-can-reduce-inner-city-gun-violence

Boyle, C. (2020, July 12). Young Chicagoans march against gun violence, remember those they lost: "I'm tired of seeing my friends in caskets." *BlockClub*. https://block clubchicago.org/2020/07/12/goodkids-madcity-activists-want-2-of- police-budget- to-be-reallocated-for-schools-violence-prevention-instead/

Braga, A. (2016). *Gun violence among serious young offenders*. Problem-Oriented Guides for Police Problem-Specific Guides Series No. 23. Washington, DC: U.S. Department of Justice.

Braga, A. A., Brunson, R. K., Cook, P. J., Turchan, B., & Wade, B. (2021). Underground gun markets and the flow of illegal guns into the Bronx and Brooklyn: A mixed methods analysis. *Journal of Urban Health*, *98*(5), 596–608.

Braga, A. A., & Cook, P. J. (2018). The association of firearm caliber with likelihood of death from gunshot injury in criminal assaults. *JAMA Network Open*, *1*(3), e180833.

Braga, A. A., Griffiths, E., Sheppard, K., & Douglas, S. (2020b). Firearm instrumentality: Do guns make violent situations more lethal? Annual Review of Criminology, 4, 147–164.

Braga, A. A., & Kennedy, D. M. (2021). *A framework for addressing violence and serious crime: Focused deterrence, legitimacy, and prevention*. New York, NY: Cambridge University Press.

Braga, A. A., Turchan, B., Papachristos, A. V., & Hureau, D. M. (2019). Hot spots policing of small geographic areas effects on crime. *Campbell Systematic Reviews*, *15*(3), e1046.

Braga, A. A., & Weisburd, D. L. (2015). Focused deterrence and the prevention of violent gun injuries: Practice, theoretical principles, and scientific evidence. *Annual Review of Public Health*, *36*, 55–68.

Braithwaite, R., & Warren, R. (2020). The African American petri dish. *Journal of Health Care for the Poor and Underserved*, *31*(2), 491–502.

Branas, C. C., Reeping, P. M., & Rudolph, K. E. (2020). Beyond gun laws—Innovative interventions to reduce gun violence in the United States. *JAMA Psychiatry*. https:// jamanetwork.com/journals/jamapsychiatry/article-abstract/2769625

Brantingham, P. J., Yuan, B., & Herz, D. (2020). Is gang violent crime more contagious than non-gang violent crime? *Journal of Quantitative Criminology*, 1–25.

Braun, E. (2019). *Never again: The Parkland shooting and the teen activists leading a movement*. Minneapolis, MN: Lerner.

Brewer Jr, J. W., Cox, C. S., Fletcher, S. A., Shah, M. N., Sandberg, M., & Sandberg, D. I. (2019). Analysis of pediatric gunshot wounds in Houston, Texas: A social perspective. *Journal of Pediatric Surgery*, *54*(4), 783–791.

Brice, J. M., & Boyle, A. A. (2020). Are ED-based violence intervention programmes effective in reducing revictimisation and perpetration in victims of violence? A systematic review. *Emergency Medicine Journal*, *37*(8), 489–495.

Bridgeri, H. (2020, January 15). The burden of firearm injuries. *Harvard Medical School News & Research*. https://hms.harvard.edu/news/burden-firearm- injuries

Brisson, J., Pekelny, I., & Ungar, M. (2020). Methodological strategies for evaluating youth gang prevention programs. *Evaluation and Program Planning*, *79*.

Brito, S. A, Gugala, Z., Tan, A., & Lindsey, M. (2013). Injury classification: Statistical validity and clinical merits of a new civilian gunshot injury classification. *Clinical Orthopedics Related Research*, *471*, 3981–3987.

Brooks, A. (2019, July 29). Survivors of violence: A life of pain and deep wounds that don't heal. *WBUR*. https://www.wbur.org/news/2019/07/29/shooting-injur ies- trauma-ptsd-farm-recovery

Brown, M. E., & Barthelemy, J. J. (2019). The aftermath of gun violence: Implications for social work in communities. *Health & Social Work, 44*(4), 271–275.

Bruce, M. M., Ulrich, C. M., Kassam-Adams, N., & Richmond, T. S. (2016). Seriously injured urban black men's perceptions of clinical research participation. *Journal of Racial and Ethnic Health Disparities, 3*(4), 724–730.

Brunson, R. K., & Wade, B. A. (2019). "Oh hell no, we don't talk to police": Insights on the lack of cooperation in police investigations of urban gun violence. *Criminology & Public Policy, 18*(3), 623–648.

Brunson, S. D. (2019). Paying for gun violence. *Minnesota Law Review, 104*, 605–610.

Bubolz, B. F., & Lee, S. (2019). Putting in work: The application of identity theory to gang violence and commitment. *Deviant Behavior, 40*(6), 690–702.

Buchanan, C. (2014). *Gun violence, disability and recovery*. Bloomington, IN: Xlibris Corporation.

Bulger, E. M., Kuhls, D. A., Campbell, B. T., Bonne, S., Cunningham, R. M., Betz, M., . . . Sakran, J. V. (2019). Proceedings from the Medical Summit on Firearm Injury Prevention: A public health approach to reduce death and disability in the US. *Journal of the American College of Surgeons, 229*(4), 415–430.

Bureau of Justice Statistics. (2020). *NCVS Victimization Analysis Too. Number of violent victimizations by injury and weapon category, 2010–2018*. Washington, DC.

Burgason, K. A., DeLisi, M., Heirigs, M. H., Kusow, A., Erickson, J. H., & Vaughn, M. G. (2020). The code of the street fights back! Significant associations with arrest, delinquency, and violence withstand psychological confounds. *International Journal of Environmental Research and Public Health, 17*(7), 2432.

Burnham, M., & Lee, J. (2020). Guns and kids: Treatment of pediatric firearm and air gun missile injuries in the Emergency Department. *Pediatrics, 144*(2).

Burrell, M., White, A. M., Frerichs, L., Funchess, M., Cerulli, C., DiGiovanni, L., & Lich, K. H. (2021). Depicting "the system": How structural racism and disenfranchisement in the United States can cause dynamics in community violence among males in urban Black communities. *Social Science & Medicine*, March.

Busch, J. (2019, January 28). Trauma surgery: What a gunshot wound patient can expect in the operating room. *AffirmResearch*. https://affirmresearch.org/2019/01/28/tra uma-surgery-what-a-gun-shot-wound- patient-can-expect-in-the-operating-room- and-intensive-care-unit/

Butkus, R., Doherty, R., & Bornstein, S. S. (2018). Reducing firearm injuries and deaths in the United States: A position paper from the American College of Physicians. *Annals of internal medicine, 169*(10), 704–707.

Butkus, R., Rapp, K., Cooney, T. G., & Engel, L. S. (2020). Envisioning a better US health care system for all: Reducing barriers to care and addressing social determinants of health. *Annals of Internal Medicine, 172*(2 Supplement), S50–S59.

Butts, J. A., Roman, C. G., Bostwick, L., & Porter, J. R. (2015). Cure violence: A public health model to reduce gun violence. *Annual Review of Public Health, 36*, 39–53.

Byrdsong, T. R., Devan, A., & Yamatani, H. (2016). A ground-up model for gun violence reduction: A community-based public health approach. *Journal of Evidence-Informed Social Work, 13*(1), 76–86.

Cacciatori, A., Godino, M., & Mizraji, R. (2018, March). Does traumatic brain injury by firearm injury accelerates the brain death cascade? Preliminary results. *Transplantation Proceedings, 50*(2), 400–404.

Calabrese, E. (2017, June 20). Journey of a bullet. *NBC.* https://www.nbcnews.com/specials/journey-of-a-bullet/

Calhoun, T. L. (2019). *Engaging with African American youth following gunshot wound trauma: The Calhoun cultural competency course* (Doctoral dissertation, Boston University, Boston, MA).

Calvert, C. M., Brady, S. S., & Jones-Webb, R. (2020). Perceptions of violent encounters between police and young Black men across stakeholder groups. *Journal of Urban Health, 97,* 279–295.

Campbell, S., & Nass, D. (2019a, August 13). How one hospital dramatically skewed CDC's estimate of nonfatal gun injuries. *The Trace.* https://www.thetrace.org/2019/08/cdc-gun-injury-estimate-hospital-selection/

Campbell, S., & Nass, D. (2019b). *The CDC's gun injury data is becoming even more unreliable.* Center for Victim Research Depository.

Campbell, S., Nass, D., & Nguyen, M. (2018, October 8). The CDC says gun injuries are on the rise. But there are big problems with its data. *The Trace.* https://www.thetrace.org/2018/10/cdc-nonfatal-gun-injury-data-estimate- problems/

Carlson, J. (2020a). Police warriors and police guardians: Race, masculinity, and the construction of gun violence. *Social Problems, 67*(3), 399–417.

Carlson, J. (2020b). Gun studies and the politics of evidence. *Annual Review of Law and Social Science, 16,* 183–202.

Carman, M. (2019). Leading the effort to promote bleeding control in our communities. *American Journal of Nursing, 119*(5), 51–53.

Carmichael, H., Steward, L., Peltz, E. D., Wright, F. L., & Velopulos, C. G. (2019). Preventable death and interpersonal violence in the United States: Who can be saved? *Journal of Trauma and Acute Care Surgery, 87*(1), 200–204.

Carmona, R. (2020). Life on both sides of the gun: A Surgeon General's call to action. *Current Trauma Reports, 6*(1), 1–4.

Carpenter, S., Borrup, K., & Campbell, B. T. (2020). Gun buyback programs in the United States. In M. Crandall, S. Bonne, J. Bronson, & W. Kessel (Eds.), *Why we are losing the war on gun violence in the United States* (pp. 173–186). New York, NY: Springer.

Carroll, P., Witten, K., & Duff, C. (2020). "How can we make it work for you?" Enabling sporting assemblages for disabled young people. *Social Science & Medicine.*

Carswell, S. M. (2022). Have we surrendered to gun violence in urban America? Federal neglect stymies efforts to stop the slaughter among young black men. *Race and Justice, 12*(1), 126–140.

Carter, P. M., & Cunningham R. M. (2021). Firearm homicide and assaults. In L. K. Lee & E. W. Fleegler (Eds.), *Pediatric firearm injuries and fatalities* (pp. 31–52). New York, NY: Springer.

Carter, P. M., Mouch, C. A., Goldstick, J. E., Walton, M. A., Zimmerman, M. A., Resnicow, K., & Cunningham, R. M. (2020). Rates and correlates of risky firearm behaviors among adolescents and young adults treated in an urban emergency department. *Preventive Medicine, 130.*

Carter, P. M., Walton, M. A., Roehler, D. R., Goldstick, J., Zimmerman, M. A., Blow, F. C., & Cunningham, R. M. (2015). Firearm violence among high-risk emergency department youth after an assault injury. *Pediatrics, 135*(5), 805–815.

Casey, S. (2018, December 3). Spinal cord injury. *Affirm.* https://affirmresearch.org/2018/12/03/spinal-cord-injury/

Castro, M., Schober, D., De Maio, F., & Ahmed, C. (2018, November). A multi-method community based approach to assessing gun violence in high-risk Chicago communities. In *APHA's 2018 Annual Meeting & Expo* (Nov. 10–Nov. 14). American Public Health Association.

Castro, H. M., Gross, A., Chuang, A., Mankiewicz, K. A., Richani, K., & Crowell, E. L. (2020). Characteristics of open globes secondary to gunshot wounds presenting at a level 1 trauma center. *Investigative Ophthalmology & Visual Science, 61*(7), 2106.

Ceccato, V., & Assiago, J. (2020). Responding to crime and fear in public places. In V. Ceccato & J. Assiago (Eds.), *Crime and fear in public places* (pp. 433–440). New York, NY: Routledge.

Ceccato, V., Canabarro, A., & Vazquez, L. (2020). Do green areas affect crime and safety? In V. Caccato & M. K. Nalla (Eds.), *Crime and fear in public places* (75–107). New York, NY: Routledge.

Cenk, S. C. (2019). An analysis of the exposure to violence and burnout levels of ambulance staff. *Turkish Journal of Emergency Medicine, 19*(1), 21–25.

Center for American Progress. (2019, October 7). Bullet control: How lax regulations on ammunition contribute to America's gun violence epidemic. https://www.americanprogress.org/issues/guns- crime/reports/2019/10/07/475538/bullet-control/

Center for American Progress. (2020, October). CAP analysis of Uniform Crime Reporting program. FBI, "Crime in the U.S." https://ucr.fbi.gov/crime-in-the- U.S.

Centers for Disease Control and Prevention. (2020, May 22). *Firearm violence prevention.* Atlanta, GA: National Center for Health Statistics.

Cerdá, M., Tracy, M., & Keyes, K. M. (2018). Reducing urban violence: A contrast of public health and criminal justice approaches. *Epidemiology, 29*(1), 142–150.

Chamberlin, V. (2020, October 30). We don't know enough about gun injuries: That's hurting local economies. WAMU *88*(5). https://gunsandamerica.org/story/20/10/28/firearm-injury-research-economic- impact/

Chan, M. (2019, May 31). They survived mass shootings. Years later, the bullets are still trying to kill them. *Time Magazine.* https://time.com/longform/gun-violence- survivors-lead-poisoning/

Chandran, A., Long, A., Price, A., Murray, J., Fields, E. L., Schumacher, C. M., . . . IMPACT Partner Collaborative. (2020). The association between social support, violence, and social service needs among a select sample of urban adults in Baltimore City. *Journal of Community Health, 45*, 987–996.

Chaudhary, M. A., Sharma, M., Scully, R. E., Sturgeon, D. J., Koehlmoos, T., Haider, A. H., & Schoenfeld, A. J. (2018). Universal insurance and an equal access healthcare system eliminate disparities for Black patients after traumatic injury. *Surgery, 163*(4), 651–656.

Cheon, C., Lin, Y., Harding, D. J., Wang, W., & Small, D. S. (2020). Neighborhood racial composition and gun homicides. *JAMA Network Open, 3*(11), e2027591.

Chernock, B., Anjaria, D., Traba, C., Chen, S., Nasser, W., Fox, A., . . . Lamba, S. (2020). Integrating the bleeding control basic course into medical school curriculum. *The American Journal of Surgery, 219*(4), 660–664.

Chien, L. C., Gakh, M., Coughenour, C., & Lin, R. T. (2020). Temporal trend of research related to gun violence from 1981 to 2018 in the United States: A bibliometric analysis. *Injury Epidemiology, 7*(1), 1–9.

Chiu, R. G., Fuentes, A. M., & Mehta, A. I. (2019). Gunshot wounds to the head: Racial disparities in inpatient management and outcomes. *Neurosurgical Focus, 47*(5), E11.

Choi, J., Carlos, G., Nassar, A. K., Knowlton, L. M., Spain, D. A., & Gregg, D. L. (2021). The impact of trauma systems on patient outcomes. *Current Problems in Surgery, 58*(1).

Choi, P. M., Dekonenko, C., Aguayo, P., & Juang, D. (2020). Pediatric firearm injuries: Midwest experience. *Journal of Pediatric Surgery, 55*(10), 2140–2143.

Choi, P. M., Hong, C., Bansal, S., Lumba-Brown, A., Fitzpatrick, C. M., & Keller, M. S. (2016). Firearm injuries in the pediatric population: A tale of one city. *Journal of Trauma and Acute Care Surgery, 80*(1), 64–69.

Chong, V. E., Smith, R., Garcia, A., Lee, W. S., Ashley, L., Marks, A., . . . Victorino, G. P. (2015). Hospital-centered violence intervention programs: A cost-effectiveness analysis. *The American Journal of Surgery, 209*(4), 597–603.

Chopra, N., Gervasio, K. A., Kalosza, B., & Wu, A. Y. (2018). Gun trauma and ophthalmic outcomes. *Eye, 32*(4), 687–692.

Chopra, T., Kaye, K., & Sobel, J. (2017). Gunshot injury paraplegics—a population dying a slow, irreversible, and expensive death—a viewpoint on preventing pressure ulcers. *Infection Control & Hospital Epidemiology, 38*(6), 759–760.

Christensen, A. J., Cunningham, R., Delamater, A., & Hamilton, N. (2019). Introduction to the special issue on gun violence: Addressing a critical public health challenge. *Journal of Behavioral Medicine, 42*, 581–583.

Christensen, J. (2016, June 17). For gunshot survivors, recovery can last a lifetime. *CNN.* https://www.cnn.com/2016/06/17/health/gunshot-wound-long- recovery/index.html

Ciomek, A. M., Braga, A. A., & Papachristos, A. V. (2020). The influence of firearms trafficking on gunshot injuries in a co-offending network. *Social Science & Medicine, 259*, 255–264.

Ciraulo, L. A., Ciraulo, N. A., Ciraulo, R. S., Robaczewski, G. D., Andreasen, K. P., Falank, C. R., . . . Ciraulo, D. L. (2020). American College of Surgeons Committee on Trauma "Stop the Bleed Program": Quantifying the impact of training upon public school educators readiness. *The American Surgeon.*

Circo, G. M. (2019). Distance to trauma centres among gunshot wound victims: Identifying trauma "deserts" and "oases" in Detroit. *Injury Prevention, injuryprev- 2019.*

Cirone, J., Bendix, P., & An, G. (2020). A system dynamics model of violent trauma and the role of violence intervention programs. *Journal of Surgical Research, 247*, 258–263.

Claiborne, S., & Martin, M. (2019). *Beating guns: Hope for people who are weary of violence.* Ada, MI: Brazos Press.

Clayton, A. (2018, April 28). Shot in their friend's car: Survivors on learning how live after a shooting. *The Guardian.* https://www.theguardian.com/us- news/2020/apr/28/gun-violence-survivors-oakland-bay-area

Clery, M. J., Dworkis, D. A., Sonuyi, T., Khaldun, J. S., & Abir, M. (2020). Location of violent crime relative to trauma resources in Detroit: Implications for community interventions. *Western Journal of Emergency Medicine, 21*(2), 291–294.

Cogan, R. (2019). Creating partnerships that reflect the collective will of healthcare professionals: An interview with Megan Ranney. *Nursing Economic, 37*(3), 140–143.

Colbert, K. J., Barghouth, U., Moore, D. A., & Johnson, J. (2019). *Firearm injury in Detroit: Examining seasonal variability and outcomes.* Henry Ford Health System Scholarly Commons.

Coles, Z. J., Tufariello, A., & Bonne, S. (2020). Unpacking the causes of PTSD in violently injured patients. *Journal of Surgical Research, 256*, 43–47.

Collins, J. W. (2019). Achieving engagement in injury and violence prevention research. *Injury Prevention, 25*(5), 472–475.

Collins, R. (2019). Preventing violence: Insights from micro-sociology. *Contemporary Sociology, 48*(5), 487–494.

Colson, D. (2019, October 21). Shooter's ear: Hearing loss caused by gunfire. *Healthy Hearing.* https://www.healthyhearing.com/report/7904-Shooting-sports-and-hearing

Cone, J., Williams, B., Hampton, D., Prakash, P., Bendix, P., Wilson, K., . . . Zakrison, T. (2021). The ethics and politics of gun violence research. *Journal of Laparoendoscopic & Advanced Surgical Techniques, 31*(9), 983–987.

Congiusta, D. V., Oettinger, J. P., Merchant, A. M., Vosbikian, M. M., & Ahmed, I. H. (2021). Epidemiology of orthopaedic fractures due to firearms. *Journal of Clinical Orthopaedics and Trauma, 12*(1), 45–49.

Connerton, C., & St Clair, J. (2020). *Using a mix of strategies to prepare nursing students for disaster response.* University of Southern Indiana. http://hdl.handle.net/20.500.12419/491

Cook, A., Hosmer, D., Glance, L., Kalesan, B., Weinberg, J., Rogers, A., . . . Rogers, F. (2019). Population-based analysis of firearm injuries among young children in the United States, 2010–2015. *The American Surgeon, 85*(5), 449–455.

Cook, E. A. (2021). *Family activism in the aftermath of fatal violence.* New York, NY: Routledge.

Cook, K. (2015). *Gone too soon: The effects of Philadelphia's urban gun violence crisis* (Doctoral dissertation, University of Missouri).

Cook, P. J. (2018). Expanding the public health approach to gun violence prevention. *Annals of Internal Medicine, 169*(10), 723–724.

Cook, P. J. (2020). Thinking about gun violence. *Criminology & Public Policy, 19*(4), 1371–1393.

Cook, P. J., Braga, A. A., Turchan, B. S., & Barao, L. M. (2019). Why do gun murders have a higher clearance rate than gunshot assaults? *Criminology & Public Policy. 18*(3), 525–551.

Cook, P. J., Pollack, H. A., & White, K. (2018*). Results of the Chicago inmate survey of gun access and use.* Urban Crime Lab, University of Chicago. https://urbanlabs.uchicago.edu/attachments/ec4f519cf18f8ba65e70f361d7 4b9ff4a767c9be/store/a29f7714dd69 87d282735d9d6744c343df61654c687f698df e584bc1928c/Gun+Offender+Survey+Report_9.20.19.pdf

Cooper, H. L., & Fullilove, M. T. (2020). *From enforcers to guardians: A public health primer on ending police violence.* Baltimore, MD: Johns Hopkins University Press.

Cooper, J. J., Stock, R. C., & Wilson, S. J. (2020). Emergency department grief support: A multidisciplinary intervention to provide bereavement support after death in the emergency department. *The Journal of Emergency Medicine, 58*(1), 141–147.

Cooper, R. (2016). *Reclaiming one's life after paralysis: A narrative inquiry exploring the role of the child life specialist in providing psychosocial support to adolescents with spinal cord injury* (Doctoral dissertation, Mills College, Oakland, CA).

Corazon, A. (2019). *Literature review on urban trauma and applying a trauma-informed approach* (Doctoral dissertation, Lesley University, Cambridge, MA).

Coupet, E., Huang, Y., & Delgado, M. K. (2019). US emergency department encounters for firearm injuries according to presentation at trauma vs. nontrauma centers. *JAMA Surgery, 154*(4), 360–362.

Coupet Jr, E., Karp, D., Wiebe, D. J., & Delgado, M. K. (2018). Shift in US payer responsibility for the acute care of violent injuries after the Affordable Care Act: Implications for prevention. *The American Journal of Emergency Medicine, 36*(12), 2192–2196.

Cox, K. S. (2018). A public health crisis: Recommendations to reduce gun violence in America. *Nursing Outlook, 66*(3), 219–220.

Crane's Chicago Business. (2019, August 28). Hidden costs push price of city's gun violence in the billions. https://www.chicagobusiness.com/crains-forum-gun- violence/ hidden-costs-push-price-citys-gun-violence-billions

Crifasi, C. K., Buggs, S. A., Booty, M. D., Webster, D. W., & Sherman, S. G. (2020). Baltimore's underground gun market: Availability of and access to guns. *Violence and Gender, 7*(2), 78–83.

Crifasi, C. K., McCourt, A., & Webster, D. W. (2019). *Policies to reduce gun violence in Illinois.* Baltimore, MD: Johns Hopkins Bloomberg School of Public Health.

Cromer, K. D., D'Agostino, E. M., Hansen, E., Alfonso, C., & Frazier, S. L. (2019). After-school poly-strengths programming for urban teens at high risk for violence exposure. *Translational Behavioral Medicine, 9*(3), 541–548.

Crosby, R. A., & Salazar, L. F. (2020). *Essentials of public health research methods.* Burlington, MA: Jones & Bartlett Learning.

Crutcher, C. L., Fannin, E. S., & Wilson, J. D. (2016). Racial disparities in cranial gunshot wounds: Intent and survival. *Journal of Racial and Ethnic Health Disparities, 3*(4), 687–691.

Cuevas, C. (2019). *The pushes and pulls toward desistance from gangs: As told by the lived experiences of ten former gang members* (Doctoral dissertation, California State University, Fullerton, CA).

Cukier, W., & Eagen, S. A. (2018). Gun violence. *Current Opinion in Psychology, 19*, 109–112.

Culyba, A. J., Branas, C. C., Guo, W., Miller, E., Ginsburg, K. R., & Wiebe, D. J. (2021). Route choices and adolescent–adult connections in mitigating exposure to environmental risk factors during daily activities. *Journal of Interpersonal Violence, 36*, 15–21.

Culyba, A. J., & Sigel, E. (2020). Firearms and substance use: Bringing synergy to counseling and intervention. *The American Journal of Drug and Alcohol Abuse, 46*(3), 263–265.

Cunningham, A. C. (2016). *Critical perspectives on gun control.* Berkeley Heights, NJ: Enslow.

Cunningham, R. M., Carter, P. M., Ranney, M. L., Walton, M., Zeoli, A. M., Alpern, E. R., . . . Goldstick, J. E. (2019). Prevention of firearm injuries among children and adolescents: consensus-driven research agenda from the Firearm Safety Among Children and Teens (FACTS) Consortium. *JAMA Pediatrics, 173*(8), 780–789.

Cure Violence Global. (Undated). *Who we are.* https://cvg.org/who-we-are/

D'Onofrio, J., & Wall, C. (2020, July 6). 87 shot, 17 fatally, in Chicago July 4th weekend violence, police say. *ABC News.* https://abc7chicago.com/chicago-shooting- shootings-this-weekend-violence-how-many-shot-in/6301523/

Dabash, S., Gerzina, C., Simson, J. E., Elabd, A., & Abdelgawad, A. (2018). Pediatric gunshot wounds of the upper extremity. *International Journal of Orthopaedics, 5*(2), 910–915.

Daley, B. J. (2021, January 13). Considerations in pediatric trauma. *Medscape.* https://emedicine.medscape.com/article/435031-overview

Dalve, K., Gause, E., Mills, B., Floyd, A. S., Rivara, F. P., & Rowhani-Rahbar, A. (2021). Neighborhood disadvantage and firearm injury: Does shooting location matter? *Injury Epidemiology, 8*(1), 1–9.

Danner, O. K., Hudak, M. D., Bayakly, R., Koplan, C., Kelly, A., Sharon, L., . . . Sheryl, L. (2020). Redefining our understanding of the impact of firearm-related injury in the state of Georgia: A white paper by the Violence Prevention Task Force of IPRCE. *Journal of the Georgia Public Health Association, 8*(1), 90–100.

DaViera, A. L., & Roy, A. L. (2020). Chicago youths' exposure to community violence: Contextualizing spatial dynamics of violence and the relationship with psychological functioning. *American Journal of Community Psychology, 65*(3–4), 332–342.

Davis, A. B., Gaudino, J. A., Soskolne, C. L., & Al-Delaimy, W. K. (2018). The role of epidemiology in firearm violence prevention: A policy brief. *International Journal of Epidemiology, 47*(4), 1015–1058.

de Anda, H., Dibble, T., Schlaepfer, C., Foraker, R., & Mueller, K. (2018). A cross- sectional study of firearm injuries in emergency department patients. *Missouri Medicine, 115*(5), 456–462.

de Freytas-Tamura, K., Hu, W., & Cook, L. R. (2020, May 27). In hard-hit Bronx, high-rises have become "death towers." *The New York Times*, pp. A1, A12.

de Witte, A. M., Hoozemans, M. J., Berger, M. A., van der Woude, L. H., & Veeger, D. (2016). Do field position and playing standard influence athlete performance in wheelchair basketball? *Journal of Sports Sciences, 34*(9), 811–820.

Decker, H. C., Hubner, G., Nwabuo, A., Johnson, L., Texada, M., Marquez, R., . . . Juillard, C. (2020). "You don't want anyone who hasn't been through anything telling you what to do, because how do they know?": Qualitative analysis of case managers in a hospital-based violence intervention program. *PloS One, 15*(6), e0234608.

Dedel, K. (2007). The problem of drive-by shootings. Center for Problem-Oriented Policing. Population Center Arizona State University. Guide 47. https://popcenter. asu.edu/content/drive-shootings-0

Dedel, K. (2016). *Drive-by shootings.* Washington, DC: U.S. Department of Justice.

Degeneffe, C. E. (2019). The phenomenological experience of family caregiving following traumatic brain injury. In C. M. Hayre & D. J. Muller (Eds.), *Enhancing healthcare and rehabilitation: The impact of qualitative research* (pp. 65–78). Boca Raton, FL: CRC Press.

Delgado, M. (1999). *Social work practice in nontraditional urban settings.* New York, NY: Oxford University Press.

Delgado, M. (2000). *Community social work practice in an urban context: The potential of a capacity enhancement perspective.* New York, NY: Oxford University Press.

Delgado, M. (2015). *Urban youth and photovoice: Visual ethnography in action.* New York, NY: Oxford University Press.

Delgado, M. (2016a). *Community practice and urban youth: Social justice service- learning and civic engagement.* New York, NY: Routledge.

Delgado, M. (2016b). *Celebrating urban community life: Fairs, festivals, parades and community practice.* Toronto: University of Toronto Press.

Delgado, M. (2018). *Music, song, dance, and theatre: Broadway meets youth community practice.* New York, NY: Oxford University Press.

Delgado, M. (2020a). *Urban youth trauma: Using community interventions to overcome gun violence.* Lanham, MD: Rowman & Littlefield.

Delgado, M. (2020b). *Community health workers in action: The efforts of "Promotores de Salud" in bringing health care to marginalized communities.* New York, NY: Oxford University Press.

Delgado, M. (2020c). *The silent epidemic of gun injuries: Challenges and opportunities for treating and preventing gun injuries.* New York, NY: Oxford University Press.

Delgado, M. (2020d). *State sanctioned violence: Advancing a social work social justice agenda.* Series on Interpersonal Violence. New York, NY: Oxford University Press.

Delgado, M. (2021). *Urban gun violence: Self-help organizations as healing sites, catalysts for change, and collaborative partners.* New York, NY: Oxford University Press.

Delgado, M., & Humm-Delgado, D. (2013). *Asset assessments and community social work practice.* New York, NY: Oxford University Press.

Delgado, M., & Staples, L. (2008). *Youth-led community organizing: Theory and action.* New York, NY: Oxford University Press.

Dell'Aria, A. (2020). Loaded objects: Addressing gun violence through art in the gallery and beyond. *Palgrave Communications, 6*(1), 1–11.

DeMario, V. M., Sikorski, R. A., Efron, D. T., Serbanescu, M. A., Buchanan, R. M., Wang, E. J., . . . Ken Lee, K. H. (2018). Blood utilization and mortality in victims of gun violence. *Transfusion, 58*(10), 2326–2334.

Deng, H., Yue, J. K., Winkler, E. A., Dhall, S. S., Manley, G. T., & Tarapore, P. E. (2019a). Pediatric firearm-related traumatic brain injury in United States trauma centers. *Journal of Neurosurgery: Pediatrics, 24*(5), 481–610.

Denne, S. C., Baumberger, J., & Mariani, M. (2020). Funding for gun violence research: The importance of sustained advocacy by academic pediatricians. *Pediatric Research, 87*, 800–801.

Deng, H., Yue, J. K., Winkler, E. A., Dhall, S. S., Manley, G. T., & Tarapore, P. E. (2019b). Adult firearm-related traumatic brain injury in United States trauma centers. *Journal of Neurotrauma, 36*(2), 322–337.

Denver Public Health. (2019, September 9). 700 Denver youth harmed by gun violence each year. http://www.denverpublichealth.org/news/2019/09/700- denver-youth-harmed-by-gun-violence-each-year

Diaz, M. F., & Shepard, B. (Eds.). (2019). *Narrating practice with children and adolescents.* New York, NY: Columbia University Press.

Dicker, R. A. (2016). Hospital-based violence intervention: An emerging practice based on public health principles. *Trauma Surgery Acute Care Open, 1*(1), 1–2.

Dicker, R. A., Gaines, B. A., Bonne, S., Duncan, T., Violano, P., & Aboutanous, M. (2017). Violence intervention programs: A primer for developing a comprehensive program for trauma centers. *Bulletin of American College of Surgeons, 102*(10), 20–29.

Dicker, R., & Juillard, C. (2020). Hospital-based interventions to reduce violence and recidivism: Wraparound programs. In M. Siegler & S. O. Rogers Jr (Eds.), *Violence, trauma, and trauma surgery* (pp. 3–15). New York, NY: Springer.

Dicker, R. A., & Punch, L. J. (2020). Long-term consequences in trauma: At the center of the public health approach is the survivor's voice. *JAMA Surgery, 155*(1), 59–60.

DiClemente, C. M., & Richards, M. H. (2021). Community violence in early adolescence: Assessing coping strategies for reducing delinquency and aggression. *Journal of Clinical Child & Adolescent Psychology*, 1–15.

Diebel, A. M., Robertson, B., Nesiama, J. A., & Alder, A. (2018). Pediatric firearm injuries: Demographics and context of injuries in urban and rural communities. *Pediatrics, 141*(1), 75. https://doi.org/10.1542/peds.141.1_MeetingAbstract.75

Dierkhising, C. B., Sánchez, J. A., & Gutierrez, L. (2021). "It changed my life": Traumatic loss, behavioral health, and turning points among gang-involved and justice-involved youth. *Journal of Interpersonal Violence, 39*(5), 859–978.

Dill, L. J., & Ozer, E. J. (2016). "I'm not just runnin' the streets": Exposure to neighborhood violence and violence management strategies among urban youth of color. *Journal of Adolescent Research, 31*(5), 536–556.

DiMaio, V. J. M. (2015). *Gunshot wounds: Practical aspects of firearms, ballistics, and forensic techniques (Practical aspects of criminal and forensic investigations)* (3rd ed.). Boca Raton, FL: CRC Press.

Disability Visibility Project. (2020, December 14). Q & A with Andres "Jay" Molina and Alexis Neophytides. https://disabilityvisibilityproject.com/2020/12/16/qa-with-andres-jay-molina-and-alexis-neophytides/

DiZazzo-Miller, R. (2015). Spinal cord injury induced by gun shot wounds: Implications for occupational therapy. *The Open Journal of Occupational Therapy, 3*(1), 7. https://scholarworks.wmich.edu/cgi/viewcontent.cgi?article=1127&context=ojot

Dobaria, V., Aguayo, E., Sanaiha, Y., Tran, Z., Hadaya, J., Sareh, S., . . . Benharash, P. (2020). National trends and cost burden of surgically treated gunshot wounds in the US. *Journal of the American College of Surgeons, 231*(4), 448–459.

Dodington, J., Violano, P., Baum, C. R., & Bechtel, K. (2017). Drugs, guns and cars: How far we have come to improve safety in the United States; yet we still have far to go. *Pediatric Research, 81*(1), 227–232.

Dodson, N. A., & Hemenway, D. (2020). Toward a deeper understanding of gun violence. *Pediatrics, 146*(1), e20193333.

Dodson, N. A., Talib, H. J., Gao, Q., Choi, J., & Coupey, S. M. (2021). Pediatricians as child health advocates: The role of advocacy education. *Health Promotion Practice, 22*(1), 13–17.

Doering, J. (2020). *Us versus them: Race, crime, and gentrification in Chicago neighborhoods.* New York, NY: Oxford University Press.

Dong, B., Morrison, C. N., Branas, C. C., Richmond, T. S., & Wiebe, D. J. (2020). As violence unfolds: A space–time study of situational triggers of violent victimization among urban youth. *Journal of Quantitative Criminology, 36*(1), 119–152.

Donnelly, K. A., Kafashzadeh, D., Goyal, M. K., Badolato, G. M., Patel, S. J., Bhansali, P., . . . Cohen, J. S. (2020). Barriers to firearm injury research. *American Journal of Preventive Medicine, 58*(6), 825–831.

Dowd, B., McKenney, M., Boneva, D., & Elkbuli, A. (2020). Disparities in National Institute of Health trauma research funding: The search for sufficient funding opportunities. *Medicine, 99*(6), e19027.

Drane, K. (2020, February 4). America's gun violence epidemic persists, according to new CDC Data. *Giffords.* https://giffords.org/blog/2020/02/americas-gun-viole nce- epidemic-persists- according-to-new-cdc-data-blog/

Drayton, T. (2018). *Coping with gun violence.* New York, NY: The Rosen Publishing Group.

Drinkard, A. M., Schell, C. G., & Adams, R. (2019). Fear of violence, family support, and well-being among urban adolescents. *Open Journal of Social Sciences, 7*(9), 86–105.

Duda, T., Sharma, A., Ellenbogen, Y., Martyniuk, A., Kasper, E., Engels, P. T., & Sharma, S. (2020). Outcomes of civilian pediatric craniocerebral gunshot wounds: A systematic review. *Journal of Trauma and Acute Care Surgery, 89*(6), 1239–1247.

Duffey, T., Haberstroh, S., & Del Vecchio-Scully, D. (2020). In T. Duffey & S. Haberstroh (Eds.), *Introduction to crisis and trauma counseling: Responding to community violence and community trauma* (pp. 227–246). Alexandra, VA: American Counseling Association.

Dukes, K., & Gaither, S. (2017). Black racial stereotypes and victim blaming: Implications for media coverage and criminal proceedings in cases of police violence against racial and ethnic minorities. *Journal of Social Issues, 73*(4), 789–807.

Durando, S. (2018). *Under the gun: A children's hospital on the front line of an American crisis.* Self-published.

Dycus, Y. D. (2020, July 24). I've been on a mission to prevent other mothers from the tragedy that changed our family's lives. *Everytown.* https://everytown.org/deandra-yates-dycus/

Dzau, V. J., & Leshner, A. I. (2018). Public health research on gun violence: Long overdue. *Annals of Internal Medicine, 168*(12), 876–877.

Eastman, A. L., Acevedo, A., & McDonnell, J. (2020). Law enforcement: A vital partnership in the public health approach to gun violence. *Current Trauma Reports, 6*(1), 51–55.

Educational Fund to Stop Gun Violence. (2019). *Nonfatal gun violence.* https://efsgv.org/learn/type-of-gun-violence/nonfatal-firearm-violence/

Educational Fund to Stop Gun Violence. (2020). *Community gun violence.* https://efsgv.org/learn/type-of-gun-violence/community-gun-violence/

Eisenman, D. P., & Flavahan, L. (2017). Canaries in the coalmine: Interpersonal violence, gang violence, and violent extremism through a public health prevention lens. *International Review of Psychiatry, 29*(4), 341–349.

Eklund, K., Rossen, E., Koriakin, T., Chafouleas, S. M., & Resnick, C. (2018). A systematic review of trauma screening measures for children and adolescents. *School Psychology Quarterly, 33*(1), 30–43.

El-Menyar, A., Mekkodathil, A., Abdelrahman, H., Latifi, R., Galwankar, S., Al-Thani, H., & Rizoli, S. (2019). Review of existing scoring systems for massive blood transfusion in trauma patients: Where do we stand? *Shock, 52*(3), 288–299.

Eng, M. (2020, July 14). Chicago funeral homes are on edge after recent mass-shootings. *WBEZ.* https://www.wbez.org/stories/chicago-funeral-homes-are-on-edge-after-recent-mass-shooting/8e198fa9-c6bd-4ed7-9474-e9ea5f4b803b

Erickson, B. P., Feng, P. W., Ko, M. J., Modi, Y. S., & Johnson, T. E. (2020). Gun-related eye injuries: A primer. *Survey of Ophthalmology, 65*(1), 67–78.

Erickson, J. H., Hochstetler, A., & Dorius, S. F. (2020). Code in transition? The evolution of code of the street adherence in adolescence. *Deviant Behavior, 41*(3), 329–347.

Esparaz, J. R., Waters, A. M., Mathis, M. S., Deng, L., Xie, R., Chen, M. K., . . . Russell, R. T. (2021). The disturbing findings of pediatric firearm injuries from the National Trauma Data Bank: 2010–2016. *Journal of Surgical Research, 259*, 224–229.

Evans, E. J., & Thompson, M. (2019). Questions and answers from research centers on gun violence. *Health & Social Work, 44*(4), 221–223.

Evans, P. T., Pennings, J. S., Samade, R., Lovvorn III, H. N., & Martus, J. E. (2020). The financial burden of musculoskeletal firearm injuries in children with and without concomitant intra-cavitary injuries. *Journal of Pediatric Surgery, 55*(9), 1754–1760.

Evans, W. N., Garthwaite, C., & Moore, T. J. (2018). *Guns and violence: The enduring impact of crack cocaine markets on young black males* (No. w24819). Washington, DC: National Bureau of Economic Research.

Everytown for Gun Surgery. (2016, June 16). *Strategies for reducing gun violence in American cities.* https://everytownresearch.org/report/strategies-for-reducing-gun-violence-in-american-cities/

Everytown USA. (2019a, October 17). *Guns and violence against women.* https://everyt ownresearch.org/report/guns-and-violence-against-women-americas-uniquely-lethal-intimate-partner-violence-problem/

Everytown USA. (2019b, November 11). *A more complete picture: The contours of gun injury in the United States.* https://everytownresearch.org/a-more-complete-picture-the-contours-of-gun-injury-in-the-united-states/

Everytown USA (2020a, June 17). *Gun violence and (COVID-19: Colliding public health issues.* https://everytownresearch.org/reports/covid- gun-violence/

Everytown USA. (2020b, October 15). *Community-led public safety strategies.* https://everytownresearch.org/report/community-led-public-safety-strategies/

Everytown USA. (2020c, December 4). *A more complete picture: The contours of gun injury in the United States.* https://everytownresearch.org/report/nonfatals- in-the-us/

Everytown USA. (2020d, August 7). *Impact of gun violence on Black Americans.* https://everytown.org/issues/gun-violence-black-americans/

Ezeonu, I. (2008). Dudes, let's talk about us: The black "community" construction of gun violence in Toronto. *Journal of African American Studies, 12*(3), 193–214.

Fagel, M. J., & Benson, G. (2020). Saving a life, it's in your hands. In S. J. Davies (Ed.), *The professional protection officer* (pp. 139–140). Waltham, MA: Butterworth-Heinemann.

Fahimi, J., Larimer, E., Hamud-Ahmed, W., Anderson, E., Schnorr, C. D., Yen, I., & Alter, H. J. (2016). Long-term mortality of patients surviving firearm violence. *Injury Prevention, 22*(2), 129–134.

Fahmy, C., Jackson, D. B., Pyrooz, D. C., & Decker, S. H. (2020). Head injury in prison: Gang membership and the role of prison violence. *Journal of Criminal Justice, 67 .*

Fairchild, H. H. (2016). *Solving violence in America.* New Delhi, India: Indo American Books.

Fast, J. (2020). New medical studies, same old politics: The effects of anti-Black racism on penetrative trauma research. *Journal of Critical Race Inquiry, 7*(1), 23–45.

Feder, L., & Angel, S. (2020). Criminal justice research: Incorporating a public health approach. In B. A. Fiedler (Ed.), *Three facets of public health and paths to improvements* (pp. 295–316). New York, NY: Academic Press.

Feldman, K. A., Tashiro, J., Allen, C. J., Perez, E. A., Neville, H. L., Schulman, C. I., & Sola, J. E. (2017). Predictors of mortality in pediatric urban firearm injuries. *Pediatric Surgery International, 33*(1), 53–58.

Felson, R. B., & Lantz, B. (2016). When are victims unlikely to cooperate with the police? *Aggressive Behavior, 42*(1), 97–108.

Fernandez, M. (2020, April 20). U.S. still has a gun violence problem despite coronavirus lockdowns. *Axios.* https://www.axios.com/coronavirus-gun-violence- shootings-crime-rate-11e273ce-e474-4c29-b92a-391446bbb91d.html

Fields, L., Valdez, C. E., Richmond, C., Murphy, M. J., Halloran, M., Boccellari, A., & Shumway, M. (2020). Communities healing and transforming trauma (CHATT): A trauma-informed speakers' bureau for survivors of violence. *Journal of Trauma & Dissociation, 21*(4), 437–451.

Finlay, J., Esposito, M., Kim, M. H., Gomez-Lopez, I., & Clarke, P. (2019). Closure of "third places"?: Exploring potential consequences for collective health and wellbeing. *Health & Place, 60.*

Fischer, K. R., Bakes, K. M., Corbin, T. J., Fein, J. A., Harris, E. J., James, T. L., & Melzer-Lange, M. D. (2019). Trauma-informed care for violently injured patients in the emergency department. *Annals of Emergency Medicine, 73*(2), 193–202.

Fischer, K. R., Cooper, C., Marks, A., & Slutkin, G. (2020). Prevention professional for violence intervention: A newly recognized health care provider for population health programs. *Journal of Health Care for the Poor and Underserved, 31*(1), 25–34.

Fisher, R. M. (2019). *Gun violence problem is a fear violence problem.* Werklund School of Education, University of Calgary, Calgary, Canada.

Fitzpatrick, V., Castro, M., Jacobs, J., Sebro, N., Gulmatico, J., Shields, M., & Homan, S. M. (2019). Nonfatal firearm violence trends on the Westside of Chicago between 2005 and 2016. *Journal of Community Health, 44*(5), 866–873.

Flaherty, M. R., & Klig, J. E. (2020). Firearm-related injuries in children and adolescents: An emergency and critical care perspective. *Current Opinion in Pediatrics, 32*(3), 349–353.

Fletcher, Y. (2020). Application of a racial equity framework for gun violence prevention. *Injury Prevention, 26* (Suppl. 1), A51–A52.

Fleurant, M. (2019). Trauma-informed care: A focus on African American men. In M. Gerber (Ed.), *Trauma-informed healthcare approaches* (pp. 69–83). New York, NY: Springer.

Flynn, K., Mathias, B., & Wiebe, D. J. (2021). Investigating the (in) visible: Piloting an integrated methodology to explore multiple forms of violence. *Journal of Community Psychology, 49*(4), 947–961.

Follman, M., Lee, J., Lurie, J., & West, J. (2018). *The true cost of gun violence in America.* Center for Victim Research Repository.

Follman, M., Lurie, J., & West, J. (2015, April 15). True cost of gun violence in America. *Mother Jones.* https://www.motherjones.com/politics/2015/04/ true- cost- of-gun-violence-in-america/

Foran, C. P., Clark, D. H., Henry, R., Lalchandani, P., Kim, D. Y., Putnam, B. A., . . . Demetriades, D. G. (2019). Current burden of gunshot wound injuries at two Los Angeles County Level I trauma centers. *Journal of the American College of Surgeons, 229*(2), 141–149.

Formica, M. K. (2021). An eye on disparities, health equity, and racism—The case of firearm injuries in urban youth in the United States and globally. *Pediatric Clinics, 68*(2), 389–399.

Fountain, A. J., Corey, A., Malko, J. A., Strozier, D., & Allen, J. W. (2021). Imaging appearance of ballistic wounds predicts bullet composition: Implications for MRI safety. *American Journal of Roentgenology, 216*(2), 542–551.

Fowler, K. A., Dahlberg, L. L., Haileyesus, T., & Annest, J. L. (2015). Firearm injuries in the United States. *Preventive Medicine, 79*, 5–14.

Fowler, K. A., Dahlberg, L. L., Haileyesus, T., Gutierrez, C., & Bacon, S. (2017). Childhood firearm injuries in the United States. *Pediatrics, 140*(1), e20163486.

Francis, M. (2018). A narrative inquiry into the experience of being a victim of gun violence. *Journal of Trauma Nursing, 25*(6), 381–388.

Franke, P. (2019). Reducing firearm injuries and deaths in the United States. *Annals of Internal Medicine, 170*(12), 910.

Fransdottir, E., & Butts, J.A. (2020, May 11). Who pays for gun violence? You do. *John Jay Research Brief.* https://johnjayrec.nyc/2020/05/11/whopays/

Frazer, E., Mitchell Jr, R. A., Nesbitt, L. S., Williams, M., Mitchell, E. P., Williams, R. A., & Browne, D. (2018). The violence epidemic in the African American community: A call by the National Medical Association for comprehensive reform. *Journal of the National Medical Association, 110*(1), 4–15.

Fredenburg, B., & Warner, H. (2020). *Stop the Bleed: A community event.* https://dune.
une.edu/cgi/viewcontent.cgi?article=1006&context=cecespring2020

Free, J. L. (2020). "We're brokers": How youth violence prevention workers intervene in
the lives of at-risk youth to reduce violence. *Criminal Justice Review, 45*(3), 281–302.

Free, J. L., & MacDonald, H. Z. (2019). "I've had to bury a lot of kids over the
years . . .": Violence prevention streetworkers' exposure to trauma. *Journal of
Community Psychology, 47*(5), 1197–1209.

Freire-Vargas, L. (2018). Violence as a public health crisis. *AMA Journal of Ethics,
20*(1), 25–28.

Friedman, J., Hoof, M., Smith, A., Tatum, D., Ibraheem, K., Guidry, C., . . . McGrew, P.
(2019). Pediatric firearm incidents: It is time to decrease on-scene mortality. *Journal
of Trauma and Acute Care Surgery, 86*(5), 791–796.

Friedman, J. K., Mytty, E., Ninokawa, S., Reza, T., Kaufman, E., Raza, S., . . . Duchesne,
J. (2021). A tale of two cities: What's driving the firearm mortality difference in two
large urban centers? *The American Surgeon.*

Frisby, J. C., Kim, T. W. B., Schultz, E. M., Adeyemo, A., Lo, K. W., Hazelton, J. P., &
Miller, L. S. (2019). Novel policing techniques decrease gun-violence and the cost to
the healthcare system. *Preventive Medicine Reports, 16*, 100995.

Fu, C. Y., Bajani, F., Tatebe, L., Butler, C., Starr, F., Dennis, A., . . . Poulakidas, S. (2019).
Right hospital, right patients: Penetrating injury patients treated at high-volume
penetrating trauma centers have lower mortality. *Journal of Trauma and Acute Care
Surgery, 86*(6), 961–966.

Furman, L. (2018). Firearm violence: Silent victims. *Pediatrics, 142*(4), 1–2. https://pedi-
atrics.aappublications.org/content/pediatrics/142/4/e20182060.full.pdf

Galea, S., & Abdalla, S. M. (2019). The public's health and the social meaning of guns.
Palgrave Communications, 5(1), 1–4.

Galea, S., Branas, C. C., Flescher, A., Formica, M. K., Hennig, N., Liller, K. D., . . . Ying, J.
(2018). Priorities in recovering from a lost generation of firearms research. *American
Journal of Public Health, 108*(7), 858–861.

Galea, S., & Vaughan, R. D. (2019). Tendrils of hope in the gun epidemic: A public
health of consequence, November 2019. *American Journal of Public Health, 109*(11),
1490–1491.

Galiatsatos, P., Cudjoe, T. K., Bratcher, J., Heikkinen, P., Leaf, P., & Golden, S. H. (2021).
Second victims: Aftermath of gun violence and faith-based responses. *Journal of
Religion and Health, 60*(3), 1832–1838.

Gallagher, A., & Hodge Sr., D. A. (2018). Gun violence: Care ethicists making the invisible visible. *Nursing Ethics, 25*(1), 3–5.

Galvagno Jr, S. M., Massey, M., Bouzat, P., Vesselinov, R., Levy, M. J., Millin, M.
G., . . . Hirshon, J. M. (2019). Correlation between the revised trauma score and in-
jury severity score: Implications for prehospital trauma triage. *Prehospital Emergency
Care, 23*(2), 263–270.

Gambacorta, D., & Ubiñas, H. (2018, November 28). Shot and forgotten America's hidden
toll of gun violence: Shooting victims face lifelong disabilities and financial burdens.
Philadelphia Inquirer. https://www.inquirer.com/news/gun-violence-philadelphia-
shooting-victims-columbine-wheelchair-jalil-frazier-ralph-brooks-20181127.html

Gani, F. (2017, November 2). The price of gun violence. *Health Affairs.* https://www.
healthaffairs.org/do/10.1377/hblog20171031.874550/full/

Gani, F., Sakran, J. V., & Canner, J. K. (2017). Emergency department visits for firearm-related injuries in the United States, 2006–14. *Health Affairs, 36*(10), 1729–1738.

Garbarino, J. (2017). Gun violence in Chicago. *Violence and gender, 4*(2), 45–47.

Garcia, S., Myers, R., Vega, L., & Feske-Kirby, K. (2020). 175 variable trajectories of posttraumatic stress symptoms among violently injured youth. *Injury Prevention, 26* (Suppl. 1).

Gardner, K. (2020, November 19). Gunshot survivors study sheds light on devastating long-lasting effects of firearm violence in the U.S. *Newsweek.* https://www.newsweek.com/gunshot-survivors-study-sheds-light-devastating-long0lasting-effects-firearm-violence-us-1472909

Garo, L. A., & Lawson, T. (2019). My story, my way: Conceptualization of narrative therapy with trauma-exposed black male youth. *Urban Education Research & Policy Annuals, 6*(1), 47–59.

Garrett, B. (2018). *Stakeholder perceptions of gun violence perpetrated by young men* (Doctoral dissertation, Walden University, Minneapolis, MN).

Gastineau, K. A., & Andrews, A. L. (2020). Pediatric hospitalists at the front line of gun violence prevention: Every patient encounter is an opportunity to promote safe gun storage. *Hospital Pediatrics, 10*(1), 98–100.

Gavine, A., MacGillivray, S., & Williams, D. J. (2017). Universal community-based social development interventions for preventing community violence by young people 12 to 18 years of age. *The Cochrane Database of Systematic Reviews, 2017*(1), 1–18.

Gebo, E., & Bond, B. J. (2020). Improving interorganizational collaborations: An application in a violence reduction context. *The Social Science Journal,* 1–12.

Gibbs, E. (2020). *Violence as an infectious disease: Using a public health approach to mitigate the United States' proliferation of police violence.*(Doctoral dissertation, Vassar College, Poughkeepsie, New York, NY).

Gibson, P. D., Ippolito, J. A., Shaath, M. K., Campbell, C. L., Fox, A. D., & Ahmed, I. (2016). Pediatric gunshot wound recidivism: Identification of at-risk youth. *Journal of Trauma and Acute Care Surgery, 80*(6), 877–883.

Giffords Center. (2016, March 10). *Healing communities in crisis: Live saving solutions to the urban gun violence epidemic.* Giffords Center. https://giffords.org/lawcenter/report/healing-communities-in-crisis-lifesaving-solutions-to-the-urban-gun-violence-epidemic/

Giffords Center. (2018, September). *Confronting the inevitable myth.* https://giffords.org/wp-content/uploads/2018/09/Giffords-Law-Center-Confronting-The-Inevitability-Myth_Factsheet.pdf

Giffords Center. (2020a, September 10). *Community violence.* https://giffords.org/issues/community-violence/

Glffords Law Center. (2020b, October 23). *Investing in local intervention strategies in New York.* https://giffords.org/lawcenter/state-laws/investing-in-local-intervention-strategies-in-new-york/

Giffords, G., & McCann, C. (2017). *Bullets into bells: Poets & citizens respond to gun violence.* Boston, MA: Beacon Press.

Giles, C. (2020, March 17). A Baltimore barber who's with his clients through life, and death. *The New York Times.* https://www.nytimes.com/2020/03/17/us/baltimore-barber-gun-violence.html

Giles, D. N. (2020). *An examination of the rhetoric surrounding gun violence in the United States through the voices of student activists* (Doctoral dissertation, Iowa State University, Ames, Iowa).

Giordano, R. (2019, November 20). Gun violence scars victims long after the wounds have healed, Penn study finds. *The Philadelphia Inquirer.* https://www.inquirer.com/health/gunshot-victims-study-ptsd-psychological-harm-   penn-hospital-20191120.html

Giovanelli, A., Hayakawa, M., Englund, M. M., & Reynolds, A. J. (2018). African-American males in Chicago: Pathways from early childhood intervention to reduced violence. *Journal of Adolescent Health, 62*(1), 80–86.

Giran, G., Bertin, H., Koudougou, C., Sury, F., Croisé, B., & Laure, B. (2019). About a pediatric facial trauma. *Journal of Stomatology, Oral and Maxillofacial Surgery, 120*(2), 154–156.

Gokhale, P., Young, M. R., Williams, M. N., Reid, S. N., Tom, L. S., O'Brian, C. A., & Simon, M. A. (2020). Refining trauma-informed perinatal care for urban prenatal care patients with multiple lifetime traumatic exposures: A qualitative study. *Journal of Midwifery & Women's Health, 65*(2), 224–230.

Golden, T. (2019). *Innovation and equity in public health research: Testing arts-based methods for trauma-informed, culturally-responsive inquiry* (Doctoral dissertation, University of Louisville, Louisville, KY).

Goldenberg, A., Rattigan, D., Dalton, M., Gaughan, J. P., Thomson, J. S., Remick, K., . . . Hazelton, J. P. (2019). Use of ShotSpotter detection technology decreases pre-hospital time for patients sustaining gunshot wounds. *Journal of Trauma and Acute Care Surgery, 87*(6), 1253–1259.

Goldman, T. R. (2020). Interrupting violence from within the trauma unit and well beyond: A report on a hospital-based program to provide short-and long-term support to victims of violence. *Health Affairs, 39*(4), 556–561.

Goldstick, J. E., Carter, P. M., Heinze, J. E., Walton, M. A., Zimmerman, M., & Cunningham, R. M. (2019). Predictors of transitions in firearm assault behavior among drug-using youth presenting to an urban emergency department. *Journal of Behavioral Medicine, 42*(4), 635–645.

González, S., & Beauthier, J. P. (2020). Injury crimes and the temporary incapacity for work: A critique. *Aggression and Violent Behavior, 51* (March-April).

Goodwin, A. K., & Grayson, T. J. (2020). Investing in the frontlines: Why trusting and supporting communities of color will help address gun violence. *The Journal of Law, Medicine & Ethics, 48*(4 Suppl.), 164–171.

Goolsby, C., Strauss-Riggs, K., Rozenfeld, M., Charlton, N., Goralnick, E., Peleg, K., . . . Hurst, N. (2019). Equipping public spaces to facilitate rapid point-of-injury hemorrhage control after mass casualty. *American Journal of Public Health, 109*(2), 236–241.

Gomez, M. B. (2016). Policing, community fragmentation, and public health: Observations from Baltimore. *Journal of Urban Health, 93*(1), 154–167.

Gonzalez, E. P. (2020). *Southern culture and defensive gun ownership* (Doctoral dissertation, The University of Texas at San Antonio, TX).

Goolsby, C., Strauss-Riggs, K., Rozenfeld, M., Charlton, N., Goralnick, E., Peleg, K., . . . Hurst, N. (2019). Equipping public spaces to facilitate rapid point-of-injury hemorrhage control after mass casualty. *American Journal of Public Health, 109*(2), 236–241.

Goralnick, E., Ezeibe, C., Chaudhary, M. A., McCarty, J., Herrera-Escobar, J. P., Andriotti, T., . . . Weissman, J. S. (2020). Defining a research agenda for layperson prehospital hemorrhage control: A consensus statement. *JAMA Network Open, 3*(7), e20939.

Green, B., Horel, T., & Papachristos, A. V. (2017). Modeling contagion through social networks to explain and predict gunshot violence in Chicago, 2006 to 2014. *JAMA Internal Medicine, 177*(3), 326–333.

Green, C. (2019). Desistance and disabled masculine identity: Exploring the role of serious violent victimization in the desistance process. *Journal of Developmental and Life-course Criminology, 5*(3), 287–309.

Green, C. (2020). *American gun: A poem by 100 Chicagoans.* Chicago, IL: Depaul University.

Green, N. (2016, May 26). At apartment complex where 6-year-old was killed volunteers try to make a difference. *VRN.* https://www.wlrn.org/post/apartment-complex-where-6-year-old-was-killed-volunteers-try-make-difference#stream/0

Green, N. (2017, August 14). After getting shot, survivors of gun violence struggle to find resources to deal with trauma. *LRN.* https://www.wlrn.org/post/after-getting-shot-survivors-gun-violence-struggle-find-resources-deal-trauma#stream/0

Grigorian, A., Nahmias, J., Chin, T., Allen, A., Kuncir, E., Dolich, M., . . . Lekawa, M. (2019). Patients with gunshot wounds to the torso differ in risk of mortality depending on treating hospital. *Updates in Surgery, 71*(3), 561–567.

Gross, B. W., Cook, A. D., Rinehart, C. D., Lynch, C. A., Bradburn, E. H., Bupp, K. A., . . . Rogers, F. B. (2017). An epidemiologic overview of 13 years of firearm hospitalizations in Pennsylvania. *Journal of Surgical Research, 210,* 188–195.

Grossman, D. C., & Choucair, B. (2019). Violence and the US health care sector: Burden and response. *Health Affairs, 38*(10), 1638–1645.

Grossman, D. C., Mueller, B. A., Riedy, C., Dowd, M. D., Villaveces, A., Prodzinski, J., . . . Harruff, R. (2005). Gun storage practices and risk of youth suicide and unintentional firearm injuries. *JAMA, 293*(6), 707–714.

Grossman, L. S., & Clear, T. R. (2021). Combatting gun violence in Newark, New Jersey. In M. Crandall, S. Bonne, J. Bronson & W. Kessel (Eds.), *Why we are losing the war on gun violence in the United States* (pp. 151–162). New York, NY: Springer.

Gunn, J. F., & Boxer, P. (2021). Gun laws and youth gun carrying: Results from the Youth Risk Behavior Surveillance System, 2005–2017. *Journal of Youth and Adolescence, 50*(3), 446–458.

Gunter, H. (2018). Independent study finds almost 60% reduction in gun injury in NYC community due to Cure Violence. *Cure Violence.* http://cureviolence.org/post/indep endentstudy-finds-almost-60-reduction-in-gun-   injury-in-nyc-community-due-to-cure-violence/.

Gurney, O., Menaker, J., & Springer, B. L. (2020). Penetrating torso trauma. *Trauma Reports, 21*(2).

Gushue, K., & Wong, J. S. (2018). "When you choose to be a gangbanger, you deserve everything you get": Victim dichotomization, fear, and the problem frame. *Journal of Contemporary Criminal Justice, 34*(4), 364–382.

Gutierrez, A., Su, Y. S., Vaughan, K. A., Miranda, S., Chen, H. I., Petrov, D., . . . Schuster, J. M. (2020). Penetrating spinal column injuries (pSI): An institutional experience with 100 consecutive cases in an urban trauma center. *World Neurosurgery, 138,* e551–e556.

Haberman, C. P., & Link, N. W. (2020). Broken windows, hot spots, and focused deterrence: The state and impact of the "Big Three" in policing innovations. In C. Chouhy, J. C. Cochran, & C. L. Jonson, C. L (Eds.), *Criminal justice theory, Volume 26: Explanations and effects* (pp. 221–243). New York, NY: Routledge.

Hachem, L. D., Ahuja, C. S., & Fehlings, M. G. (2017). Assessment and management of acute spinal cord injury: From point of injury to rehabilitation. *The Journal of Spinal Cord Medicine, 40*(6), 665–675.

Hagan, J., & Foster, H. (2020). Imprisonment, opioids and health care reform: The failure to reach a high-risk population. *Preventive Medicine, 130,* 105897.

Hai, S., McKenney, M., & Elkbuli, A. (2020). Use of ShotSpotter detection technology decreases prehospital time for patients sustaining gunshot wounds. *Journal of Trauma and Acute Care Surgery, 89*(2), e16–e17.

Haider, A. H. (2017). Yes. This is a missed opportunity. Let's turn traumatic injury into a life-saving event. *Annals of Surgery, 265*(5), 854–855.

Hakki, L., Smith, A., Babin, J., Hunt, J., Duchesne, J., & Greiffenstein, P. (2019). Effects of a fragmenting handgun bullet: Considerations for trauma care providers. *Injury, 50*(5), 1143–1146.

Halimeh, B. N., Hughes, D., Evans, B., Emberson, J., Turco, L., Zakrison, T. L., & Winfield, R. D. (2021). Empowering the affected—Informing community-based solutions through interviews with survivors of interpersonal firearm violence. *Journal of Trauma and Acute Care Surgery, 90*(6), 980–986.

Hamby, S. (2020). Strengths-based approaches to conducting research with low income and other marginalized populations. In K. C. McLean (Ed.), *Cultural methods in psychology: Describing and transforming cultures* (pp. 76–108). New York, NY: Oxford University Press.

Hameed, S. M., Knebel, K., & Rogers, S. O. (2020). The future of injury control is precise. In M. Siegler & S. Rogers Jr. S (Eds.), *Violence, trauma, and trauma surgery* (pp. 183–198). New York, NY: Springer.

Hamilton, D. (2020, November 2). A new curriculum helps surgical trainees prepare to comprehensively treat victims of firearm violence. *American College of Surgeons.* https://www.facs.org/media/press-releases/2020/firearm-violence- 110220

Hammack, A. Y. (2019). *Assault-related, penetrating trauma and gunshot wound recidivism in New Orleans* (Doctoral dissertation, Louisiana State University Health Sciences Center).

Han, H. S., & Helm, S. (2020). Does demolition lead to a reduction in nearby crime associated with abandoned properties? *Housing Policy Debate,* 1–24.

Hansen, A.J. (2019, January 23). Costs of gun-related hospitalizations, readmissions examined in study. *SCOPE.* https://scopeblog.stanford.edu/2019/01/23/costs-of-gun-related-hospitalizations-readmissions-examined-in-study/

Harcke, H. T., Lawrence, L. L., Gripp, E. W., Kecskemethy, H. H., Kruse, R. W., & Murphy, S. G. (2019). Adult tourniquet for use in school-age emergencies. *Pediatrics, 143*(6). https://pediatrics.aappublications.org/content/143/6/e20183447.full \

Hardiman, E. R., Jones, L. V., & Cestone, L. M. (2019). Neighborhood perceptions of gun violence and safety: Findings from a public health-social work intervention. *Social Work in Public Health, 34*(6), 492–504.

Harper, A. W. (2020). Strengthening congregational communities: Social justice engagement through deliberative dialogue. *Social Work & Christianity, 47*(3), 85–122.

Hatchimonji, J. S., Swendiman, R. A., Goldshore, M. A., Blinman, T. A., Nance, M. L., Allukian III, M., & Nace Jr, G. W. (2020a). Pediatric firearm mortality in the United States, 2010 to 2016: A National Trauma Data Bank analysis. *Journal of Trauma and Acute Care Surgery*, *88*(3), 402–407.

Hatchimonji, J. S., Swendiman, R. A., Seamon, M. J., & Nance, M. L. (2020b). Trauma does not quarantine: Violence during the Covid-19 pandemic. *Annals of Surgery*, *272*(2), e53–e54.

Hatten, D. N., & Wolff, K. T. (2020). Rushing gunshot victims to trauma care: The influence of first responders and the challenge of the geography. *Homicide Studies*, *24*(4), 377–397.

Haywood, C., Pyatak, E., Leland, N., Henwood, B., & Lawlor, M. C. (2019). A qualitative study of caregiving for adolescents and young adults with spinal cord injuries: Lessons from lived experiences. *Topics in Spinal Cord Injury Rehabilitation*, *25*(4), 281–289.

He, K., & Sakran, J. V. (2019). Elimination of the moratorium on gun research is not enough: The need for the CDC to set a budgetary agenda. *JAMA Surgery*, *154*(3), 195–196.

Healthline. (2021, January 4). Why addressing gun violence as a health crisis is crucial for change. https://www.healthline.com/health-news/why-addressing-gun- violence-as-a-health-crisis-is-crucial-for-change#Gun-violence-a-serious- problem-in-America

Heath, S. (2020, August 10). What is the difference between health disparities, equity? *Patient Engagement*. https://patientengagementhit.com/news/what-is-the-difference-between-health-disparities-equity

Held, M., Engelmann, E., Dunn, R., Ahmad, S. S., Laubscher, M., Keel, M. J., . . . Hoppe, S. (2017). Gunshot induced injuries in orthopaedic trauma research: A bibliometric analysis of the most influential literature. *Orthopaedics & Traumatology: Surgery & Research*, *103*(5), 801–807.

Hemenway, D. (2018). Firearms data, and an ode to data systems. *Chance*, *31*(1), 7–11.

Hemenway, D. (2020). Let's make it a priority to improve injury data. *Injury Prevention*, *26*(4), 395–396.

Hemenway, D., & Nelson, E. (2020). The scope of the problem: Gun violence in the USA. *Current Trauma Reports*, *6*(1), 29–35.

Hersh, E. (2019, February 4). Exploring the hidden burden of firearm injuries. *Health City*. https://www.bmc.org/healthcity/population-health/exploring-hidden-burden-firearm-injuries

Hickner, J. (2018). We need to treat gun violence like an epidemic. *Journal of Family Practice*, *67*(4), 198.

Hills-Evans, K., Mitton, J., & Sacks, C. A. (2018). Stop posturing and start problem solving: A call for research to prevent gun violence. *AMA Journal of Ethics*, *20*(1). https://journalofethics.ama-assn.org/article/stop-posturing-and- start-problem-solving-call-research-prevent-gun-violence/2018- 01?Effort%2BCode=FBB007

Hink, A. B., Bonne, S., Levy, M., Kuhls, D. A., Allee, L., Burke, P. A., . . . Stewart, R. M. (2019). Firearm injury research and epidemiology: A review of the data, their limitations, and how trauma centers can improve firearm injury research. *Journal of Trauma and Acute Care Surgery*, *87*(3), 678–689.

Hinton, A. (2017). "And so I bust back" violence, race, and disability in hip hop. *CLA Journal*, *60*(3), 290–304.

Hipple, N. K., Huebner, B. M., Lentz, T. S., McGarrell, E. F., & O'Brien, M. (2020). The case for studying criminal nonfatal shootings: Evidence from four Midwest cities. *Justice Evaluation Journal*, *3*(1), 94–113.

Hipple, N. K., & Magee, L. A. (2017). The difference between living and dying: victim characteristics and motive among nonfatal shootings and gun homicides. *Violence and Victims*, *32*(6), 977–997.

Hipple, N. K., Thompson, K. J., Huebner, B. M., & Magee, L. A. (2019). Understanding victim cooperation in cases of nonfatal gun assaults. *Criminal Justice and Behavior*, *46*(12), 1793–1811.

Hirner, K. (2019). *Fear and awe: Social construction of fear in Detroit.* Institut Fur Ethnologie. Munich, Germany. https://epub.ub.uni- muenchen.de/60886/1/26_ Hirner_Detroit.pdf

Hofmann, L. J., Keric, N., Cestero, R. F., Babbitt-Jonas, R., Khoury, L., Panzo, M., & Cohn, S. M. (2018). Trauma surgeons' perspective on gun violence and a review of the literature. *Cureus*, *10*(11), E3599.

Hohl, B. C., Kondo, M. C., Kajeepeta, S., MacDonald, J. M., Theall, K. P., Zimmerman, M. A., & Branas, C. C. (2019). Creating safe and healthy neighborhoods with place-based violence interventions. *Health Affairs*, *38*(10), 1687–1694.

Hohl, B. C., Stegal, M. B., & Webster, D. W. (2021). A new narrative on gun violence. In A. L. Plough (Ed.), *Community resilience: Equitable practices for an uncertain future* (pp. 30–45). New York, NY: Oxford University Press.

Holzer, K. J., Carbone, J. T., DeLisi, M., & Vaughn, M. G. (2019). Traumatic brain injury and coextensive psychopathology: New evidence from the 2016 Nationwide Emergency Department Sample (NEDS). *Journal of Psychiatric Research*, *114*, 149–152.

Hong, J. S., & Espelage, D. L. (2020). An introduction to the special issue: Firearms homicide and perceptions of safety in American schools post-Columbine. *Journal of School Violence*, *19*(1), 1–5.

Hooker, L. (2021, February 1). Firearm injury study and violence prevention, intervention program aim to break cycle for at-risk populations. *Medical University of South Carolina.* https://web.musc.edu/about/news- center/2021/02/01/firearm-injury-study-and-violence-prevention-intervention- program-aim-to-break-cycle

House, S. (2018). Addressing gun violence in the United States. *Gettysburg Social Sciences Review*, *2*(2), 6–191.

Howard, A., & Rawsthorne, M. (2020). *Everyday community practice: Principles and practice.* New York, NY: Routledge.

Howell, E. M., & Gangopadhyaya, A. (2017). *State variation in the hospital costs of gun violence, 2010 and 2014.* Washington, DC: The Urban Institute.

Hsu, Y. T., Chang, D. C., Perez, N. P., Westfal, M. L., Kelleher, C. M., Sacks, C. A., & Masiakos, P. T. (2020). Civilian firearm-related injuries: How often is a tourniquet beneficial? *Annals of Surgery*, *271*(2), e12–e13.

Hu, L. (2019, April 4). How poetry is helping gun violence survivors find purpose in life. *Spectrum News NY.* https://www.ny1.com/nyc/all- boroughs/news/2019/04/05/how-poetry-is-helping-gun-violence-survivors-find- new-purpose-in-life

Huebner, B. M., Lentz, T. S., & Schafer, J. A. (2020). Heard shots–Call the police? An examination of citizen responses to gunfire. *Justice Quarterly*, *46*(3), 1–24.

Huebner, B. M., Martin, K., Moule Jr, R. K., Pyrooz, D., & Decker, S. H. (2016). Dangerous places: Gang members and neighborhood levels of gun assault. *Justice Quarterly*, *33*(5), 836–862.

Hughes, B. D., Cummins, C. B., Shan, Y., Mehta, H. B., Radhakrishnan, R. S., & Bowen-Jallow, K. A. (2020). Pediatric firearm injuries: Racial disparities and predictors of healthcare outcomes. *Journal of Pediatric Surgery*, *55*(8), 1596–1603.

Humar, P., Goolsby, C. A., Forsythe, R. M., Reynolds, B., Murray, K. M., Bertoty, D., . . . Neal, M. D. (2020). Educating the public on hemorrhage control: Methods and challenges of a public health initiative. *Current Surgery Reports, 8*, 1–7.

Hunter-Pazzara, B. (2020). The possessive investment in guns: Towards a material, social, and racial analysis of guns. *Palgrave Communications, 6*(1), 1–10.

Hurlbert, R. J. (2019). Penetrating spinal trauma: Snapshot of the American epidemic. *Neurosurgical Focus, 46*(3), E5.

Hutchinson, A. J., Kusnezov, N. A., Dunn, J. C., Rensing, N., Prabhakar, G., & Pirela-Cruz, M. A. (2019). Epidemiology of gunshot wounds to the hand. *Hand Surgery and Rehabilitation, 38*(1), 14–19.

Hutson, H. R., Anglin, D., & Eckstein, M. (1996). Drive-by shootings by violent street gangs in Los Angeles: A five-year review from 1989 to 1993. *Academic Emergency Medicine, 3*(4), 300–303.

Hyak, J. M., Todd, H., Rubalcava, D., Vogel, A. M., Fallon, S., & Naik-Mathuria, B. (2020). Barely benign: The dangers of BB and other nonpowder guns. *Journal of Pediatric Surgery, 55*(8), 1604–1609.

Hylton, A., & Eggers, C. (2020, September 30). Victims of gun violence in Chicago offer an alternative to traditional policing. *NBC News.* https://www.nbcnews.com/news/us-news/victims-gun-violence-chicago-offer- alternative-traditional-policing-n1241444

Ingram, L., & Drew-Branch, V. (2017). Taking what we have and making what we need: Utilizing natural helping support networks to decrease self directed violence among adolescents of color. *Journal of Family Strengths, 17*(1), 9–23.

Irizarry, C. R., Hardigan, P. C., Mc Kenney, M. G., Holmes, G., Flores, R., Benson, B., & Torres, A. M. (2017). Prevalence and ethnic/racial disparities in the distribution of pediatric injuries in South Florida: Implications for the development of community prevention programs. *Injury Epidemiology, 4*(1), 12.

Irwin-Erickson, Y., Bai, B., Gurvis, A., & Mohr, E. (2016). *The effect of gun violence on local economies.* Washington, DC: Urban Institute.

Iturricastillo, A., Yanci, J., Granados, C., & Goosey-Tolfrey, V. (2016). Quantifying wheelchair basketball match load: a comparison of heart-rate and perceived-exertion methods. *International Journal of Sports Physiology and Performance, 11*(4), 508–514.

Izadi, S. N., Patel, N., Fofana, D., Paredes, A. Z., Snyder, S. K., Torres-Reveron, A., & Skubic, J. J. (2020). Racial inequality in the trauma of women: A disproportionate decade. *Journal of Trauma and Acute Care Surgery, 89*(1), 254–262.

Jacob, H., Travers, C., & Hann, G. (2021). Hospital youth workers for violence reduction. *Archives of Disease in Childhood, 106*(5).

Jacobs, L. M., McSwain Jr, N. E., Rotondo, M. F., Wade, D., Fabbri, W., Eastman, A. L., . . . Sinclair, J. (2013). Improving survival from active shooter events: The Hartford Consensus. *Journal of Trauma and Acute Care Surgery, 74*(6), 1399–1400.

Jacoby, S. F., Reeping, P. M., & Branas, C. C. (2020a). Police-to-hospital transport for violently injured individuals: A way to save lives? *The ANNALS of the American Academy of Political and Social Science, 687*(1), 186–201.

Jacoby, S. F., Rich, J. A., Webster, J. L., & Richmond, T. S. (2020c). "Sharing things with people that I don't even know": Help-seeking for psychological symptoms in injured Black men in Philadelphia. *Ethnicity & Health, 25*(6), 777–795.

Jacoby, S. F., Richmond, T. S., Holena, D. N., & Kaufman, E. J. (2018). A safe haven for the injured? Urban trauma care at the intersection of healthcare, law enforcement, and race. *Social Science & Medicine, 199*, 115–122.

Jacoby, S., Webster, J., Twomey, C., & Gebreyesus, A. (2020b). "The toll that it's taken": Perspectives of black men recovering from traumatic injury in Philadelphia. *BMJ Injury Prevention 26* (Suppl. 1). Online.

Jaffe, S. (2020). Decisions to be made on US gun violence research funds. *The Lancet, 395*(10222), 403–404.

James, J., Fitzgibbon, J., & Blackford, M. (2016). Nausea, vomiting, and weight loss in a young adult patient with a history of a gunshot wound. *Pediatric Emergency Care, 32*(9), 616–618.

Jamrozik, A., Oraa Ali, M., Sarwer, D. B., & Chatterjee, A. (2019). More than skin deep: Judgments of individuals with facial disfigurement. *Psychology of Aesthetics, Creativity, and the Arts, 13*(1), 117–129.

January, A. M., Kirk, S., Zebracki, K., Chlan, K. M., & Vogel, L. C. (2018). Psychosocial and health outcomes of adults with violently acquired pediatric spinal cord injury. *Topics in Spinal Cord Injury Rehabilitation, 24*(4), 363–370.

Jarrett, A. L., & Devers, J. J. (2020). Impact of access on trauma mortality. *Emergency Medicine Trauma. EMTCJ-100036, 2*(4).

Javanbakht, A. (2020, July 1). Fireworks can torment veterans and survivors of gun violence. *Psychology Today.* https://www.psychologytoday.com/us/blog/the- many-faces-anxiety-and-trauma/202007/fireworks-can-torment-veterans-and- survivors-gun

Jay, J., Miratrix, L. W., Branas, C. C., Zimmerman, M. A., & Hemenway, D. (2019). Urban building demolitions, firearm violence and drug crime. *Journal of Behavioral Medicine, 42*(4), 626–634.

Jiang, T., Webster, J. L., Robinson, A., Kassam-Adams, N., & Richmond, T. S. (2018). Emotional responses to unintentional and intentional traumatic injuries among urban black men: A qualitative study. *Injury, 49*(5), 983–989.

Jocson, R. M., Alers-Rojas, F., Ceballo, R., & Arkin, M. (2020). Religion and spirituality: Benefits for Latino adolescents exposed to community violence. *Youth & Society, 52*(3), 349–376.

Johnson, J., & Lecci, L. (2020). How caring is "nullified": Strong racial identity eliminates White participant empathy effects when police shoot an unarmed Black male. *Psychology of Violence, 10*(1), 58–67.

Joseph, B., Bible, L., & Hanna, K. (2020). Developing data-driven Solutions to Firearm violence. *Current Trauma Reports, 6*, 44–50.

Joseph, B., Hanna, K., Callcut, R. A., Coleman, J. J., Sakran, J. V., & Neumayer, L. A. (2019). The hidden burden of mental health outcomes following firearm-related injures. *Annals of Surgery, 270*(4), 593–601.

John Jay Research and Evaluation Center. (2020). Reducing violence without the police: A review of research evidence. https://johnjayrec.nyc/2020/11/09/av2020/

Johnson, D., & Barsky, A. E. (2020). Preventing gun violence in schools: Roles and perspectives of social workers. *School Social Work Journal, 44*(2), 26–48.

Johnson, N. J. (2016). Us v. them: Remnants of urban war zones. *Penn GSE Perspectives on Urban Education, 13*(1), 49–55.

Johnston, E. R. (2020). Pathologizing the wounded?: Post-traumatic stress disorder in an era of gun violence. *Rhetoric of Health & Medicine, 3*(1), 1–33.

Joint Economic Committee of the U.S. Congress. (2019, September 18). A state-by-state examination of the economic costs of gun violence. https://www.jec.senate.gov/pub lic/_cache/files/b2ee3158-aff4-4563-8c3b- 0183ba4a8135/economic-costs-of-gun-violence.pdf

Joint Economic Committee Democrats. (2017). *America can't afford gun violence.* Washington, DC. https://www.jec.senate.gov/public/_cache/files/a8c89469- 30a1-4b88-b3f5-0c5e54ad5df0/economic-impact-of-gun-violence-final.pdf

Jones, K. A. (2020). A seat at the table: Centering the voices of gun violence survivors. *Pediatrics, 146*(4).

Jordan, E., & Harper, K. (2020). Supporting policy making with research: Challenges, opportunities, and lessons learned. *Child Trends.* https://www.childtrends.org/wp-content/uploads/2020/01/Duke-bridges- brief_ChildTrends_Jan2020-1.pdf

Joseph, B., Bible, L., & Hanna, K. (2020). Developing data-driven solutions to firearm violence. *Current Trauma Reports, 6*, 44–50.

Joseph, K., & Reese, C. (2020). Primum non nocere: When is it our moral duty to do more for our trauma patients in need? In M. Siegler & S. Rogers Jr (Eds.), *Violence, trauma, and trauma surgery* (pp. 137–144). New York, NY: Springer.

Joseph, K., Turner, P., Barry, L., Cooper, C., Danner, O., Enumah, S., ... Stanford, A. (2018). Reducing the impact of violence on the health status of African-Americans: Literature review and recommendations from the Society of Black Academic Surgeons. *The American Journal of Surgery, 216*(3), 393–400.

The Joyce Foundation. (2020, December). *The next one hundred questions: A research agenda for ending gun violence.* Chicago, IL. http://www.joycefdn.org/assets/images/TJF-The-Next-100-Questions-A- Research-Agenda-for-Ending-Gun-Violence.pdf

Juette, M., & Berger, R. J. (2008). *Wheelchair warrior: Gangs, disability, and basketball.* Philadelphia, PA: Temple University Press.

Jundoria, A. K., Grant, B., Olufajo, O. A., De La Cruz, E., Metcalfe, D., Williams, M., ... Hughes, K. (2020). Assessment of the "weekend effect" in lower extremity vascular trauma. *Annals of Vascular Surgery, 66*, 233–241.

Juzang, I. (2020). Moving young black men beyond survival mode: Protective factors for their mental health. In R. Majors, K. Carberry, & T. S. Ransaw (Eds.), *The international handbook of black community mental health* (257–273). Bingley, UK: Emerald.

Kahn, J. S. (2018). What can I do as a physician to prevent firearm injury? *Annals of Internal Medicine, 169*(10), 725–726.

Kalra, M. (2019). A ricochet of pain—The long echo of gun violence. *The New England Journal of Medicine, 381*(18), 1704–1705.

Kalesan, B. (2017). The cost of firearm violence survivorship. *American Journal of Public Health, 107*(5), 638–639.

Kalesan, B., Siracuse, J. J., Cook, A., Prosperi, M., Fagan, J., & Galea, S. (2021). Prevalence and hospital charges from firearm injuries treated in US emergency departments from 2006 to 2016. *Surgery, 169*(5), 1188–1198.

Kalesan, B., Vyliparambil, M. A., Bogue, E., Villarreal, M. D., Vasan, S., Fagan, J., ... Firearm Injury Research Group. (2016). Race and ethnicity, neighborhood poverty and pediatric firearm hospitalizations in the United States. *Annals of Epidemiology, 26*(1), 1–6.

Kalesan, B., Weinberg, J., & Galea, S. (2016). Gun violence in Americans' social network during their lifetime. *Preventive Medicine, 93*, 53–56.

Kalesan, B., Zuo, Y., Vasan, R. S., & Galea, S. (2019). Risk of 90-day readmission in patients after firearm injury hospitalization: A nationally representative retrospective cohort study. *Journal of Injury and Violence Research, 11*(1), 65–80.

Kalesan, B., Zuo, Y., Xuan, Z., Siegel, M. B., Fagan, J., Branas, C., & Galea, S. (2018). A multi-decade joinpoint analysis of firearm injury severity. *Trauma Surgery & Acute Care Open, 3*(1), e000139.

Kalra, M. (2019). A ricochet of pain—The long echo of gun violence. *The New England Journal of Medicine, 381*(18), 1704–1705.

Kalyanaraman, N. (2020, October). A countywide public health approach to gun violence. In *APHA's 2020 virtual annual meeting and expo* (October 24–28). American Public Health Association.

Kamat, P. P., Santore, M. T., Hoops, K. E., Wetzel, M., McCracken, C., Sullivan, D., ... Grunwell, J. R. (2020). Critical care resource use, cost, and mortality associated with firearm-related injuries in US children's hospitals. *Journal of Pediatric Surgery, 55*(11), 2475–2479.

Kang, D., & Swaroop, M. (2019). Community-based approach to trauma and violence: guns, germs, and bystanders. *JAMA Surgery, 154*(3), 196–197.

Kang, D., & Swaroop, M. (2020). Empowerment: The ethical dilemma. In M. Siegler & S. Rogers Jr (Eds.), *Violence, trauma, and trauma surgery* (pp. 173–181). New York, NY: Springer.

Kao, A. M., Schlosser, K. A., Arnold, M. R., Kasten, K. R., Colavita, P. D., Davis, B. R., ... Heniford, B. T. (2019). Trauma recidivism and mortality following violent injuries in young adults. *Journal of Surgical Research, 237*, 140–147.

Kaplan, M. S., & Mueller-Williams, A. C. (2019). The hidden epidemic of firearm suicide in the United States: challenges and opportunities. *Health & Social Work, 44*(4), 276–279.

Kaufman, E., & Delgado, K. (2020). 210 Characteristics of hospitals that care for patients with firearm injuries: evidence from the nationwide emergency department sample. *Injury Prevention, 26* (Suppl. 1).

Kaufman, E., Holena, D. N., Yang, W. P., Morrison, C. N., Jacoby, S. F., Seamon, M., ... Beard, J. H. (2019). Firearm assault in Philadelphia, 2005–2014: A comparison of police and trauma registry data. *Trauma Surgery & Acute Care Open, 4*(1), e000316.

Kaufman, E. J., Passman, J. E., Jacoby, S. F., Holena, D. N., Seamon, M. J., MacMillan, J., & Beard, J. H. (2021). Making the news: Victim characteristics associated with media reporting on firearm injury. *JAMA Internal Medicine, 181*(2), 237–244.

Kaufman, E. J., & Richmond, T. S. (2020). Beyond band-aids for bullet holes: Firearm violence as a public health priority. *Critical Care Medicine, 48*(3), 391–397.

Kaufman, S. B. (2020, April 20). Post-traumatic growth: Finding meaning and creativity in adversity. *Scientific American.* https://blogs.scientificamerican.com/beautiful- minds/post-traumatic-growth-finding-meaning-and-creativity-in-adversity/

Keil, S., Beardslee, J., Schubert, C., Mulvey, E., & Pardini, D. (2020). Perceived gun access and gun carrying among male adolescent offenders. *Youth Violence and Juvenile Justice, 18*(2), 179–195.

Kelly, J. R., Levy, M. J., Reyes, J., & Anders, J. (2020). Effectiveness of the combat application tourniquet for arterial occlusion in young children. *Journal of Trauma and Acute Care Surgery, 88*(5), 644–647.

Kennedy, D. M. (2019). Policing and the lessons of focused deterrence. In D. Weisburd & A. A. Braga (Eds.), *Police innovation: Contrasting perspectives* (pp. 205–226). New York, NY: Cambridge University Press.

Kent, A. J., Sakran, J. V., Efron, D. T., Haider, A. H., Cornwell III, E. E., Haut, E. R., ... Garen Wintemute, M. D. (2017). Understanding increased mortality after gunshot injury: Cook et al. respond. *American Journal of Public Health, 107*(12), E22–E23.

Kersten, L., Vriends, N., Steppan, M., Raschle, N. M., Praetzlich, M., Oldenhof, H., ... Martinelli, A. (2017). Community violence exposure and conduct problems in children and adolescents with conduct disorder and healthy controls. *Frontiers in Behavioral Neuroscience, 11*, 219, 1–14.

Khubchandani, J., & Price, J. H. (2018). Violent behaviors, weapon carrying, and firearm homicide trends in African American adolescents, 2001–2015. *Journal of Community Health, 43*(5), 947–955.

Kieltyka, J., Kucybala, K., & Crandall, M. (2016). Ecologic factors relating to firearm injuries and gun violence in Chicago. *Journal of Forensic and Legal Medicine, 37*, 87–90.

Kim, D. H., Michalopoulos, L. M., & Voisin, D. R. (2021). Validation of the Brief Symptom Inventory–18 among low-income African American adolescents exposed to community violence. *Journal of Interpersonal Violence, 36*(1-2), NP984–NP1002.

Kim, S. J., Ramirez-Valles, J., Watson, K., Allen-Mears, P., Matthews, A., Martinez, E., . . . Winn, R. A. (2020). Fostering health equity research: Development and implementation of the Center for Health Equity Research (CHER) Chicago. *Journal of Clinical and Translational Science, 4*(1), 53–60.

King, K. (2021, February 17). Gun violence crisis: With revenge and retaliation on the rise, how police are responding. *ABC News.* https://wlos.com/news/local/gun-violence-crisis-with-revenge-retaliation-on-the-rise-how-asheville-police-are-responding

King, M. (2021). *Battlefield medics: How warfare changed the history of medicine.* London, UK: Arcturus.

King, T. (2021). Youth gun violence prevention organizing. In M. Crandall, S. Bonne, J. Bronson, & W. Kessel (Eds.), *Why we are losing the war on gun violence in the United States* (pp. 233–247). New York, NY: Springer.

Kingsland, J. (2020, August 14). Gunshot wounds needing surgery in the US increase in frequency, severity. *Medical News Today.* https://www.medicalnewstoday.com/articles/gunshot-wounds-needing-surgery-in-the-us-increase-in-frequency-severity#Major-surgery

Kircher, J. C., Stilwell, C., Talbot, E. P., & Chesborough, S. (2011, October). Academic bullying in social work departments: The silent epidemic. In *Annual Meeting of the National Association of Christians in Social Work.* Pittsburgh, PA.

Kirsch, T. D., Prytz, E., Hunt, R. C., & Jonson, C. O. (2020). Recommended process outcome measures for Stop the Bleed education programs. *AEM Education and Training, 5*(1), 139–142.

Klassen, A. B., Core, S. B., Lohse, C. M., & Sztajnkrycer, M. D. (2018). A descriptive analysis of care provided by law enforcement prior to EMS arrival in the United States. *Prehospital and Disaster Medicine, 33*(2), 165–170.

Klofas, J., Altheimer, I., & Petitti, N. (2020). *Retaliatory violent disputes.* Strategies for policing innovation: Problem-oriented guides for police. Series no. 74. Phoenix: Arizona State University.

Knight, C., & Borders, L. D. (2018). Trauma-informed supervision: Core components and unique dynamics in varied practice contexts. *The Clinical Supervisor, 37*(1), 1–6.

Kondo, M. C., South, E. C., Branas, C. C., Richmond, T. S., & Wiebe, D. J. (2017). The association between urban tree cover and gun assault: A case-control and case-crossover study. *American Journal of Epidemiology, 186*(3), 289–296.

Kondo, M. C., Andreyeva, E., South, E. C., MacDonald, J. M., & Branas, C. C. (2018). Neighborhood interventions to reduce violence. *Annual Review of Public Health, 39*, 253–271.

Konstam, M. A., & Konstam, A. D. (2019). Gun violence and cardiovascular health: We need to know. *Circulation, 139*(22), 2499–2501.

Koper, C. S. (2020). Assessing the potential to reduce deaths and injuries from mass shootings through restrictions on assault weapons and other high-capacity semiautomatic firearms. *Criminology & Public Policy, 19*(1), 147–170.

Koper, C. S., Johnson, W. D., Nichols, J. L., Ayers, A., & Mullins, N. (2018). Criminal use of assault weapons and high-capacity semiautomatic firearms: An updated examination of local and national sources. *Journal of Urban Health, 95*(3), 313–321.

Koper, C. S., Johnson, W. D., Stesin, K., & Egge, J. (2019). Gunshot victimisations resulting from high-volume gunfire incidents in Minneapolis: Findings and policy implications. *Injury Prevention, 25*(Suppl 1), i9–i11.

Koper, C. S., Woods, D. J., & Isom, D. (2016). Evaluating a police-led community initiative to reduce gun violence in St. Louis. *Police Quarterly, 19*(2), 115–149.

Korten, T. (2016, June 25). The other victims of gun violence. *Modern Health Care.* https://www.modernhealthcare.com/reports/other-victims-of-gun-violence/#!/

Kotlowitz, A. (2019). *An American summer: Love and death in Chicago.* New York, NY: Nan A. Talese.

Kravitz-Wirtz, N., Pallin, R., Schleimer, J., Tomsich, E., Buggs, S., Charbonneau, A., . . . Wintemute, G. (2019, November). Socio-emotional consequences of exposure to violence: Does firearm involvement exacerbate the impact of a violent event? In *APHA's 2019 Annual Meeting and Expo* (Nov. 2–Nov. 6). American Public Health Association.

Krisberg, K. (2018). Gun violence research hurt by shortage of funding, data. *American Journal of Public Health, 108*(8), 967.

Kuehn, B. M. (2019). Growing evidence linking violence, trauma to heart disease. *Circulation, 139*(7), 981–982.

Kuhls, D. A., Stewart, R. M., & Bulger, E. M. (2020). Building consensus to decrease firearm injuries and death in the USA: Engaging medical and public health organizations to find common ground. *Current Trauma Reports, 6,* 56–61.

Kuo, E. C., Harding, J., Ham, S. W., & Magee, G. A. (2019). Successful treatment and survival after gunshot wound to the aortic arch with bullet embolism to superficial femoral artery. *Journal of Vascular Surgery Cases and Innovative Techniques, 5*(3), 283–288.

Kurek, N., Darzi, L. A., & Maa, J. (2020). A worldwide perspective provides insights into why a US Surgeon General Annual Report on Firearm Injuries is needed in America. *Current Trauma Reports, 6,* 36–43.

Kurtenbach, S. (2021). "I fear only the neighbourhood and the Lord!" Youth violence in marginalized spaces. *European Journal of Criminology,*

Kurtenbach, S., & Rauf, A. (2019). Violence-related norms and the "code of the street." In W. Heitmeyer, S. Howell, S. Kurtenbach, A. Rauf, M. Zaman, & S. Zdun (Eds.), *The codes of the street in risky neighborhoods* (pp. 21–38). New York, NY: Springer.

Kwak, H., Dierenfeldt, R., & McNeeley, S. (2019). The code of the street and cooperation with the police: Do codes of violence, procedural injustice, and police ineffectiveness discourage reporting violent victimization to the police? *Journal of Criminal Justice, 60,* 25–34.

Kwong, J. Z., Gray, J. M., Rein, L., Liu, Y., & Melzer-Lange, M. D. (2019). An educational intervention for medical students to improve self-efficacy in firearm injury prevention counseling. *Injury Epidemiology, 6*(1), 27.

Lale, A., Krajewski, A., & Friedman, L. S. (2017). Undertriage of firearm-related injuries in a major metropolitan area. *JAMA Surgery, 152*(5), 467–474.

Landeo-Gutierrez, J., Forno, E., Miller, G. E., & Celedón, J. C. (2020). Exposure to violence, psychosocial stress, and asthma. *American Journal of Respiratory and Critical Care Medicine, 201*(8), 917–922.

Lane, S. D., Rubinstein, R. A., Bergen-Cico, D., Jennings-Bey, T., Fish, L. S., Larsen, D. A., . . . Robinson, J. A. (2017). Neighborhood trauma due to violence: A multilevel analysis. *Journal of Health Care for the Poor and Underserved, 28*(1), 446–462.

Lauger, T. R. (2019). Group processes within gangs. In C. Haney (Ed.), *Oxford research encyclopedia of criminology and criminal justice*. New York, NY: Oxford University Press.

Laurencin, C. T., & Walker, J. M. (2020). Racial profiling is a public health and health disparities issue. *Journal of Racial and Ethnic Health Disparities, 7*, 393–397.

Law, T., Seiver, S., Papachristos, A. V., & Violano, P. (2017). *Age of gunshot would victims in New Haven, 2003–2015.* New Haven, CT: Institution for Social and Policy Studies, Yale University.

Leasy, M., O'Gurek, D. T., & Savoy, M. L. (2019). Unlocking clues to current health in past history: Childhood trauma and healing. *Family Practice Management, 26*(2), 5–10.

Leavitt, S. (2018). Firearm-related injury and death. *Annals of Internal Medicine, 169*(1), 64–66.

Lee, D. (2018). Gun violence and health equity. *Diverse Issues in Higher Education, 35*(5), 40.

Lee, D. B., Hsieh, H. F., Stoddard, S. A., Heinze, J. E., Carter, P. M., Goldstick, J. E., . . . Zimmerman, M. A. (2020). Longitudinal pathway from violence exposure to firearm carriage among adolescents: The role of future expectation. *Journal of Adolescence, 81*, 101–113.

Lee, J. (2013). The pill hustle: Risky pain management for a gunshot victim. *Social Science & Medicine, 99*, 162–168.

Lee, J. R. S., Hunter, A. G., Priolli, F., & Thornton, V. J. (2020). "Pray that I live to see another day": Religious and spiritual coping with vulnerability to violent injury, violent death, and homicide bereavement among young Black men. *Journal of Applied Developmental Psychology, 70*, 101–180.

Lee, L. K., & Schaechter, J. (2019). No silver bullet: Firearm laws and pediatric death prevention. *Pediatrics, 144*(2), e20191300.

Lee, V. J., Meloche, A., Grant, A., Neuman, D., & Tecce DeCarlo, M. J. (2019). "My thoughts on gun violence": An urban adolescent's display of agency and multimodal literacies. *Journal of Adolescent & Adult Literacy, 63*(2), 157–168.

Leeolou, M. C., & Takooshian, H. (2020). Homicide activism: A call for research on a neglected phenomenon. In R. Javier, E. Owen, & J. Maddux (Eds.), *Assessing trauma in forensic contexts* (pp. 139–150). New York, NY: Springer, Cham.

Leibbrand, C., Hill, H., Rowhani-Rahbar, A., & Rivara, F. (2020). Invisible wounds: Community exposure to gun homicides and adolescents' mental health and behavioral outcomes. *SSM-Population Health, 12*. https://www.ncbi.nlm.nih.gov/pmc/articles/PMC7653279/

Lenart, E. K., Lewis Jr, R. H., Sharpe, J. P., Fischer, P. E., Croce, M. A., & Magnotti, L. J. (2020). They only come out at night: Impact of time of day on outcomes after penetrating abdominal trauma. *Surgery Open Science, 2*(4), 1–4.

Lennon, T. (2020). Factors associated with perception of life expectancy in assault-injured urban youth: An emergency department sample. *Injury Prevention, 26*(Suppl. 1), A41.

Leonard, D. J. (2017). Illegible black death, legible white pain: Denied media, mourning, and mobilization in an era of "post-racial" gun violence. *Cultural Studies↔ Critical Methodologies, 17*(2), 101–109.

Levas, M. N., Melzer-Lange, M., Tarima, S., Beverung, L. M., & Panepinto, J. (2020). Youth victims of violence report worse quality of life than youth with chronic diseases. *Pediatric Emergency Care, 36*(2), e72–e78.

Levine, M., Philpot, R., & Kovalenko, A. G. (2020). Rethinking the bystander effect in violence reduction training programs. *Social Issues and Policy Review, 14*(1), 273–296.

Lewis, J. A., Hamilton, J. C., & Elmore, J. D. (2019). Describing the ideal victim: A linguistic analysis of victim descriptions. *Current Psychology*, July, 1–9.

Lewis, C. Y., Carmona, R. H., & Roberts, C. S. (2020). Should every physician be ready to act as a community first responder? *Injury, 51*(12), 2731–2733.

Lewis, S., & de Mesquita, B. B. (2020). Racial differences in hospital evaluation after the use of force by police: A tale of two cities. *Journal of Racial and Ethnic Health Disparities, 7*, 1178–1187.

Lewis, J. A., Hamilton, J. C., & Elmore, J. D. (2021). Describing the ideal victim: A linguistic analysis of victim descriptions. *Current Psychology, 40*(9), 4324–4332.

Lipscomb, A. E., Emeka, M., Bracy, I., Stevenson, V., Lira, A., Gomez, Y. B., & Riggins, J. (2019). Black male hunting! A phenomenological study exploring the secondary impact of police induced trauma on the Black man's psyche in the United States. *Journal of Sociology, 7*(1), 11–18.

Lo, C. C., Ash-Houchen, W., Gerling, H. M., & Cheng, T. C. (2020). From childhood victim to adult criminal: Racial/ethnic differences in patterns of victimization— Offending among Americans in early adulthood. *Victims & Offenders, 15*(4), 430–456.

Loeffler, C., & Flaxman, S. (2016). Is gun violence contagious? *arXiv:1611.06713*. https://arxiv.org/pdf/1611.06713.pdf

LoFaso, C. A. (2020). Solving homicides: The influence of neighborhood characteristics and investigator caseload. *Criminal Justice Review, 45*(1), 84–103.

Loggini, A., Vasenina, V. I., Mansour, A., Das, P., Horowitz, P. M., Goldenberg, F. D., . . . Lazaridis, C. (2020). Management of civilians with penetrating brain injury: A systematic review. *Journal of Critical Care, 56*(April), 159–166.

Lopez, G. (2018). America's unique gun violence problem, explained in 17 maps and charts. *Vox*. https://www. vox. com/policy-and-politics/2017/10/2/16399418/us- gun-violence-statistics-maps-charts.

Lotfollahzadeh, S., & Burns, B. (2020). Penetrating abdominal trauma. *StatPearls*. https://www.ncbi.nlm.nih.gov/books/NBK459123/

Loveland-Jones, C., Ferrer, L., Charles, S., Ramsey, F., Van Zandt, A., Volgraf, J., . . . Rappold, J. (2016). A prospective randomized study of the efficacy of "Turning Point," an inpatient violence intervention program. *Journal of Trauma and Acute Care Surgery, 81*(5), 834–842.

Lowenstein, J., & Dharmawardene, M. (2020). A violent thread: How violence cuts across the generations on Chicago's South Side. In M. Siegler & S. Rogers Jr. (Eds.), *Violence, trauma, and trauma surgery* (pp. 25–50). New York, NY: Springer, Cham.

Lu, S. W., & Spain, D. A. (2020). The research agenda for stop the bleed: Beyond focused empiricism in prehospital hemorrhage control. *JAMA Network Open, 3*(7), e209465.

Lynch, H. F., Bateman-House, A., & Rivera, S. M. (2020). Academic advocacy: Opportunities to influence health and science policy under US lobbying law. *Academic Medicine, 95*(1), 44–51.

Lyons, V. H., Benson, L. R., Griffin, E., Floyd, A. S., Kiche, S. W., Haggerty, K. P., . . . Rowhani-Rahbar, A. (2020a). Fidelity assessment of a social work–led intervention among patients with firearm injuries. *Research on Social Work Practice*, *30*(6), 678–687.

Lyons, V. H., Rowhani-Rahbar, A., Adhia, A., & Weiss, N. S. (2020b). Selection bias and misclassification in case–control studies conducted using the National Violent Death Reporting System. *Injury Prevention*, *26*(6).

Maa, J., & Darzi, A. (2018). Firearm injuries and violence prevention—The potential power of a surgeon general's report. *New England Journal of Medicine*, *379*(5), 408–410.

MacFarquhar, N., & Charito, R. (2020, July 6). Violence in Chicago spikes, taking toll on young. *The New York Times*, pp. A1, A16–A17.

Magee, L. (2020a). Identifying victims of firearm assault in Indianapolis through police and clinical data. *Injury Prevention*, *26*(1), A7–A8.

Magee, L. A. (2020b). Community-level social processes and firearm shooting events: A multilevel analysis. *Journal of Urban Health*, *97*, 296–305.

Magee, L. A., Dir, A. L., Clifton, R. L., Wiehe, S. E., & Aalsma, M. C. (2020). Patterns of over an 11-year time period. *Preventive Medicine*, *139*, 106199.

Magruder, K. M., Kassam-Adams, N., Thoresen, S., & Olff, M. (2016). Prevention and public health approaches to trauma and traumatic stress: A rationale and a call to action. *European Journal of Psychotraumatology*, *7*(1), 29715.

Maldonado, K. L. (2019). *Formerly gang-involved Chicana mothers resisting trails of violence in the barrio* (Doctoral dissertation, UC Riverside, Riverside, CA).

Malina, D., Morrissey, S., Campion, E. W., Hamel, M. B., & Drazen, J. M. (2016). Rooting out gun violence. *New England Journal of Medicine*, *374* (January), 175–176.

Maloney, C. (2020, February 4). *Maloney and Warren request GAO review of health care costs of gun violence.* House Committee Maloney.

Mangual, R. A. (2021, January 20). The homicide spike is real. *The New York Times*, p. A27.

Manhattan, L. H. (2019, April 4). How poetry is helping gun violence survivors find new purpose in life. *Spectrum News NY.* https://www.ny1.com/nyc/all- boroughs/news/2019/ 04/05/how-poetry-is-helping-gun-violence-survivors-find- new-purpose-in-life

Manley, N. R., Croce, M. A., Fischer, P. E., Crowe, D. E., Goines, J. H., Sharpe, J. P., . . . Magnotti, L. J. (2019). Evolution of firearm violence over 20 years: integrating law enforcement and clinical data. *Journal of the American College of Surgeons*, *228*(4), 427–434.

Manley, N. R., Fabian, T. C., Sharpe, J. P., Magnotti, L. J., & Croce, M. A. (2018). Good news, bad news: An analysis of 11,294 gunshot wounds (GSWs) over two decades in a single center. *Journal of Trauma and Acute Care Surgery*, *84*(1), 58–65.

Manley, N. R., Fischer, P. E., Sharpe, J. P., Stranch, E. W., Fabian, T. C., Croce, M. A., & Magnotti, L. J. (2020). Separating truth from alternative facts: 37 years of guns, murder, and violence across the US. *Journal of the American College of Surgeons*, *230*(4), 475–481.

Maqungo, S., Fegredo, D., Brkljac, M., & Laubscher, M. (2020). Gunshot wounds to the hip. *Journal of Orthopaedics*, *29*(3), 135–140.

Maqungo, S., Kauta, N., Held, M., Mazibuko, T., Keel, M. J., Laubscher, M., & Ahmad, S. S. (2020). Gunshot injuries to the lower extremities: Issues, controversies and algorithm of management. *Injury*, *51*(7), 1425–1431.

Marantidis, J., & Biggs, G. (2020). Migrated bullet in the bladder presenting 18 years after a gunshot wound. *Urology Case Reports, 28*, 101016.

Marshall, W. A., Egger, M. E., Pike, A., Bozeman, M. C., Franklin, G. A., Nash, N. A.,... Miller, K. R. (2020). Recidivism rates following firearm injury as determined by a collaborative hospital and law enforcement database. *Journal of Trauma and Acute Care Surgery, 89*(2), 371–376.

Martin Luther, Jr. 1958. *Stride toward freedom.* New York, NY: Harper and Row.

Martínez-Alés, G., & Keyes, K. M. (2019). Fatal and non-fatal self-injury in the USA: critical review of current trends and innovations in prevention. *Current Psychiatry Reports, 21*(10), 104.

Marx, G. (2020). Violence interrupters. In R. Kupers (Ed.), *A climate policy revolution: What the science of complexity reveals about saving our planet* (pp. 68–73).Cambridge, MA: Harvard University Press.

Masho, S. W., Schoeny, M. E., Webster, D., & Sigel, E. (2016). Outcomes, data, and indicators of violence at the community level. *Journal of Primary Prevention, 37*(2), 121–139.

Masiakos, P. T., & Warshaw, A. L. (2017). Stopping the bleeding is not enough. *Annals of Surgery, 265*(1), 37–38.

Masiakos, P. T., & Warshaw, A. L. (2018). Addressing the public health epidemic of firearm-related violence in America. *Annals of Surgery, 268*(1), e15.

Mason, T., Campbell, K., Battle, A., Ho, L., & Cheng, S. (2019, November). Public health impact of slavery on identity: The cultural self-identity destruction and gun violence in Chicago. *APHA's 2019 Annual Meeting and Expo* (Nov. 2–Nov. 6). American Public Health Association.

Matoba, N., Reina, M., Prachand, N., Davis, M. M., & Collins, J. W. (2019). Neighborhood gun violence and birth outcomes in Chicago. *Maternal and Child Health Journal, 23*, 1251–1259.

Mattson, S. A., Sigel, E., & Mercado, M. C. (2020). Risk and protective factors associated with youth firearm access, possession or carrying. *American Journal of Criminal Justice, 45*, 844–864.

Max, W., & Rice, D. P. (1993). Shooting in the dark: Estimating the cost of firearm injuries. *Health Affairs, 12*(4), 171–185.

Mazeika, D. M., & Uriarte, L. (2019). The near repeats of gun violence using acoustic triangulation data. *Security Journal, 32*(4), 369–389.

McCann, H. (2020, July 21). Police: 15 injured after shooting outside funeral home. *WHSV3.* https://www.whsv.com/2020/07/22/police-14-being-treated-after-gunfire-erupts-in-chicago-funeral/

McCarthy, E. (2019). *Personal trauma, community healing: Reimagining sanctuary for survivors of gun violence* (Doctoral dissertation, Rhode Island School of Design, Providence, RI).

McCarty, J. C., Caterson, E. J., & Goralnick, E. (2020). Bleeding control training for the lay public: Keep it simple—Reply. *Jama Surgery, 155*(2), 176.

McClure, J., & Leah, C. (2021). Is independence enough? Rehabilitation should include autonomy and social engagement to achieve quality of life. *Clinical Rehabilitation, 35*(1), 3–12.

McCoy, H. (2020, April 1). Violence in urban communities must get the same attention as suburban school shootings. *The Hill.* https://thehill.com/opinion/civil- rights/490669-gun-violence-in-urban-communities-must-get-the-same-attention- as-white

McCrea, K. T., Richards, M., Quimby, D., Scott, D., Davis, L., Hart, S., . . . Hopson, S. (2019). Understanding violence and developing resilience with African American youth in high-poverty, high-crime communities. *Children and Youth Services Review*, *99*, 296–307.

McGaha, P., Stewart, K., Garwe, T., Johnson, J., Sarwar, Z., & Letton, R. W. (2020). Is it time for firearm injury to be a separate activation criteria in children? An assessment of penetrating pediatric trauma using need for surgeon presence. *The American Journal of Surgery*, *221*(1), 21–24.

McGarrell, E. F. (2019). Focused deterrence violence prevention at community and individual levels. *Marquette Law Review*, *103*, 963–981.

McGee, Z., Alexander, C., Cunningham, K., Hamilton, C., & James, C. (2019). Assessing the linkage between exposure to violence and victimization, coping, and adjustment among urban youth: Findings from a research study on adolescents. *Children*, *6*(3), 36.

McLean, R. M. (2019). Firearm-related injury and death in the United States. *Annals of Internal Medicine*, *171*(8), 573–577.

McLean, R. M., Harris, P., Cullen, J., Maier, R. V., Yasuda, K. E., Schwartz, B. J., & Benjamin, G. C. (2019). Firearm-related injury and death in the United States: A call to action from the nation's leading physician and public health professional organizations. *Annals of Internal Medicine*, *171*(8), 573–577.

McLean, S. J., Worden, R. E., Wilkens, C., Reynolds, D. L., Cochran, H., & Worden, K. (2019). *Police interactions with victims of violence*. Washington, DC: Office of Justice Programs, National Criminal Justice Reference Service.

McLively, M. (2019). Gun violence prevention 2.0: A new framework for addressing America's enduring epidemic. *Washington University Journal of Law & Policy*, *60*, 235–254.

McLone, S., Mason, M., Arunkumar, P., Zakariya, E., & Sheehan, K. (2017). 113 homicides in the city of Chicago, 2008–2016. *Injury Prevention*, *23*(Suppl. 1), A42–A43.

McManus, H. D., Engel, R. S., Cherkauskas, J. C., Light, S. C., & Shoulberg, A. M. (2020). *Street violence crime reduction strategies: A review of the evidence*. Cincinnati, OH: Center for Police Research and Policy.

McMillan, J. A. (2020). *Epidemiology and criminology: Managing youth firearm homicide violence in urban areas* (Doctoral dissertation, Walden University, Minneapolis, MN).

McNamara, M., Cane, R., Hoffman, Y., Reese, C., Schwartz, A., & Stolbach, B. (2021). Training hospital personnel in trauma-informed care: Assessing an inter-professional workshop with patients as teachers. *Academic Pediatrics*, *27*(1), 158–164.

Meizoso, J. P., Ray, J. J., Karcutskie IV, C. A., Allen, C. J., Zakrison, T. L., Pust, G. D., . . . Livingstone, A. S. (2016). Effect of time to operation on mortality for hypotensive patients with gunshot wounds to the torso: The golden 10 minutes. *Journal of Trauma and Acute Care Surgery*, *81*(4), 685–691.

Mencken, F. C., & Froese, P. (2019). Gun culture in action. *Social Problems*, *66*(1), 3–27.

Menger, R., Kalakoti, P., Hanif, R., Ahmed, O., Nanda, A., & Guthikonda, B. (2017). A political case of penetrating cranial trauma: The injury of James Scott Brady. *Neurosurgery*, *81*(3), 545–551.

Merritt, C. H., Taylor, M. A., Yelton, C. J., & Ray, S. K. (2019). Economic impact of traumatic spinal cord injuries in the United States. *Neuroimmunol Neuroinflammation*, *6*, 9.

Merry, M. K. (2019). Angels versus devils: The portrayal of characters in the gun policy debate. *Policy Studies Journal*, *47*(4), 882–904.

Merry, M. K. (2020). *Warped narratives: Distortion in the framing of gun policy.* Ann Arbor: University of Michigan Press.

Metzl, J., McKay, T., & Piemonte, J. (2021). Structural competency and the future of firearm research. *Social Science & Medicine, 277*(May).

Metzl, J. M. (2019). *What guns mean: The symbolic lives of firearms.* Palgrave Communications. Center for Medicine, Health, and Science, Vanderbilt University.

Mikhail, J. N., & Nemeth, L. S. (2016). Trauma center based youth violence prevention programs: An integrative review. *Trauma, Violence, & Abuse, 17*(5), 500–519.

Mikhail, J. N., Nemeth, L. S., Mueller, M., Pope, C., & NeSmith, E. G. (2018). The social determinants of trauma: A trauma disparities scoping review and framework. *Journal of Trauma Nursing, 25*(5), 266–281.

Milam, A. J., Furr-Holden, C. D., Leaf, P., & Webster, D. (2018). Managing conflicts in urban communities: Youth attitudes regarding gun violence. *Journal of Interpersonal Violence, 33*(24), 3815–3828.

Miller, A. D. (2020). *Preventing community violence: A case study of Metro Detroit and interfaith activism* (Doctoral dissertation, Virginia Tech University, Blacksburg, Virginia).

Miller, G. F., Kegler, S. R., & Stone, D. M. (2020). Traumatic brain injury–related deaths from firearm suicide: United States, 2008–2017. *American Journal of Public Health, 110*(6), 897–899.

Miller, K. C., & Dahmen, N. (2020). "This is still their lives": Photojournalists' ethical approach to capturing and publishing graphic or shocking images. *Journal of Media Ethics, 35*(1), 17–30.

Milliff, A. (2019). *Facts shape feelings: An information-based framework for emotional responses to violence.* Cambridge, MA: MIT. https://www.aidanmilliff.com/pdf/Facts_shape_feelings_v10Dec2019.pdf

Mills, B. (2017). *Firearm-related morbidity and mortality by injury intent: Analysis of medical, criminal, and vital records in Seattle, WA* (Doctoral dissertation, University of Washington, Seattle, WA).

Minnesota Coalition for Common Sense. (2016). *The economic costs of gun violence.* https://protectmn.org/wp-content/uploads/2016/12/The-Economic-Cost-of-Gun-Violence_FINAL-1.pdf

Mitchell, K. J., Jones, L. M., Turner, H. A., Beseler, C. L., Hamby, S., & Wade Jr, R. (2021). Understanding the impact of seeing gun violence and hearing gunshots in public places: Findings from the Youth Firearm Risk and Safety Study. *Journal of Interpersonal Violence, 36*(17-18), 8835–8851.

Mitchell, R. J., & Ryder, T. (2020). Rethinking the public health model for injury prevention. *Injury Prevention, 26*(1), 2–4.

Moise, I. K. (2020). Youth and weapons: Patterns, individual and neighborhood correlates of violent crime arrests in Miami-Dade County, Florida. *Health & Place, 65.*

Moms Demand Action. (2021, January 26). Take it from me: Gun violence can happen to anyone. https://momsdemandaction.org/take-it-from-me-gun-violence-can-happen-to-anyone/

Monopoli, W. J., Myers, R. K., Paskewich, B. S., Bevans, K. B., & Fein, J. A. (2021). Generating a core set of outcomes for hospital-based violence intervention programs. *Journal of Interpersonal Violence, 36*(9-10), 4771–4786.

Moore, M., Whiteside, L. K., Dotolo, D., Wang, J., Ho, L., Conley, B., . . . Zatzick, D. F. (2016). The role of social work in providing mental health services and care coordination in an urban trauma center emergency department. *Psychiatric Services*, *67*(12), 1348–1354.

Morral, A. R. (2019). *Reducing disagreements on gun policy through scientific research and an improved data infrastructure*. Santa Monica, CA: RAND.

Morrow, K. D., Podet, A. G., Spinelli, C. P., Lasseigne, L. M., Crutcher, C. L., Wilson, J. D., ... DiGiorgio, A. M. (2019). A case series of penetrating spinal trauma: Comparisons to blunt trauma, surgical indications, and outcomes. *Neurosurgical Focus*, *46*(3), E4.

Moselle, A. (2020, August 20). When it comes to its gun epidemic Philly is struggling to control the contagion. *WHHY*. https://whyy.org/articles/when-it-comes-to-its-gun-violence-epidemic-philly-is-struggling-to-control-the-contagion/

Moser, W. (2014, October 8). Q/A: Lawrence Ralph on injury and resilience in gangland Chicago. *Politics & City Life*. https://www.chicagomag.com/city-life/October- 2014/Q-A-Laurence-Ralph-on-Injury-and-Resilience-in-Gangland-Chicago/

Motley, R., & Banks, A. (2018). Black males, trauma, and mental health service use: A systematic review. *Perspectives on Social Work*, *14*(1), 4–19.

Moton, R., Baus, C., Brandt, C., Coleman, A., Kennedy, K., Swank, S., . . . Evoy, K. E. (2020). Stop the Bleed: An interprofessional community service learning project assessing the efficacy of pharmacist-led hemorrhage control education for laypersons. *Disaster Medicine and Public Health Preparedness*, 1–6.

Moule Jr, R. K., & Fox, B. (2021). Belief in the Code of the Street and individual involvement in offending: A meta-analysis. *Youth Violence and Juvenile Justice*, *19*(2), 227–247.

Moyer, R., MacDonald, J. M., Ridgeway, G., & Branas, C. C. (2019). Effect of remediating blighted vacant land on shootings: A citywide cluster randomized trial. *American Journal of Public Health*, *109*(1), 140–144.

Mueller, K. L., Trolard, A., Moran, V., Landman, J. M., & Foraker, R. (2020). Positioning public health surveillance for observational studies and clinical trials: The St. Louis region-wide hospital-based violence intervention program data repository. *Contemporary Clinical Trials Communications*, *21*, 100683.

Muret-Wagstaff, S. L., Faber, D. A., Gamboa, A. C., & Lovasik, B. P. (2020). Increasing the effectiveness of "Stop the Bleed" training through stepwise mastery learning with deliberate practice. *Journal of Surgical Education*, *77*(5), 1146–1153.

Nacasio, C. (2015). What really happens when you get shot. *Wired*. https://www.wired.com/2015/12/what-really-happens-when-you-get-shot/

Nagengast, A. K., Benns, M. V., Bozeman, M. C., Nash, N. A., Smith, J. W., Harbrecht, B. G., . . . Miller, K. R. (2017). Firearm injuries in women at an urban trauma center. *Journal of the American Academy of Surgeons*, *225*(4), e178–e178.

Naik-Mathuria, B., & Gill, A. C. (2020). Firearm injuries in children: Prevention. *UpToDate*. https://www.uptodate.com/contents/firearm-injuries-in-children-prevention

Nally, E., Jelinek, J., & Bunning, R. D. (2017). Quadriparesis caused by lead poisoning nine years after a gunshot wound with retained bullet fragments: A case report. *PM&R*, *9*(4), 411–414.

Nanassy, A. D., Graf, R. L., Budziszewski, R., Thompson, R., Zwislewski, A., Meyer, L., & Grewal, H. (2020). Stop the bleed: The impact of a basic bleeding control course on high school personnel's perceptions of self-efficacy and school preparedness. *Workplace Health & Safety*, *68*(12), 552–559.

Nass, D. (2020, March 4). You still can't rely on CDC's gun injury numbers. But this may be the first step toward a fix. *The Trace*. https://www.thetrace.org/2020/03/cdc-gun- injury-estimates-cpsc-hospital-database/

National Academies of Sciences, Engineering, and Medicine. (2017). *Community violence as a population health issue: Proceedings of a workshop.* Washington, DC: National Academies Press.

National Academies of Sciences, Engineering, and Medicine. (2019). *Health systems interventions to prevent firearm injuries and death: Proceedings of a workshop.* Washington, DC: National Academies Press.

National Institute for Criminal Justice Reform. (2020a). *The true cost of gun violence.* https://costofviolence.org/

National Institute for Criminal Justice Reform. (2020b). *Stockton California: The cost of gun violence: The direct cost to taxpayers.* https://nicjr.org/wp- content/themes/nicjr-child/assets/Stockton.pdf

National Spinal Cord Injury Statistical Center. (2017). *Facts and figures at a glance.* Birmingham: University of Alabama at Birmingham.

Ndikum, C. M. (2018). *Perceptions of people's experiences regarding gun violence* (Doctoral dissertation, Walden University, Minneapolis, MN).

Nebbitt, V., Lombe, M., Pitzer, K. A., Foell, A., Enelamah, N., Chu, Y., . . . Gaylord-Harden, N. (2021). Exposure to violence and posttraumatic stress among youth in public housing: Do community, family, and peers matter? *Journal of Racial and Ethnic Health Disparities, 8,* 264–277.

Nelson, E., Sawyer, L., & McKinney, M. (2019, November 23). As gunfire continues in St. Paul, so does the ShotSpotter debate. *Star Tribune.* https://www.startribune.com/as-gunfire-continues-in-st-paul-so-does-shotspotter- debate/565382652/

Nelson, E. W. (2017). Confronting the firearm injury plague. *Pediatrics, 140*(1), e20171300.

Nelson-Arrington, K. (2020). *Community violence, protective factors, and resilience: Gender differences in African American youth* (Doctoral dissertation, National Louis University, Chicago, IL).

Nesoff, E. D., Milam, A. J., Barajas, C. B., & Furr-Holden, C. D. M. (2020). Expanding tools for investigating neighborhood indicators of drug use and violence: Validation of the NIfETy for virtual street observation. *Prevention Science, 21*(2), 203–210.

Newgard, C. D., Sanchez, B. J., Bulger, E. M., Brasel, K. J., Byers, A., Buick, J. E., . . . Minei, J. P. (2016). A geospatial analysis of severe firearm injuries compared to other injury mechanisms: Event characteristics, location, timing, and outcomes. *Academic Emergency Medicine, 23*(5), 554–565.

Newman, D. M. (2020). *Sociology: Exploring the architecture of everyday life.* Thousand Oaks, CA: Sage.

Ngo, Q. M., Sigel, E., Moon, A., Stein, S. F., Massey, L. S., Rivara, F., . . . FACTS Consortium. (2019). State of the science: a scoping review of primary prevention of firearm injuries among children and adolescents. *Journal of Behavioral Medicine, 42*(4), 811–829.

Nickerson, A. B., Shisler, S., Eiden, R. D., Ostrov, J. M., Schuetze, P., Godleski, S. A., & Delmerico, A. M. (2020). A longitudinal study of gun violence attitudes: role of childhood aggression and exposure to violence, and early adolescent bullying perpetration and victimization. *Journal of School Violence, 19*(1), 62–76.

Niemi, L., & Young, L. (2016, June 24). Who blames the victim? *The New York Times.* https://moralitylab.bc.edu/wp-content/uploads/2011/10/niemi_young_nyt.pdf

Nieto, B., & Mclively, M. (2020, December 17). *America at the cross-roads: Reimagining federal funding to end community violence.* Giffords Law Center. https://giffords.

org/lawcenter/report/america-at-a-crossroads-reimagining-federal-funding-to-end-community-violence/

Nina Vinik, N. (2019, June 19). Solutions to America's gun violence epidemic—academic researchers weight in. *Joyce Foundation*. http://www.joycefdn.org/news/solutions-to-americas-gun-violence- epidemic-academic-researchers-weigh-in

Nishimura, L., & Robledo, G. (2019, February 29). Works of justice: The reality poets on "wheeling and healing". *Pen America*. https://pen.org/works-of-justice-reality- poets/

Nordin, A., Coleman, A., Shi, J., Wheeler, K., Xiang, H., & Kenney, B. (2018). In harm's way: Unintentional firearm injuries in young children. *Journal of Pediatric Surgery*, *53*(5), 1020–1023.

Novak, K. J., & King, W. R. (2020). *Evaluation of the Kansas City Crime Gun Intelligence Center*. https://crimegunintelcenters.org/wp- content/uploads/2020/10/KC-CGIC-Final-Report-2020.pdf

Nugent, N. R., Sumner, J. A., & Amstadter, A. B. (2014). Resilience after trauma: From surviving to thriving. *European Journal of Psychotraumatology*, *5*. 10.3402/ejpt.v5.25339.

Nunn, S. D. (2020). *The impact of community violence on student motivation to attend college* (Doctoral dissertation, Missouri Baptist University, St. Louis, MO).

Nusbaum, S., Medina, A. G., Kim, B., Torosyan, S., & Narine, S. R. (2020). The effectiveness and costs associated with hospital-based violence intervention programs. http://www.thewagnerreview.org/wp-content/uploads/2020/05/Effectiveness-and-Costs-Associated-with-Hospital-based-Violence-Intervention-Programs_The-Wagner-Review-1.pdf

Oakes, K. (2019, December 4). Gunshot wound victims are at high risk for readmission. *MDedge*. https://www.mdedge.com/surgery/article/213505/trauma/gunshot- wound-victims-are-high-risk-readmission

O'Brien, W. J., Gupta, K., & Itani, K. M. (2020). Association of postoperative infection with risk of long-term infection and mortality. *JAMA Surgery*, *155*(1), 61–68.

Odak, S. (2021). *Religion, conflict, and peacebuilding*. New York, NY: Springer.

Odom-Forren, J. (2016). Gun violence: A public health and nursing concern. *Journal of Perianesthesia Nursing*, *31*(4), 285–288.

Ohmer, M. L., Teixeira, S., Booth, J., Zuberi, A., & Kolke, D. (2016). Preventing violence in disadvantaged communities: Strategies for building collective efficacy and improving community health. *Journal of Human Behavior in the Social Environment*, *26*(7–8), 608–621.

O'Leary, A., White, L. D., Anderson, L., & Bishop, L. (2017). Descriptions of gun violence. In A. O'Leary & P. Frew (Eds.), *Poverty in the United States* (pp. 207–220). New York, NY: Springer.

Oliphant, S. N., Mouch, C. A., Rowhani-Rahbar, A., Hargarten, S., Jay, J., Hemenway, D., . . . FACTS Consortium. (2019). A scoping review of patterns, motives, and risk and protective factors for adolescent firearm carriage. *Journal of Behavioral Medicine*, *42*(4), 763–810.

Olufajo, O. A., Zeineddin, A., Nonez, H., Okorie, N. C., De La Cruz, E., Cornwell III, E. E., & Williams, M. (2020). Trends in firearm injuries among children and teenagers in the United States. *Journal of Surgical Research*, *245*, 529–536.

O'Malley, N. (2019). *Bringing bleeding control (STOP THE BLEED) education to the community*. Cleveland, OH: Cleveland State University. https://reducedisparity.org/

site/wp- content/uploads/2019/03/PowerPntCommunity-Based-Research-Network-STB- presenation-002.pdf

Omid, R., Stone, M. A., Zalavras, C. G., & Marecek, G. S. (2019). Gunshot wounds to the upper extremity. *Journal of the American Academy of Orthopaedic Surgeons, 27*(7), e301–e310.

O'Neill, K. M., Vega, C., Saint-Hilaire, S., Jahad, L., Violano, P., Rosenthal, M. S., . . . Dodington, J. (2020). Survivors of gun violence and the experience of recovery. *Journal of Trauma and Acute Care Surgery, 89*(1), 29–35.

Onufer, E. J., Andrade, E., Cullinan, D. R., Kramer, J., Leonard, J., Stewart, M., . . . Punch, L. J. (2020). Anatomy of gun violence: Contextualized curriculum to train surgical residents in both technical and non-technical skills in the management of gun violence. *Journal of the American College of Surgeons, 231*(6), 628–637.

Opara, I., Lardier, D. T., Metzger, I., Herrera, A., Franklin, L., Garcia-Reid, P., & Reid, R. J. (2020). "Bullets have no names": A qualitative exploration of community trauma among Black and Latinx youth. *Journal of Child and Family Studies, 29*, 2117–2129.

Oppel, Jr., R. A., Gebeloff, R., Lai, R., Wright, W., & Smith, M. (2020, July 6). Racial disparity in cases stretches all across board. *The New York Times*, pp. A1, A6.

Orejuela-Davila, A. I. (2020). *Posttraumatic growth and race-based trauma among African Americans* (Doctoral dissertation, The University of North Carolina, Charlotte, NC).

Orthopedist, H. S. C. (2016). Spinal cord injury and its impact on the patient, family, and the society. *International Journal of Recent Surgical and Medical Sciences, 2*(1), 1–4.

Otto, F. (2020, August 7). Trauma centers weathered increase in gun violence from Philadelphia's hardest-hit COVID areas. *Medical News.* https://medicalxpress.com/news/2020–08-trauma-centers-weathered-gun- violence.html

Outland, R. L. (2019). Symbolic meaning of violence: Urban African-American adolescent males' perspectives. *Journal of African American Studies, 23*(3), 233–255.

Owusu, J. A., Stewart, C. M., & Boahene, K. (2018). Facial nerve paralysis. *Medical Clinics of North America, 102*(6), 1135–1143.

Ozuna, L., Champion, C., & Yorkgitis, B. K. (2020). Partnering with patients to reduce firearm-related death and injury. *Journal of the American Osteopathic Association, 120*(6), 413–417.

Paddock, E., Jetelina, K. K., Bishopp, S. A., Gabriel, K. P., & Gonzalez, J. M. R. (2020). Factors associated with civilian and police officer injury during 10 years of officer-involved shooting incidents. *Injury Prevention, 26*(6), 509–515.

Paddock, E., Samuels, J., Vinik, N., & Overton, S. (2017). *Federal actions to engage communities in reducing gun violence.* Washington, DC: Justice Policy Center, Urban Institute.

Pallin, R., Spitzer, S. A., Ranney, M. L., Betz, M. E., & Wintemute, G. J. (2019). Preventing firearm-related death and injury. *Annals of Internal Medicine, 170*(11), ITC81–ITC96.

Palumbo, A. J., Wiebe, D. J., Kassam-Adams, N., & Richmond, T. S. (2019). Neighborhood environment and health of injured urban Black men. *Journal of Racial and Ethnic Health Disparities, 6*(6), 1068–1077.

Papachristos, A. V., Brazil, N., & Cheng, T. (2018). Understanding the crime gap: Violence and inequality in an American city. *City & Community, 17*(4), 1051–1074.

Parham-Payne, W. (2014). The role of the media in the disparate response to gun violence in America. *Journal of Black Studies, 45*(8), 752–768.

Parikh, K., Silver, A., Patel, S. J., Iqbal, S. F., & Goyal, M. (2017). Pediatric firearm-related injuries in the United States. *Hospital Pediatrics, 7*(6), 303–312.

Parker, S. T. (2020). Estimating nonfatal gunshot injury locations with natural language processing and machine learning models. *JAMA Network Open, 3*(10).

Parnell, W. (2020, November 29). 'What am I going to do at home?': After losing a leg in Queens shooting, amputee talks about recovery. *Daily News.* https://www.nydailynews.com/new-york/nyc-crime/ny-gun-violence-leg-amputee-20201130-j2rnp7schzfwpkfel2svsqm2lu-story.html

Parreco, J., Sussman, M. S., Crandall, M., Ebler, D. J., Lee, E., Namias, N., & Rattan, R. (2020). Nationwide outcomes and risk factors for reinjury after penetrating trauma. *Journal of Surgical Research, 250,* 59–69.

Parsons, C., Vargas, E. W., & Bhatia, R. (2020, August 6). The gun industry in America: The overlooked player in the national crisis. *American Progress.* https://www.americanprogress.org/issues/guns- crime/reports/2020/08/06/488686/gun-industry-america/

Passman, J., Xiong, R., Hatchimonji, J., Kaufman, E., Sharoky, C., Yang, W., . . . Holena, D. (2020). Readmissions after injury: Is fragmentation of care associated with mortality? *Journal of Surgical Research, 250,* 209–215.

Patel, S. J., Badolato, G. M., Parikh, K., Iqbal, S. F., & Goyal, M. K. (2018). Geographic regions with stricter gun laws have fewer emergency department visits for pediatric firearm-related injuries: A five-year national study. *Pediatrics.* https://pediatrics.aappublications.org/content/142/1_MeetingAbstract/108?sso=1   &sso_redirect_count=1&nfstatus=401&nftoken=00000000–0000-0000–0000- 000000000000&nfstatusdescription=ERROR%3A%20No%20local%20token&ut m_source=TrendMD&utm_medium=TrendMD&utm_campaign=Pediatrics_Trend MD_0

Patton, D. U., & Roth, B. J. (2016). Good kids with ties to "deviant" peers: Network strategies used by African American and Latino young men in violent neighborhoods. *Children and Youth Services Review, 66,* 123–130.

Pavoni, A., & Tulumello, S. (2020). What is urban violence? *Progress in Human Geography, 44*(1), 49–76.

Payán, D. D., Sloane, D. C., Illum, J., & Lewis, L. B. (2019). Intrapersonal and environmental barriers to physical activity among Blacks and Latinos. *Journal of Nutrition Education and Behavior, 51*(4), 478–485.

Pear, V. A., McCort, C. D., Kravitz-Wirtz, N., Shev, A. B., Rowhani-Rahbar, A., & Wintemute, G. J. (2020). Risk factors for assaultive reinjury and death following a nonfatal firearm assault injury: A population-based retrospective cohort study. *Preventive Medicine, 139.*

Pearl, B. (2020, October 15). Beyond policing: Investing in offices of neighborhood safety. Washington, DC: Center for American Progress. https://www.americanprogress.org/issues/criminal-justice/reports/2020/10/15/491545/beyond-policing-investing-offices- neighborhood-safety/

Peek-Asa, C., Butcher, B., & Cavanaugh, J. E. (2017). Cost of hospitalization for firearm injuries by firearm type, intent, and payer in the United States. *Injury Epidemiology, 4*(1), 1–9.

Peluso, H., Cull, J. D., & Abougergi, M. S. (2020a). The effect of opioid dependence on firearm injury treatment outcomes: A nationwide analysis. *Journal of Surgical Research, 247,* 241–250.

Peluso, H., Cull, J. D., & Abougergi, M. S. (2020b). Race impacts outcomes of patients with firearm injuries. *The American Surgeon, 86*(9), 1113–1118.

Perkins, A. (2020). Spinal cord injury: A lifelong condition. *Nursing Made Incredibly Easy, 18*(5), 34–43.

Perkins, C., Scannell, B., Brighton, B., Seymour, R., & Vanderhave, K. (2016). Orthopaedic firearm injuries in children and adolescents: An eight-year experience at a major urban trauma center. *Injury, 47*(1), 173–177.

Petty, J. K., Henry, M. C., Nance, M. L., & Ford, H. R. (2019). Firearm injuries and children: position statement of the American Pediatric Surgical Association. *Journal of Pediatric Surgery, 54*(7), 1269–1276.

Peetz, A. B., & Haider, A. (2018). Gun violence research and the profession of trauma surgery. *AMA Journal of Ethics, 20*(5), 475–482.

Phalen, P., Bridgeford, E., Gant, L., Kivisto, A., Ray, B., & Fitzgerald, S. (2020). Baltimore Ceasefire 365: Estimated impact of a recurring community-led ceasefire on gun violence. *American Journal of Public Health, 110*(4), 554–559.

Phan, J. (2019). *Examining the pathologic adaptation model of community violence exposure in adolescent offenders: The moderating and mediating effects of moral disengagement* (Doctoral dissertation, Loyola University Chicago, Chicago, IL).

Philadelphia Center for Gun Violence Reporting. (2021). https://www.pcgvr.org/

Phillips, R., Shahi, N., Bensard, D., Meier, M., Shirek, G., Goldsmith, A., . . . Moulton, S. (2020). Guns, scalpels, and sutures: The cost of gun shot wounds in children and adolescents. *Journal of Trauma and Acute Care Surgery, 89*(3), 558–564.

Pica, E., Sheahan, C. L., Pozzulo, J., & Bennell, C. (2020). Guns, gloves, and tasers: Perceptions of police officers and their use of weapon as a function of race and gender. *Journal of Police and Criminal Psychology, 35*, 348–359.

Pino, E. C., Fontin, F., James, T. L., & Dugan, E. (2021). Boston violence intervention advocacy program: Challenges and opportunities for client engagement and goal achievement. *Academic Emergency Medicine, 28*(3), 281–291.

Pirelli, G., & Gold, L. (2019). Leaving Lake Wobegon: Firearm-related education and training for medical and mental health professionals is an essential competence. *Journal of Aggression, Conflict and Peace Research, 11*(2).

Pizarro, J. M., Sadler, R. C., Goldstick, J., Turchan, B., McGarrell, E. F., & Zimmerman, M. A. (2020). Community-driven disorder reduction: Crime prevention through a clean and green initiative in a legacy city. *Urban Studies, 57*(14), 2956–2972.

Pizarro, J. M., Zgoba, K. M., & Pelletier, K. R. (2020). Firearm use in violent crime: Examining the role of premeditation and motivation in weapon choice. *The Journal of Primary Prevention, 42*, 77–91.

Post, L. A., Balsen, Z., Spano, R., & Vaca, F. E. (2019). Bolstering gun injury surveillance accuracy using capture–recapture methods. *Journal of Behavioral Medicine, 42*(4), 674–680.

Powell, E. (2020). Unlawful silence: St. Louis families' fight for records after the killing of a loved one by police. *American Criminal Law Review Online, 57*.

Powell, R. E., & Sacks, C. A. (2020). A national research strategy to reduce firearm-related injury and death: Recommendations from the Health Policy Research Subcommittee of the Society of General Internal Medicine (SGIM). *Journal of General Internal Medicine, 35*(7), 2182–2185.

Powers, D. B., & Rodriguez, E. D. (2020). Characteristics of ballistic and blast injuries. In *Facial Trauma Surgery* (pp. 261–272).

Preidt, R. (2021, February 18). Tougher state gun laws, less gun violenceamong teens study. *USNEWS.* https://www.usnews.com/news/health-news/articles/2021-02- 18/tougher-state-gun-laws-less-gun-violence-among-teens-study

Preston, R. (2020). *Community violence exposure: Experiences of African American families* (Doctoral dissertation, Capella University, Minneapolis, MN).

Przybyla, H. (2021, March 12). Gun violence grows during coronavirus pandemic groups' data shows. *NBC News.* https://www.nbcnews.com/politics/meet-the- press/ blog/meet-press-blog-latest-news-analysis-data-driving-political-   discussion- n988541/ncrd1223551#blogHeader

Purtle, J., Carter, P. M., Cunningham, R., & Fein, J. A. (2016). Treating youth violence in hospital and emergency department settings. *Adolescent Medicine: State of the Art Reviews, 27*(2), 351–363.

Quimby, D., Dusing, C. R., Deane, K., DiClemente, C. M., Morency, M. M., Miller, K. M., . . . Richards, M. (2018). Gun exposure among Black American youth residing in low-income urban environments. *Journal of Black Psychology, 44*(4), 322–346.

Quiroz, H. J., Casey, L. C., Parreco, J. P., Willobee, B. A., Rattan, R., Lasko, D. S., . . . Thorson, C. M. (2020). Human and economic costs of pediatric firearm injury. *Journal of Pediatric Surgery, 55*(5), 944–949.

Radiology Society of North America. (2019, November 29). Gunshot injuries have long-term medical consequences. *RSNA.* https://press.rsna.org/timssnet/media/ pressreleases/14_pr_target.cfm?ID=2125

Rahamim, S. (2018). From dream to nightmare: Gun violence in America. *Interdisciplinary Journal of Partnership Studies, 5*(2), 7.

Raja, A., & Zane, R. D. (2016). Initial management of trauma in adults. *UpToDate.* Dostopno januarja.

Rajan, S., Branas, C. C., Hargarten, S., & Allegrante, J. P. (2018). Funding for gun violence research is key to the health and safety of the nation. *AJPH Perspectives, 108*(2), 194–195.

Rajan, S., Branas, C. C., Myers, D., & Agrawal, N. (2019). Youth exposure to violence involving a gun: Evidence for adverse childhood experience classification. *Journal of Behavioral Medicine, 42*(4), 646–657.

Ralph, L. (2012). What wounds enable: The politics of disability and violence in Chicago. *Disability Studies Quarterly, 32*(3).

Ralph, L. (2014). *Renegade dreams: Living through injury in gangland Chicago.* Chicago, IL: University of Chicago Press.

Ramji, Z., & Laflamme, M. (2017). Ankle lead arthropathy and systemic lead toxicity secondary to a gunshot wound after 49 years: A case report. *The Journal of Foot and Ankle Surgery, 56*(3), 648–652.

RAND Corporation. (2018, March 2). Mass shootings: Definitions and trends. RAND Corporation Gun Policy in America. https://www.rand.org/research/gun- policy/ analysis/essays/mass-shootings.html.

Range, B., Gutierrez, D., Gamboni, C., Hough, N. A., & Wojciak, A. (2018). Mass trauma in the African American community: Using multiculturalism to build resilient systems. *Contemporary Family Therapy, 40*(3), 284–298.

Ranney, M. L., Fletcher, J., Alter, H., Barsotti, C., Bebarta, V. S., Betz, M. E., . . . Fahimi, J. (2017). A consensus-driven agenda for emergency medicine firearm injury prevention research. *Annals of Emergency Medicine, 69*(2), 227–240.

Ranney, M. L., Herges, C., Metcalfe, L., Schuur, J. D., Hain, P., & Rowhani-Rahbar, A. (2020). Increases in actual health care costs and claims after firearm injury. *Annals of Internal Medicine, 173*(12), 949–955.

Ranney, M., Karb, R., Ehrlich, P., Bromwich, K., Cunningham, R., Beidas, R. S., & FACTS Consortium. (2019). What are the long-term consequences of youth exposure to firearm injury, and how do we prevent them? A scoping review. *Journal of Behavioral Medicine, 42*(4), 724–740.

Ransford, C., & Slutkin, G. (2017). Seeing and treating violence as a health issue. In F. Brookman, E. R. Maguire, & M. Maguire (Eds.), *The handbook of gomicide* (pp. 601–625). Chichester, West Sussex, England: Wiley-Blackwell.

Rattan, R., Namias, N., & Zakrison, T. L. (2018). Response: Hidden costs of hospitalization after firearm injury. *Annals of Surgery, 268*(6), e78.

Ray, E. M. (2020). *Exposure to community violence: An adolescent perspective* (Doctoral dissertation, The Chicago School of Professional Psychology, Chicago, IL).

Raza, S., Thiruchelvam, D., & Redelmeier, D. A. (2020). Death and long-term disability after gun injury: a cohort analysis. *CMAJ Open, 8*(3), E469–E478.

Reagan, M. (2016) *Lessons my father taught me: The strength, integrity, and faith of Ronald Reagan.* New York, NY: Humanix Books.

Reardon, C. (2020). Gun violence trauma: Beyond the numbers. *Social Work Today, 20*(1), 10.

Reed, T. (2019, August 7). Gun injuries are increasingly viewed as a public health threat: Experts weigh in on what hospitals need to do. *Fierce.* https://www.fierc ehealthcare.com/hospitals-health-systems/gun-injuries-are-increasingly-viewed-as-public-health-threat-trauma

Reeping, P. M., & Hemenway, D. (2020). The association between weather and the number of daily shootings in Chicago (2012–2016). *Injury Epidemiology, 7*(1), 1–8.

Reid, J. A., Richards, T. N., Loughran, T. A., & Mulvey, E. P. (2017). The relationships among exposure to violence, psychological distress, and gun carrying among male adolescents found guilty of serious legal offenses: A longitudinal cohort study. *Annals of Internal Medicine, 166*(6), 412–418.

Reinberg, S. (2020, September 2). Gun violence costs U.S. health care system $170 billion a year. *US News.* https://www.usnews.com/news/health- news/articles/2020–09-02/gun-violence-costs-us-health-care-system-170-billion- annually

Renson, A., Schubert, F. D., Gabbe, L. J., & Bjurlin, M. A. (2019). Interfacility transfer is associated with lower mortality in undertriaged gunshot wound patients. *Journal of Surgical Research, 236*, 74–82.

Reny, D., Root, S., Chreiman, K., Browning, R., & Sims, C. (2020). A body of evidence: Barriers to family viewing after death by gun violence. *Journal of Surgical Research, 247*, 556–562.

Rhee, N. (2019, November 11). For Chicago kids shaken by gun violence, a shortage of trauma support compounds the harm. *The Trace.* https://www.thetrace.org/2019/11/for-chicago-kids-shaken-by-gun-violence-a-shortage-of-trauma-support-compounds-the-harm/

Rhine, M., McQueen, L., Slidell, M., Kane, J., & Pinto, N. (2020). 362: Characterizing critically injured pediatric gunshot would patients. *Critical Care Medicine, 48*(1), 163.

Rich, J. A., Corbin, T. J., Jacoby, S. F., Webster, J. L., & Richmond, T. S. (2020). Pathways to help-seeking among black male trauma survivors: A fuzzy set qualitative comparative analysis. *Journal of Traumatic Stress, 33*(4), 528–540.

Richardson, J. B., & Bullock, C. (2021). Hospital-based violence prevention programs: From the ground up. In M. Crandall M, S. Bonne, J. Bronson, & W. Kessel (Eds.), *Why we are losing the war on gun violence in the United States* (pp. 187–221). New York, NY: Springer.

Richardson, J. B., Vil, C. S., Sharpe, T., Wagner, M., & Cooper, C. (2016). Risk factors for recurrent violent injury among black men. *Journal of Surgical Research, 204*(1), 261–266.

Richardson, J. B., Wical, W., Kottage, N., Galloway, N., & Bullock, C. (2021). Staying out of the way: Perceptions of digital non-emergency medical transportation services, barriers, and access to care among young Black male survivors of firearm violence. *The Journal of Primary Prevention, 42*, 43–58.

Richardson, M. A. (2019). Framing community-based interventions for gun violence: A review of the literature. *Health & Social Work, 44*(4), 259–270.

Richardson Jr, J. B., Wical, W., Kottage, N., & Bullock, C. (2020). Shook ones: Understanding the intersection of nonfatal violent firearm injury, incarceration, and traumatic stress among young black men. *American Journal of Men's Health, 14*(6).

Riemann, M. (2019). Problematizing the medicalization of violence: A critical discourse analysis of the "Cure Violence" initiative. *Critical Public Health, 29*(2), 146–155.

Rijos, P., Muhammad, S., Meyers, J., Trolard, A., & Bildner, M. (2020, October). *A regional public health response to gun violence: Centering community- academic partnerships*. In APHA's 2020 VIRTUAL Annual Meeting and Expo (Oct. 24–28). American Public Health Association.

Risser, L. (2020). *The Intergenerational Healing Project: Community-academic partnerships to evaluate trauma interventions within the African American community* (Doctoral dissertation, University of Pittsburgh, Pittsburgh, PA).

Ritter, N. (2019). A public health approach to reduce shootings and killings. In F. De Maio, R. C. Shah, J. Mazzeo & D. A. Ansell (Eds.), *Community health equity: A Chicago reader* (pp. 379–385). Chicago, IL: University of Chicago Press.

Rivara, F., Adhia, A., Lyons, V., Massey, A., Mills, B., Morgan, E., . . . Rowhani- Rahbar, A. (2019). The effects of violence on health. *Health Affairs, 38*(10), 1622–1629.

Rivers, T. (2018). *Shoot or be shot: Urban America and gun violence among African American males* (Doctoral dissertation, California State University, Long Beach, CA).

Roberto, E., Braga, A. A., & Papachristos, A. V. (2018). Closer to guns: The role of street gangs in facilitating access to illegal firearms. *Journal of Urban Health, 95*(3), 372–382.

Robertson, E. L., Frick, P. J., Walker, T. M., Kemp, E. C., Ray, J. V., Thornton, L. C., . . . Cauffman, E. (2020). Callous-unemotional traits and risk of gun carrying and use during crime. *American Journal of Psychiatry, 177*(9), 827–833.

Robertson, K. (2018). *Six things we learned from young adults experiencing gun violence in Chicago*. Washington, DC: Center for Victim Research Repository.

Rodriguez, I. (2021, February 26). Can we lower urban gun violence without police? *The Crime Report.* https://thecrimereport.org/2021/02/26/can-we-lower-urban-violence-without-police/

Rodriguez, N. (2018). Expanding the evidence base in criminology and criminal justice: Barriers and opportunities to bridging research and practice. *Justice Evaluation Journal, 1*(1), 1–14.

Roman, C. (2020). "He's not helping us, so we are not helping him": The police as gatekeepers to victim services for victims of street violence. *Injury Prevention, 26*(Suppl 1), A8.

Roman, C. G., Harding, C. S., Klein, H. J., Hamilton, L., & Koehnlein, J. (2019). *The victim-offender overlap*. Washington, DC: Office of Justice Programs, National Criminal Justice Reference Service.

Roman, C. G., Klein, H. J., Harding, C. S., Koehnlein, J. M., & Coaxum, V. (2022). Postinjury engagement with the police and access to care among victims of violent street crime: Does criminal history matter? *Journal of Interpersonal Violence, 37*(3-4), 1637–1661.

Roman, C. G., Link, N. W., Hyatt, J. M., Bhati, A., & Forney, M. (2019). Assessing the gang-level and community-level effects of the Philadelphia Focused Deterrence strategy. *Journal of Experimental Criminology, 15*(4), 499–527.

Roman, J. K. (2020). *A blueprint for a U.S. firearms data infrastructure.* Chicago, IL: University of Chicago. https://craftmediabucket.s3.amazonaws.com/uploads/A- Blueprint-for-a-U.S.-Firearms-Data-Infrastructure_NORC-Expert-Panel-Final-Report_October-2020.pdf

Roman, J. K., & Van Ness, A. (2020, October 29). One answer to firearm violence: Fix our gun data infrastructure. *The Crime Report: Center on Media Crime and Justice John Jay College.* https://thecrimereport.org/2020/10/29/one-answer-to- firearm-violence-fix-our-gun-data-infrastructure/

Romo, N. D. (2020). Gone but not forgotten: Violent trauma victimization and the treatment of violence like a disease. *Hospital Pediatrics, 10*(1), 95–97.

Roochi, M. M., & Razmara, F. (2020). Maxillofacial gunshot injures and their therapeutic challenges: Case series. *Clinical Case Reports, 8*(6), 1094–1100.

Rosas-Salazar, C., Han, Y. Y., Brehm, J. M., Forno, E., Acosta-Pérez, E., Cloutier, M. M., . . . Celedón, J. C. (2016). Gun violence, African ancestry, and asthma: A case-control study in Puerto Rican children. *Chest, 149*(6), 1436–1444.

Rosbrook-Thompson, J. (2019). Legitimacy, urban violence and the public health approach. *Urbanities-Journal of Urban Ethnography, 9*(S2), 37–43.

Rosell, R. L. (2020). Guns and human suffering: A pastoral theological perspective. *Review & Expositor, 117*(3), 333–347.

Rosenberg, M. L. (2019). Let's bring the full power of science to gun violence prevention. *AJPH Perspectives, 109*(3), 396–397.

Rosenberg, M. L. (2021). Considerations for developing an agenda for gun violence prevention research. *Annual Review of Public Health, 42*, 23–41.

Rosenfeld, E. H., & Cooper, A. (2017). Organizing the community for pediatric trauma. In D. E. Wesson & B. Naik-Mathuria (Eds.), *Pediatric trauma* (pp. 7–27). New York, NY: CRC Press.

Ross, K. M., Sullivan, T., O'Connor, K., Hitti, S., & Leiva, M. N. (2021). A community-specific framework of risk factors for youth violence: A qualitative comparison of community stakeholder perspectives in a low-income, urban community. *Journal of Community Psychology, 49*, 1134–1152.

Ross, L., & Arsenault, S. (2018). Problem analysis in community violence assessment: Revealing early childhood trauma as a driver of youth and gang violence. *International Journal of Offender Therapy and Comparative Criminology, 62*(9), 2726–2741.

Rostron, A. (2018). The Dickey Amendment on federal funding for research on gun violence: A legal dissection. *American Journal of Public Health, 108*(7), 805–867.

Roth, K. R. (2019). Introduction: A special note on the heightened effects of urban marginality in the Trump Era. In M. Brug, Z. S. Ritter, & K. R. Roth (Eds.), *Marginality in the urban center* (pp. 3–10). New York, NY: Palgrave Macmillan.

Rothschild, T. (2018). *An ethnography of gun violence prevention activists: "We are thinking people."* Lanham, MD: Rowman & Littlefield.

Rowhani-Rahbar, A., Bellenger, M. A., & Rivara, F. P. (2019). Firearm violence research: Improving availability, accessibility, and content of firearm-related data systems. *JAMA, 322*(19), 1857–1858.

Rowhani-Rahbar, A., Oesterle, S., & Skinner, M. L. (2020). Initiation age, cumulative prevalence, and longitudinal patterns of handgun carrying among rural adolescents: A multistate study. *Journal of Adolescent Health, 66*(4), 416–422.

Rowhani-Rahbar, A., Zatzick, D. F., & Rivara, F. P. (2019). Long-lasting consequences of gun violence and mass shootings. *JAMA, 321*(18), 1765–1766.

Roy, A. L., Isaia, A. R., DaViera, A. L., Eisenberg, Y., & Poulos, C. D. (2021). Redefining exposure: Using mobile technology and geospatial analysis to explore when and where Chicago adolescents are exposed to neighborhood characteristics. *American Journal of Community Psychology, 68*(1-2), 18–28.

Royster, R. A. (2017). *The doll project as a liberatory art intervention for conscious raising and trauma relief in a Chicago marked by violence* (Doctoral dissertation, National Louis University, Chicago, IL).

Rubin, R. (2016). Tale of two agencies: CDC avoids gun violence research but NIH funds it. *JAMA, 315*(16), 1689–1692.

Rubinstein, R. A., Lane, S. D., Mojeed, L., Sanchez, S., Catania, E., Jennings-Bey, T., . . . Quesada, J. (2018). Blood in the Rust Belt: Mourning and memorialization in the context of community violence. *Current Anthropology, 59*(4), 439–454.

Ruderman, D., & Cohn, E. G. (2021). Predictive extrinsic factors in multiple victim shootings. *The Journal of Primary Prevention, 42,* 59–72.

Russo, R., Fury, M., Accardo, S., & Krause, P. (2016). Economic and educational impact of firearm-related injury on an urban trauma center. *Orthopedics, 39*(1), e57–e61.

Rydberg, J., Stone, R., & McGarrell, E. F. (2016). Utilizing incident-based crime data to inform strategic interventions: A problem analysis of violence in Michigan. *Justice Research and Policy, 17*(1), 3–27.

Ryley, S. (2020, January 17). What I learned from making dozens of public records requests for police data. *The Trace.* https://www.thetrace.org/2020/01/pol ice- data-documentation-public-records-requests/

Ryley, S., Singer-Vine, J., & Campbell, S. (2020, January 24). Shoot someone in a major city, and odds are you'll get away with it. *The Trace.* https://www.thetrace.org/features/ murder-solve-rate-gun-violence-baltimore- shootings/

Sachs, N. M., Veysey, B. M., & Rivera, L. M. (2020). Situational victimization cues strengthen implicit and explicit self-victim associations: An experiment with college-aged adults. *Journal of Interpersonal Violence, 37*(3-4), 1292–1310.

Sacks, B. (2018, September 6). A young anti-gun activist was shot and killed walking to the store. *BuzzFeedNews.* https://www.buzzfeednews.com/article/briannasacks/anti-gun-activist-shot-killed- chicago

Sacks, T. K., & Chow, J. C. C. (2018). A social work perspective on police violence: Evidence and interventions. *Journal of Ethnic & Cultural Diversity in Social Work, 27,* 215–218.

Said, P. Z. H., Ghosh, A., Pal, R., Poli, N., Moscote-Salazar, L. R., & Agrawal, A. (2018). Impact of traumatic brain injury on cognitive functions. *Archives of Mental Health, 19*(2), 97–101.

Saint-Hilaire, S., Jahad, L., Rosenthal, M. S., Maung, A. A., Becher, R. D., & Dodington, J. (2020). Survivors of gun violence and the experience of recovery. *Journal of Trauma and Acute Care Surgery, 89*(1), 29–35.

Sakran, J. V. (2020). The impact of bleeding control—A perspective beyond firearm injury! *Annals of Surgery, 271*(2), e14.

Sakran, J. V., Mehta, A., Fransman, R., Nathens, A. B., Joseph, B., Kent, A., . . . Efron, D. T. (2018). Nationwide trends in mortality following penetrating trauma: Are we up for the challenge? *Journal of Trauma and Acute Care Surgery, 85*(1), 160–166.

Sakran, J. V., Nance, M., Riall, T., Asmar, S., Chehab, M., & Joseph, B. (2020). Pediatric firearm injuries and fatalities: Do racial disparities exist? *Annals of Surgery, 272*(4), 556–561.

Samuels, J. T. (2020). Interest-driven sociopolitical youth engagement: Art and gun violence prevention. *Journal of Media Literacy Education, 12*(2), 80–92.

Sanchez, C., Jaguan, D., Shaikh, S., McKenney, M., & Elkbuli, A. (2020). A systematic review of the causes and prevention strategies in reducing gun violence in the United States. *The American Journal of Emergency Medicine, 38*(10), 2169–2178.

Sanchez-Jankowski, M. (2008). *Cracks in the pavement: Social change and resilience in poor neighborhoods.* Berkeley: University of California Press.

Sanders, C. (2019, June 20). 17 times this tattoo artist turned traumatic scars into works of art. *Inspire.* https://www.inspiremore.com/scars-to-works-of-art-flavia/

Sandoval, E. (2020, May 24). After battling gangs and guns, a neighborhood faces a new killer. *The New York Times,* p. 15.

Santaella-Tenorio, J., Cerdá, M., Villaveces, A., & Galea, S. (2016). What do we know about the association between firearm legislation and firearm-related injuries? *Epidemiologic Reviews, 38*(1), 140–157.

Santilli, A., O'Connor Duffany, K., Carroll-Scott, A., Thomas, J., Greene, A., Arora, A., . . . Ickovics, J. (2017). Bridging the response to mass shootings and urban violence: Exposure to violence in New Haven, Connecticut. *American Journal of Public Health, 107*(3), 374–379.

Sargent, E., Zahniser, E., Gaylord-Harden, N., Morency, M., & Jenkins, E. (2020). Examining the effects of family and community violence on African American adolescents: The roles of violence type and relationship proximity to violence. *The Journal of Early Adolescence, 40*(5), 633–661.

Sathya, C. (2020, July 14). Gun violence is killing more kids in the U.S. than COVID-19. When will we start treating it like a public health issue? *TIME Magazine.* https://time.com/5866776/gun-violence-covid/

Sauaia, A., Gonzalez, E., Moore, H. B., Bol, K., & Moore, E. E. (2016). Fatality and severity of firearm injuries in a Denver trauma center, 2000–2013. *JAMA, 315*(22), 2465–2467.

Savakar, D. G., & Kannur, A. (2016). A practical aspect of identification and classifying of guns based on gunshot wound patterns using gene expression programming. *Pattern Recognition and Image Analysis, 26*(2), 442–449.

Scarboro, M., Massetti, J., & Aresco, C. (2019). Traumatic brain injuries. In K. A. McQuillan & M. B. Maki (Eds.), *Trauma nursing e-book: From resuscitation through rehabilitation* (p. 332). Amsterdam, Holland: Elsevier.

Scarlet, S., & Rogers Jr, S. O. (2018). What is the institutional duty of trauma systems to respond to gun violence? *AMA Journal of Ethics, 20*(5), 483–491.

Schellenberg, M., Owattanapanich, N., Cremonini, C., Heindel, P., Anderson, G. A., Clark, D. H., . . . Inaba, K. (2020). Shotgun wounds: Nationwide trends in epidemiology, injury patterns, and outcomes from US trauma centers. *The Journal of Emergency Medicine, 58*(5), 719–724.

Schmidt, C. J., Rupp, L., Pizarro, J. M., Lee, D. B., Branas, C. C., & Zimmerman, M. A. (2019). Risk and protective factors related to youth firearm violence: A scoping review and directions for future research. *Journal of Behavioral Medicine, 42*(4), 706–723.

Schroll, R., Smith, A., Martin, M. S., Zeoli, T., Hoof, M., Duchesne, J., . . . Avegno, J. (2020). Stop the bleed training: Rescuer skills, knowledge, and attitudes of hemorrhage control techniques. *Journal of Surgical Research, 245*, 636–642.

Schubl, S. D., Robitsek, R. J., Sommerhalder, C., Wilkins, K. J., Klein, T. R., Trepeta, S., & Ho, V. P. (2016). Cervical spine immobilization may be of value following firearm injury to the head and neck. *The American Journal of Emergency Medicine, 34*(4), 726–729.

Schwartz, M. (2020, April 28). Nothing but death: Inside the nursing home where NYC's most vulnerable struggle to survive COVID-19. *Mother Jones.* https://www.moth erjones.com/coronavirus-updates/2020/04/nothing-but-death- inside-the-nursing-home-where-nycs-most-vulnerable-struggle-to-survive-covid- 19/

Science Daily. (2020, August 10). New study documents increasing frequency, cost, and severity of gunshot wounds: Researchers hope findings drive changes to address violence and hospital costs. *Science Daily.* https://www.sciencedaily.com/releases/2020/08/200810160147.htm

Scott, S. (2020, January 2). *A report from the NIMBioS/DySoC Investigative Workshop on the mathematics of gun violence.* https://pdfs.semanticscholar.org/2dda/1cc18255 d7206a7165a301a8b86ff63de1c 3.pdf

Scuba, S., Charles, S., & Hendrickson, M. (2020, June 8). 18 murders in 24 hours: Inside the most violent day in 60 years in Chicago. *Chicago Sun Times.* https://chicago.suntimes.com/crime/2020/6/8/21281998/chicago-deadli est-day- violence-murder-history-police-crime

Secrist, M. E., John, S. G., Harper, S. L., Edge, N. A. C., Sigel, B. A., Sievers, C., & Kramer, T. (2019). Nightmares in treatment-seeking youth: The role of cumulative trauma exposure. *Journal of Child & Adolescent Trauma, 13*, 249–256.

Sehgal, A. R. (2020). Lifetime risk of death from firearm injuries, drug overdoses, and motor vehicle accidents in the United States. *The American Journal of Medicine, 133*(10), 1162–1167.

Seim, J. (2020). *Bandage, sort, and hustle: Ambulance crews on the front lines of urban suffering.* Berkeley: University of California Press.

Sekeris, P. G., & van Ypersele, T. (2020). *An economic analysis of violent crime.* https://www.researchgate.net/profile/Petros_Sekeris/publication/340161491_An_ Economic_Analysis_of_Violent_Crime/links/5e96cd79a6fdcca78918afb0/An-Economic-Analysis-of-Violent-Crime.pdf

Serchen, J., Doherty, R., Atiq, O., & Hilden, D. (2020). Racism and health in the United States: A policy statement from the American College of Physicians. *Annals of Internal Medicine, 173*(7), 556–557.

Sertbaş, İ., & Karatay, M. (2020). The effect of the delay between injury and surgery on mortality, morbidity, and complications in craniospinal gunshot wounding. *Trauma, 22*(3), 193–200.

Shah, A. A., Zuberi, M., Cornwell, E., Williams, M., Manicone, P., Kane, T., . . . Petrosyan, M. (2019). Gaps in access to comprehensive rehabilitation following traumatic injuries in children: A nationwide examination. *Journal of Pediatric Surgery, 54*(11), 2369–2374.

Shammas, M. (2019, November 24). It's time to retire the "guns don't kill people—people kill people" argument. *Guns Do Kill People.* https://ssrn.com/abstract=3492766

Sharkey, P. (2013). *Stuck in place: Urban neighborhoods and the end of progress toward racial equality.* Chicago, IL: University of Chicago Press.

Sharkey, P. (2018). The long reach of violence: A broader perspective on data, theory, and evidence on the prevalence and consequences of exposure to violence. *Annual Review of Criminology, 1,* 85–102.

Sherrod, B., Karsy, M., Guan, J., Brock, A. A., Eli, I. M., Bisson, E. F., & Dailey, A. T. (2019). Spine trauma and spinal cord injury in Utah: A geographic cohort study utilizing the National Inpatient Sample. *Journal of Neurosurgery: Spine, 31*(1), 93–102.

Shjarback, J. A., White, M. D., & Bishopp, S. A. (2021). Can police shootings be reduced by requiring officers to document when they point firearms at citizens? *Injury Prevention, 27*(6), 508–513.

Shoptaw, S., Goodman-Meza, D., & Landovitz, R. J. (2020). Collective call to action for HIV/AIDS community-based collaborative science in the era of COVID-19. *AIDS and Behavior, 24,* 2013–2016.

Shorr, K. (2017). *SHOT: 101 survivors of gun violence in America.* New York, NY: Powerhouse Books.

Shultz, J. M., Ettman, C., & Galea, S. (2018). Insights from population health science to inform research on firearms. *The Lancet Public Health, 3*(5), e213–e214.

Siegler, M. (2020). *Violence, trauma, and trauma surgery: Ethical issues, interventions, and innovations.* New York, NY: Springer Nature.

Silas, A. B., & Akang, U. J. (2019). Lower eye lid defect repair following an ocular gunshot injury: A case study and review of literature. *Journal of Research in Basic and Clinical Sciences, 1*(2), 197–202.

Silva, K. (2019, September 16). Black boys, grief, guns in urban schools. *Physorg.* https://phys.org/news/2019-09-black-boys-grief-guns-urban.html

Silver, A. H., Andrews, A. L., Azzarone, G., Bhansali, P., Hjelmseth, E., Hogan, A. H., . . . Parikh, K. (2020). Engagement and leadership in firearm-related violence prevention: The role of the pediatric hospitalis. *Hospital Pediatrics, 10*(6), 523–530.

Singletary, G. (2020a). Beyond PTSD: Black male fragility in the context of trauma. *Journal of Aggression, Maltreatment & Trauma, 29*(5), 517–536.

Singletary, G. (2020b). Trauma and black male adolescents: A critical link. *Adolescent Psychiatry, 10*(1), 17–28.

Siracuse, J. J., Cheng, T. W., Farber, A., James, T., Zuo, Y., Kalish, J. A., . . . Kalesan, B. (2019). Vascular repair after firearm injury is associated with increased morbidity and mortality. *Journal of Vascular Surgery, 69*(5), 1524–1531.

Siracuse, J. J., Farber, A., Cheng, T. W., Jones, D. W., & Kalesan, B. (2020). Lower extremity vascular injuries caused by firearms have a higher risk of amputation and death compared with non-firearm penetrating trauma. *Journal of Vascular Surgery, 72*(4), 1298–1304.

Sivak, C. J., Pearson, A. L., & Hurlburt, P. (2021). Effects of vacant lots on human health: A systematic review of the evidence. *Landscape and Urban Planning, 208.*

Sivaraman, J. J., Marshall, S. W., & Ranapurwala, S. I. (2020). State firearm laws, race and law enforcement–related deaths in 16 US states: 2010–2016. *Injury prevention, 26*(6).

Skaggs, A. (2019, September 19). We need to talk about the economic toll of gun violence. *Giffords.* https://giffords.org/blog/2019/09/we-need-to-talk-about-the-economic-toll-of-gun-violence-blog/

Skrodzka, E., & Wicher, A. (2019). A review of gunshot noise as factor in hearing disorders. *Acta Acustica United with Acustica, 105*(6), 904–911.

Slutkin, G. (2017). Reducing violence as the next great public health achievement. *Nature Human Behaviour, 1*(1), 1.

Slutkin, G., & Ransford, C. (2020). Violence is a contagious disease: Theory and practice in the USA and abroad. In M. Siegler & S. Rogers Jr (Eds.), *Violence, trauma, and trauma surgery* (pp. 67–85). New York, NY: Springer.

Slye, N., Loux, T., Lu, Y., Mansuri, F., Brooks, S., Geissler, G., . . . Kip, K. (2019, November). Factors associated with pediatric injury hotspots (2010–2017). In *APHA's 2019 Annual Meeting and Expo* (Nov. 2–Nov. 6). Philadepphia, PA: American Public Health Association.

Smart, R., Morral, A. R., Smucker, S., Cherney, S., Schell, T. L., Peterson, S., . . . Gresenz, C. R. (2020). *The science of gun policy: A critical synthesis of research evidence on the effects of gun policies in the United States.* Santa Monica, CA: Rand Corporation.

Smith, M. E., Sharpe, T. L., Richardson, J., Pahwa, R., Smith, D., & DeVylder, J. (2020). The impact of exposure to gun violence fatality on mental health outcomes in four urban US settings. *Social Science & Medicine, 246.*

Smith, N. A., Voisin, D. R., Yang, J. P., & Tung, E. L. (2019). Keeping your guard up: hypervigilance among urban residents affected by community and police violence. *Health Affairs, 38*(10), 1662–1669.

Smith, R. N., Seamon, M. J., Kumar, V., Robinson, A., Shults, J., Reilly, P. M., & Richmond, T. S. (2018). Lasting impression of violence: Retained bullets and depressive symptoms. *Injury, 49*(1), 135–140.

Smith, R. N., Tracy, B. M., Smith, S., Johnson, S., Martin, N. D., & Seamon, M. J. (2020). Retained bullets after firearm injury: A survey on surgeon practice patterns. *Journal of Interpersonal Violence.*

Smith, V. M., Siegel, M., Xuan, Z., Ross, C. S., Galea, S., Kalesan, B., . . . Goss, K. A. (2017). Broadening the perspective on gun violence: An examination of the firearms industry, 1990–2015. *American Journal of Preventive Medicine, 53*(5), 584–591.

Smith Lee, J. R., & Robinson, M. A. (2019). "That's my number one fear in life. It's the police": Examining young black men's exposures to trauma and loss resulting from police violence and police killings. *Journal of Black Psychology, 45*(3), 143–184.

Smith-Walter, A., Peterson, H. L., Jones, M. D., & Nicole Reynolds Marshall, A. (2016). Gun stories: How evidence shapes firearm policy in the United States. *Politics & Policy, 44*(6), 1053–1088.

Snyder, K. B., Farrens, A., Raposo-Hadley, A., Tibbits, M., Burt, J., Bauman, Z. M., & Evans, C. H. (2020). Dusk to dawn: Evaluating the effect of a hospital-based youth violence prevention program on youths' perception of risk. *Journal of Trauma and Acute Care Surgery, 89*(1), 140–144.

Snyder, S. (2018). *Hospital-based violence intervention programs: A systematic review.* Hartford, CT: University of Hartford.

Sodagari, F., Katz, D. S., Menias, C. O., Moshiri, M., Pellerito, J. S., Mustafa, A., Revzin, M. V. (2020). Imaging evaluation of abdominopelvic gunshot trauma. *Radiographics, 40*(6), 1766–1788.

Sodhi, A., Ambast, S., Fitzgerald, A., & Vartak, M. (2021). Gun violence and barriers to reparation in the United States: Scars of survival. In M. Crandall, S. Bonne, J. Bronson, & W. Kessel (Eds.), *Why we are losing the war on gun violence in the United States* (pp. 267–281). New York, NY: Springer.

Sokol, R. L., Carter, P. M., Goldstick, J., Miller, A. L., Walton, M. A., Zimmerman, M. A., & Cunningham, R. M. (2020). Within-person variability in firearm carriage among high-risk youth. *American Journal of Preventive Medicine, 59*(3), 386–393.

Sondik, E. J. (2021). Data on gun violence: What do we know and how do we know it? In M. Crandall, S. Bonne, J. Bronson & W. Kessel (Eds.), *Why we are losing the war on gun violence in the United States* (pp. 15–24). New York, NY: Springer.

Søreide, K., Weber, C., & Thorsen, K. (2020). Priorities for research in trauma care: creating a bucket list. *Injury, 51*(9), 2051–2052.

South, E. C., Stillman, K., Buckler, D. G., & Wiebe, D. (2021). Association of gun violence with emergency department visits for stress-responsive complaints. *Annals of Emergency Medicine, 27*(5), 469–478.

Southall, A. (2020, May 23). City enlists violence prevention groups to encourage social distancing. *The New York Times*, p. A12.

Southall, A., & MacFarquhar, N. (2020, June 24). New York City sees its most violent start to summer 1996. *The New York Times*, p. A14.

Soyer, M. (2018). *Lost childhoods: Poverty, trauma, and violent crime in the post-welfare era*. Berkeley: University of California Press.

Spearman, A. (2020, December 16). A public health crisis: Forum seeks to reframe gun violence in D.C. *WJLA.* https://wjla.com/news/local/a-public-health-cri sis-forum- seeks-to-reframe-gun-violence-in-dc

Sperlich, M., Logan-Greene, P., Slovak, K., & Kaplan, M. S. (2019). Addressing gun violence: A social work imperative. *Health & Social Work, 44*(4), 217–220.

Spinrad, M. (2017). *A public health approach to gun violence: Evaluating strategies to improve intervention and public awareness* (Doctoral dissertation, Boston University, Boston, MA).

Spitzer, R. J. (2020a). *The politics of gun control*. New York, NY: Routledge.

Spitzer, R. J. (2020b). Gun accessories and the Second Amendment: Assault weapons, magazines, and silencers. *Law & Contemporary Problems, 83*, 231–255

Spitzer, S. A., Pear, V. A., McCort, C. D., & Wintemute, G. J. (2020). Incidence, distribution, and lethality of firearm injuries in California from 2005 to 2015. *JAMA Network Open, 3*(8), e2014736–e2014736.

Spitzer, S. A., Staudenmayer, K. L., Tennakoon, L., Spain, D. A., & Weiser, T. G. (2017). Costs and financial burden of initial hospitalizations for firearm injuries in the United States, 2006–2014. *American Journal of Public Health, 107*(5), 770–774.

Spitzer, S. A., Vail, D., Tennakoon, L., Rajasingh, C., Spain, D. A., & Weiser, T. G. (2019). Readmission risk and costs of firearm injuries in the United States, 2010–2015. *Plos One, 14*(1). https://journals.plos.org/plosone/article?id=10.1371/journal. pone.0209896

Sripong, K., Samai, W., & Liabsuetrakul, T. (2019). Feasibility and reliability of a developed and validated forensic recording form for firearm injury. *Journal of Health Science and Medical Research, 37*(3), 183–195.

Stark, D. E., & Shah, N. H. (2017). Funding and publication of research on gun violence and other leading causes of death. *JAMA, 317*(1), 84–85.

Steffen, S., & Harlow, P. (2014, July 9). Gang violence: What happens when you don't die. *CNN.* https://www.cnn.com/2014/07/08/us/paralyzed-by-gun- violence/index.html

Stein, M., & Galea, S. (2020). *Pained: Uncomfortable conversations about the public's health*. New York, NY: Oxford University Press.

Steinbrook, R., Stern, R. J., & Redberg, R. F. (2016). Firearm injuries and gun violence: Call for papers. *JAMA Internal Medicine, 176*(5), 596–597.

Stephens, J., Thorpe, E., Schrag, J., & Ramiah, K. (2017, November). Essential hospitals: Community anchors in violence prevention. In *APHA 2017 Annual Meeting & Expo* (Nov. 4–Nov. 8). Atlanta, Georgia: American Public Health Association.

Stern, J., & Zhang, S. (2017, December 14). Americans don't really understand gun violence. *The Atlantic*. https://www.theatlantic.com/politics/archive/2017/12/guns-nonfatal-shooting- newtown-las-vegas/548372/

Stern, M., & Lester, T. W. (2021). Does local ownership of vacant land reduce crime? An assessment of Chicago's large lots program. *Journal of the American Planning Association, 87*(1), 73–84.

Stevenson, D. D. (2019). Gun violence as an obstacle to educational equality. *University of Memphis Law Review*, 1092–1135.

Stewart, R. M., Kuhls, D. A., Campbell, B. T., Letton Jr, R. W., Burke, P. A., Dicker, R. A., & Gaines, B. A. (2017). The COT's consensus-based approach to firearm injury: An introduction. *Bulletin AM College of Surgeons, 102*(5), 12–13.

Stolbach, B. C., & Anam, S. (2017). Racial and ethnic health disparities and trauma-informed care for children exposed to community violence. *Pediatric Annals, 46*(10), e377–e381.

Stolbach, B. C., & Reese, C. (2020). Healing hurt people-Chicago: Supporting trauma recovery in patients injured by violence. In M. Siegler & S. Rogers Jr (Eds.), *Violence, trauma, and trauma surgery* (pp. 237–248). New York, NY: Springer.

Strong, B., Tracy, B. M., Sangji, N. F., & Barrera, K. (2018). Gun violence and firearm policy in the US: A brief history and the current status. *Bulletin of the American College of Surgeons.* https://bulletin.facs.org/2018/07/gun-violence-and-firearm- policy-in-the-u-s-a-brief-history-and-the-current-status/

Stuart, F. (2020). Code of the tweet: Urban gang violence in the social media age. *Social Problems, 67*(2), 191–207.

Su, Y. S., Gutierrez, A., Vaughan, K. A., Miranda, S., Chen, H. I., Petrov, D., ... Schuster, J. M. (2020). Penetrating spinal column injuries (pSI): An institutional experience with 100 consecutive cases in an urban trauma center. *World Neurosurgery, 38*, e551–e556.

Sutherland, M., McKenney, M., & Elkbuli, A. (2021). Gun violence during COVID-19 pandemic: Paradoxical trends in New York City, Chicago, Los Angeles and Baltimore. *The American Journal of Emergency Medicine, 39*, 225–228.

Swaroop, M. (2018). TRUE (Trauma Responders Unify to Empower) Communities: An evolution of empowerment. *The Bulletin of the Royal College of Surgeons of England, 100*(5), 201–205.

Sweeten, G., & Fine, A. D. (2021). Dynamic risk factors for handgun carrying: Are there developmental or sex differences? *Journal of Clinical Child & Adolescent Psychology, 50*(3), 311–325.

Swendiman, R. A., Hatchimonji, J. S., Allukian, M., Blinman, T. A., Nance, M. L., & Nace, G. W. (2020). Pediatric firearm injuries: Anatomy of an epidemic. *Surgery, 168*(3), 381–384.

Swendiman, R. A., Luks, V. L., Hatchimonji, J. S., Nayyar, M. G., Goldshore, M. A., Nace Jr, G. W., ... Allukian III, M. (2021). Mortality after adolescent firearm injury: Effect of trauma center designation. *Journal of Adolescent Health, 68*(5), 978–984.

Szajna, A., & Shaffer, K. (2020). Public health champions in the making: An innovative undergraduate nursing pedagogy. *Public Health Nursing, 37*(1), 130–134.

Taichman, D. B., Bauchner, H., Drazen, J. M., Laine, C., & Peiperl, L. (2017). Firearm-related injury and death: A US health care crisis in need of health care professionals. *JAMA, 318*(19), 1875.

Taichman, D. B., Bornstein, S. S., & Laine, C. (2018). Firearm injury prevention: AFFIRMing that doctors are in our lane. *Annals of Internal Medicine, 169*(12), 885–886.

Talley, C. L., Campbell, B. T., Jenkins, D. H., Barnes, S. L., Sidwell, R. A., Timmerman, G., . . . Ficke, J. (2019). Recommendations from the American College of Surgeons Committee on Trauma's firearm strategy team (FAST) workgroup: Chicago consensus I. *Journal of the American College of Surgeons, 228*(2), 198–206.

Tanga, C., Franz, R., Hill, J., Lieber, M., & Galante, J. (2018). Evaluation of experience with lower extremity arterial injuries at an urban trauma center. *The International Journal of Angiology, 27*(1), 29–34.

Tapia, M. (2019). *Gangs of the El Paso–Juárez borderland: A history.* Albuquerque: University of New Mexico Press.

Tasigiorgos, S., Economopoulos, K. P., Winfield, R. D., & Sakran, J. V. (2015). Firearm injury in the United States: An overview of an evolving public health problem. *Journal of the American College of Surgeons, 221*(6), 1005–1014.

Tatebe, L., Speedy, S., Kang, D., Barnum, T., Cosey-Gay, F., Regan, S., . . . Swaroop, M. (2019). Empowering bystanders to intervene: Trauma Responders Unify to Empower (TRUE) communities. *Journal of Surgical Research, 238*, 255–264.

Taxman, F. S. (2019). Violence reduction using the principles of risk-need-responsivity. *Marquette Law Review, 103*, 1149–1178.

Team Trace. (2018, March 23). An American crisis: 18 facts about gun violence—and 6 promising ways to reduce the suffering. https://www.thetrace.org/features/gun-violence-facts-and-solutions/

Tedesco, D., Adja, K. Y. C., Rallo, F., Reno, C., Fantini, M. P., & Hernandez-Boussard, T. (2020). Is the firearm epidemic in the US getting worse? *European Journal of Public Health, 30*(Suppl._5), ckaa165–1114.

Teixeira, P. G., Brown, C. V., Emigh, B., Long, M., Foreman, M., Eastridge, B., . . . Holcomb, J. (2018). Civilian prehospital tourniquet use is associated with improved survival in patients with peripheral vascular injury. *Journal of the American College of Surgeons, 226*(5), 769–776.

Teixeira, S. (2016). Beyond broken windows: Youth perspectives on housing abandonment and its impact on individual and community well-being. *Child Indicators Research, 9*(3), 581–607.

Teixeira, S., Hwang, D., Spielvogel, B., Cole, K., & Coley, R. L. (2020). Participatory photo mapping to understand youths' experiences in a public housing neighborhood preparing for redevelopment. *Housing Policy Debate, 30*(5), 766–782.

Telpin, L. A., Meyerson, N. S., Jakubowski, J. A., Aaby, D. A., Zheng, N., Abram, K. M., & Welty, L. J. (2021). Association of firearm access, use, and victimization during adolescence with firearm perpetration during adulthood in a 16-year longitudinal study of youth involved in the juvenile justice system. *JAMA Network Open, 4*(2), e2034208–e2034208. https://jamanetwork.com/journals/jamanetworkopen/fullarticle/2775922

Tessler, R. A., Arbabi, S., Bulger, E. M., Mills, B., & Rivara, F. P. (2019). Trends in firearm injury and motor vehicle crash case fatality by age group, 2003-2013. *JAMA Surgery*, *154*(4), 305-310.

Testa, P. A., & Legome, E. (2017, January 13). Abdominal trauma, penetrating. *Medscape*. http://emedicine.medscape.com/article/822099-overview.

Theaker, N. (2020). *Rationalized killing: How moral disengagement affects gun carrying and violence among adolescent youth* (Doctoral dissertation Wayne State University, Detroit, MI).

Thiels, C. A., Zielinski, M. D., Glasgow, A. E., & Habermann, E. B. (2018). The relative lack of data regarding firearms injuries in the United States. *Annals of Surgery*, *268*(6), e55-e56.

Thomas, P., Duffrin, M., Duffrin, C., Mazurek, K., Clay, S. L., & Hodges, T. (2020). Community violence and African American male health outcomes: An integrative review of literature. *Health & Social Care in the Community*, *28*(6), 1884-1897.

Thompson, R. A. (2017). *The hunger and thirst for justice: Barbershop Ministry Initiative— Shaping heads for the future* (Doctoral dissertation, Drew University, Madison, NJ).

Thompson, V. S. (2020). Levels and strategies for community-engaged research. *Perspective*. Washington University Brown Institute for Public Health. https://publi chealth.wustl.edu/levels-and-strategies-for-community-engaged- research/

Thurman, P. (2018). *Firearm injuries in Maryland, 2005-2014: Trends, recidivism, and costs* (Doctoral dissertation University of Maryland, MD).

Tobias, J., Miller, S., & Bermudez, A. (2020). 83 Strengthening research partnership and health care systems to prevent violence: A SWOT analysis. *Injury Prevention*, *26*(1 Suppl.).

Tobon, M., Ledgerwood, A. M., & Lucas, C. E. (2019). The urban injury severity score (UISS) better predicts mortality following penetrating gunshot wounds (GSW). *The American Journal of Surgery*, *217*(3), 573-576.

Tomberg, K. A., & Butts, J. A. (2016). Street by street. NCJRS. https://www.ncjrs.gov/ pdffiles1/ojjdp/grants/250383.pdf

Topalli, V., Dickinson, T., & Jacques, S. (2020). Learning from criminals: Active offender research for criminology. *Annual Review of Criminology*, *3*, 189-215.

Tracy, M., Braga, A. A., & Papachristos, A. V. (2016). The transmission of gun and other weapon-involved violence within social networks. *Epidemiologic Reviews*, *38*(1), 70-86.

Tripathi, P. B., Floriolli, D., & Caughlin, B. P. (2017). Delayed facial nerve paralysis following blast trauma. *Facial Plastic Surgery*, *33*(01), 116-118.

Truesdell, W., Gore, A., Primakov, D., Lieberman, H., Jankowska, D., Joshi, G., & Goyal, N. (2020). Ballistic and penetrating injuries of the chest. *Journal of Thoracic Imaging*, *35*(2), W51-W59.

Tsui, M., Carroll, S. L., Dye, D. W., Smedley, W. A., Gilbert, A. D., Griffin, R. L., . . . Jansen, J. O. (2020). Stop the bleed: Gap analysis and geographical evaluation of incident locations. *Trauma Surgery & Acute Care Open*, *5*(1).

Tuller, L. R. (2020). *Understanding psychological and collective trauma related to neighborhood violence* (Doctoral dissertation, Northeastern University, Boston, MA).

Tung, E. L., Hampton, D. A., Kolak, M., Rogers, S. O., Yang, J. P., & Peek, M. E. (2019). Race/ethnicity and geographic access to urban trauma care. *JAMA Network Open*, *2*(3), e190138.

Turco, L., Cornell, D. L., & Phillips, B. (2017). Penetrating bihemispheric traumatic brain injury: A collective review of gunshot wounds to the head. *World Neurosurgery*, *104*, 653–659.

Turner, H. A., Finkelhor, D., Mitchell, K. J., Jones, L. M., & Henly, M. (2020). Strengthening the predictive power of screening for adverse childhood experiences (ACEs) in younger and older children. *Child Abuse & Neglect*, *107*.

Turner, H. A., Mitchell, K. J., Jones, L. M., Hamby, S., Wade Jr, R., & Beseler, C. L. (2019). Gun violence exposure and posttraumatic symptoms among children and youth. *Journal of Traumatic Stress*, *32*(6), 881–889.

Turner, T., & Wise, A. (2019, September 14). Shattered: Life after being shot. *WAMU*. https://wamu.atavist.com/-

Tynan, R., Bas, N. F., & Cohen, D. (2018, June 14). Unmasking the hidden power of cities. *Medium*. https://laane.org/wp- content/uploads/2018/11/Unmasking-the-Hidden-Power-Facing.pdf

Ubiñas, H. (2019, September 10). Paralyzed gunshot survivors are coming to their support group, but they need more. *Philadelphia Inquirer*. https://www.inquirer.com/news/columnists/helen-ubinas-philadelphia-guns- violence-paralyzed-victims-temple-carousel-house-20190910.html

Ulrich, M. R. (2019). A public health approach to gun violence, legally speaking. *The Journal of Law, Medicine & Ethics*, *47*(2 Suppl.), 112–115.

UPI. (2019, January 23). Study: Gunshot patient follow-up care costs $86 million per year. https://www.upi.com/Health_News/2019/01/23/Study-Gunshot-patient- follow-up-care-costs-86M-per-year/7781548280213/

Urban CNY News. (2020, June 21). Syracuse police were busy last night handling 9 people shot; 3 additional shootings with injuries occurring between 11:15 and 11:58 PM. https://www.urbancny.com/syracuse-police-were-busy-last-night- handling-9-people-shot-3-additional-shootings-with-injuries-occurring-between- 1115-and-1158-pm/

The Urban Institute. (2016, November 3). *Is gun violence stunting business growth?* Washington, DC: Author.

U.S. National Library of Medicine. (2018, April 5). *Gunshot wounds aftercare*. https://medlineplus.gov/ency/patientinstructions/000737.htm

Urrechaga, E. M., Stoler, J., Quinn, K., Cioci, A. C., Nunez, V., Rodriguez, Y., . . . Sola, J. E. (2021). Geo-demographic analysis of pediatric firearm injuries in Miami, FL. *Journal of Pediatric Surgery*, *56*(1), 159–164.

Vakil, M. T., & Singh, A. K. (2017). A review of penetrating brain trauma: Epidemiology, pathophysiology, imaging assessment, complications, and treatment. *Emergency Radiology*, *24*(3), 301–309.

Van Brocklin, E. (2017, October 9). What it costs to treat gun violence in hospitals. *The Trace*. https://www.thetrace.org/2017/10/gun-violence-healthcare-cost-research- hospital/

Van Brocklin, E. (2018a, May 22). The wounds you can't see: Four women on the lasting trauma of gun violence. *The Trace*. https://www.thetrace.org/2018/05/gun-violence-survivors-trauma/

Van Brocklin, E. (2018b, August 2). How to report on gun violence survivors. *The Trace*. https://www.thetrace.org/2018/08/reporting-guide-gun-violence-survivors/

Van Brocklin, E. (2019a, July 10). What gun violence prevention looks like when it focuses on the communities most hurt. *The Trace*. https://www.thetrace.org/2019/07/gun-violence-prevention-communities-of-color- funding/

Van Brocklin, E. (2019b, October 8). How to live again after being shot. *VICE*. https://www.vice.com/en_us/article/43k7nw/how-to-live-again-after-being-shot

Van Brocklin, E. (2019c, October 8). The art of surviving. *The Trace*. https://www.thetrace.org/features/gunshot-survivors-new-york-art-collective/

Van Brocklin, E., & Fernandez, E. (2018, May 31). Bullets put these men in wheelchairs. They turned to poetry to process their pain. *The Trace*. https://www.thetrace.org/2018/05/gun-shot-survivors-spoken-word-poetry- performance/

Van Winkle, B., DiBrito, S. R., Amini, N., Levy, M. J., & Haut, E. R. (2021). A survey of hospitalized trauma patients in hemorrhage control education: Are trauma victims willing to stop the bleed? *Journal of Surgical Research, 264*, 469–473.

Vargas, E. W., & Bhatia, R. (2020, October 20). No shots fired. Center for American Progress. https://www.americanprogress.org/issues/guns- crime/reports/2020/10/20/491823/no-shots-fired/

Vargas, E. W., & Hemenway, D. (2021). Emotional and physical symptoms after gun victimization in the United States, 2009–2019. *Preventive Medicine, 143*.

Vargas, R. (2016). *Wounded city: Violent turf wars in a Chicago barrio*. New York, NY: Oxford University Press.

Vaughn, P. E. (2020). The effects of devaluation and solvability on crime clearance. *Journal of Criminal Justice, 68*.

Veenstra, M., Schaewe, H., Donoghue, L., & Langenburg, S. (2015). Pediatric firearm injuries: Do database analyses tell the whole story? *Current Surgery Reports, 50*(7), 1184–1187.

Vella, M. A., Warshauer, A., Tortorello, G., Fernandez-Moure, J., Giacolone, J., Chen, B., . . . Reilly, P. M. (2020). Long-term functional, psychological, emotional, and social outcomes in survivors of firearm injuries. *JAMA Surgery, 155*(1), 51–59.

Vente, T. M. (2020). The impact of gun violence on those already dying: Perspectives from a palliative care physician. *Pediatrics, 145*(2).

Vera, A. (2020, May 26). Chicago sees deadliest Memorial Day weekend with 8 killed and 24 injured in shootings. *CBS58*. https://www.cbs58.com/news/chicago-sees- deadliest-memorial-day-weekend-in-four-years-with-8-killed-and-24-injured-in- shootings

Vigil, J. D. (2020). *Multiple marginality and gangs: Through a prism darkly*. Lanham, MD: Lexington Books.

Vil, C. S., Richardson, J., & Cooper, C. (2018). Methodological considerations for research with black male victims of violent injury in an urban trauma unit. *Violence and Victims, 33*(2), 383–396.

Villarreal, A. (2020, December 7). Toll of nonfatal gun injuries among Latinos, Blacks needs to be addressed, new report finds. *NBCNews*. https://www.nbcnews.com/news/latino/toll-nonfatal-gun-injuries-among-latinos- blacks-needs-be-addressed-n1250068

Villegas, C. V., Gupta, A., Liu, S., Curren, J., Rosenberg, J., Barie, P. S., . . . Narayan, M. (2020). Stop the bleed: Effective training in need of improvement. *Journal of Surgical Research, 255*, 627–631.

Vlaszof, N. (2017). *Motivations to return to a gang after severe physical victimization* (Doctoral dissertation, Walden University, Minneapolis, MN).

Voisin, D. R. (2019). *America the beautiful and violent: Black youth and neighborhood trauma in Chicago*. Chicago, IL: University of Chicago Press.

Walker, H., Collingwood, L., & Bunyasi, T. L. (2020). White response to black death: A racialized theory of white attitudes towards gun control. *Du Bois Review: Social Science Research on Race, 17*(1), 165–188.

Walters, G. (2020, June 12). These cities replaced cops with social workers, medics, and people without guns. *VICE.* https://www.vice.com/en/article/y3zpqm/ these- cities-replaced-cops-with-social-workers-medics-and-people-without-guns

Wamser-Nanney, R., Nanney, J. T., Conrad, E., & Constans, J. I. (2019). Childhood trauma exposure and gun violence risk factors among victims of gun violence. In M. Siegler & S. Rogers Jr (Eds.), *Psychological trauma: Theory, research, practice, and policy* (Vo. *11*(1), pp. 99–106). New York: Springer.

Wamser-Nanney, R., Nanney, J. T., & Constans, J. I. (2020). Trauma exposure, posttraumatic stress symptoms, and attitudes toward guns. *Psychology of Violence, 11*(4), 376–384.

Wandling, M., Behrens, J., Hsia, R., & Crandall, M. (2016). Geographic disparities in access to urban trauma care: defining the problem and identifying a solution for gunshot wound victims in Chicago. *The American Journal of Surgery, 212*(4), 587–591.

Wang, E. A. (2016). *Building resilient neighborhoods and positive neighborhood networks to prevent gun violence.* National Institutes of Health. https://grantome.com/grant/ NIH/R01-MD010403–02

Wang, E. A., Riley, C., Wood, G., Greene, A., Horton, N., Williams, M., . . . Roy, B. (2020). Building community resilience to prevent and mitigate community impact of gun violence: Conceptual framework and intervention design. *BMJ Open, 10*(10), e040277.

Wang, J. (2020, June 9). Urban grief raises awareness of emotional trauma from gun violence. *Spectrum News.* https://www.ny1.com/nyc/all- boroughs/news/2020/06/10/ urban-grief-raises-awareness-of-emotional-trauma

Ward-Lasher, A., Messing, J. T., Cimino, A. N., & Campbell, J. C. (2020). The association between homicide risk and intimate partner violence arrest. *Policing: A Journal of Policy and Practice, 14*(1), 228–242.

Washington, M. (2017, August 31). Paralyzed Chicago gunshot victims get GEDs. *CBS Chicago Online.* https://www.sralab.org/articles/patient-story/spinal-cord- injury/ paralyzed-chicago-gunshot-victims-get-geds

Waterman, A. S. (2020). "Now what do I do?": Toward a conceptual understanding of the effects of traumatic events on identity functioning. *Journal of Adolescence, 79*, 59–69.

Watts, S. J. (2019). Gun carrying and gun victimization among American adolescents: A fresh look at a nationally representative sample. *Victims & Offenders, 14*(1), 1–14.

Webster, D. W., Crifasi, C. K., Williams, R. G., Booty, M. D., & Buggs, S. A. L. (2020). *Reducing violence and building trust: Data to guide enforcement of gun laws in Baltimore.* Baltimore, MD: Johns Hopkins Bloomberg School of Public Health, Center for Gun Policy and Research. https://www.jhsph.edu/research/centers-and-institutes/ johns-hopkins-center-for- gun-policy-and-research/_docs/reducing-violence-and- building-trust-gun-center- report-june-4–2020.pdf

Webster, J. (2019). Wanted: Local medical experts/champions to prevent gun violence. *The American Journal of Medicine, 132*(3), 276–277.

Webster, J. R. (2020). Firearm-related injury and death in the United States. *Annals of Internal Medicine, 172*(5), 367–368.

Weinman, S. (2020). Retention of tourniquet application skills following participation in a bleeding control course. *Journal of Emergency Nursing, 46*(2), 154–162.

Weisberg, D. K. (2017). Hidden victims and hidden abuse: Clients of hair salons. *Family Intimate Partner Violence Quarterly, 10*, 35. 1–2, 13–15.

Weiss, D., Lee, D., Feldman, R., & Smith, K. E. (2017a). Severe lead toxicity attributed to bullet fragments retained in soft tissue. *Case Reports.*

Weiss, D., Tomasallo, C. D., Meiman, J. G., Alarcon, W., Graber, N. M., Bisgard, K. M., & Anderson, H. A. (2017b). Elevated blood lead levels associated with retained bullet fragments—United States, 2003–2012.*Morbidity and Mortality Weekly Report, 66*(5), 130–133.

Weisser, M. (2018). Is gun violence an example of American exceptionalism?

Wen, L. S., & Goodwin, K. E. (2016). Violence is a public health issue. *Journal of Public Health Management & Practice, 22*(6), 503–505.

Wen, L. S., & Sadeghi, N. B. (2020). Treating gun violence with a public health approach. *The American Journal of Medicine, 133*(6), 883–884.

Westoby, P. (2017). *Soul, community and social change: Theorising a soul perspective on community practice.* New York, NY: Taylor & Francis.

Wetsman, N. (2018, November 9). Gun wound first aid can save a life. Here's what to do. *Popular Science.* https://www.popsci.com/gunshot-wound-first-aid-triage/

Whalen, A. N. (2020). *Compensation and victim purity: Comparing the treatment of domestic violence, gun violence and rampage shooting victims* (Doctoral dissertation, Northern Arizona University, Flagstaff, AZ).

Wheelchairs Against Gun Violence. (2020). https://wheelchairsagainstguns.org/wheelchairs-guns-support-gun-violence- awareness-month/

Wheeler, A. P., Worden, R. E., & Silver, J. R. (2019). The accuracy of the violent offender identification directive tool to predict future gun violence. *Criminal Justice and Behavior, 46*(5), 770–788.

Wheeler, T. (2019). *Advocacy education for psychology graduate students: a curriculum for professional advocacy in psychology* (Doctoral dissertation, Pepperdine University, Malibu, CA).

White, I. (2020, October). Trauma and healing among adolescents and young adults who are victims of violence. In *APHA's 2020 VIRTUAL Annual Meeting and Expo* (Oct. 24–28). American Public Health Association.

White, K., Stuart, F., & Morrissey, S. L. (2021). Whose lives matter? Race, space, and the devaluation of homicide victims in minority communities. *Sociology of Race and Ethnicity, 7*(3), 33–349.

Whitney-Snel, K., Valdez, C. E., & Totaan, J. (2020). "We break the cycle . . .": Motivations for prosocial advocacy among former gang members to end gang involvement. *Journal of Community Psychology, 48*(6), 1929–1941.

Wiard, G. (2018). *Bystander intervention to reduce mortality in gun-related trauma: An integrated literature review.* Chicago, IL: The Grace Peterson Nursing Research Colloqium, Depaul University.

Wical, W., Richardson, J., & Bullock, C. (2020). A credible messenger: The role of the violence intervention specialist in the lives of young black male survivors of violence. *Violence and Gender, 7*(2), 1–4.

Wiley, M. (2020). *Gun violence among youth: A program evaluation of training procedures for a youth mentoring program* (Doctoral dissertation, Capella University, Minneapolis, MN).

Wilkinson, D., LaMarr, F. V., Alsaada, T. F., Ahad, C., Hill, D., & Saunders Sr, J. (2018). *Building an engaged community to prevent and heal from gun violence. Community Engagement Conference.* The Ohio State University, Columbus, Ohio, January 24–25, 2018.

Willard, T. M., Khan, A., Reckard, P. E., Day, G., Leavitt, C., Roberts, S., & Schroeppel, T. J. (2020). Pediatric trauma outreach and prevention: Early data suggesting we are making an impact. *Journal of the American College of Surgeons, 231*(4), e244–e245.

Willett, J. A. (2000). *Permanent waves: The making of the American beauty shop.* New York, NY: NYU Press.

Williams, H. E., Bowman, S. W., & Jung, J. T. (2019). The limitations of government databases for analyzing fatal officer-involved shootings in the United States. *Criminal Justice Policy Review, 30*(2), 201–222.

Williams, M. A., & Bassett, M. T. (2019, July 29). Op-Ed: How do we reduce gun violence? By treating it as a disease. *Los Angeles Times.* https://www.latimes.com/opin ion/story/2019–07-29/gun-violence-gilroy-brooklyn- public-health-problem

Williams, P. (2019, April 1). Turning bystanders into first responders. *The New Yorker.* https://www.newyorker.com/magazine/2019/04/08/turning-bystanders-into-first-responders

Willoughby, M., Spittal, M. J., Borschmann, R., Tibble, H., & Kinner, S. A. (2021). Violence-related deaths among people released from prison: A data linkage study. *Journal of Interpersonal Violence, 36*(23-24), NP13229–NP13253.

Winfield, R. D., Crandall, M., Williams, B. H., Sakran, J. V., Shorr, K., & Zakrison, T. L. (2019). Firearm violence in the USA: A frank discussion on an American public health crisis—The Kansas City Firearm Violence Symposium. *Trauma Surgery & Acute Care Open, 4*(1).

Winker, M., Rowhani-Rahbar, A., & Rivara, F. P. (2020). US gun violence and deaths. *BMJ, 368*, 1–2.

Wintemute, G. J., Claire, B. E., McHenry, V. S., & Wright, M. A. (2012). Epidemiology and clinical aspects of stray bullet shootings in the United States. *Journal of Trauma and Acute Care Surgery, 73*(1), 215–223.

Wise, A. (2019, October 14). Shattered: Paralyzed shooting victims find new life in wheelchairs. *NPR.* https://www.npr.org/local/305/2019/10/14/770046019/shattered-paralyzed- shooting-victims-find-new-life-in-wheelchairs

Wise, A., & Turner, T. (2019, October 14). Shattered: Finding life in a wheelchair after being shot. *Guns in America.* https://wamu.org/story/19/10/14/shattered-find ing- life-in-a-wheelchair-after-being-shot/

Wolf, A. (2019). *Comparison of injury severity and resource utilization in pediatric firearm and sharp force injuries* (Doctoral dissertation, University of Washington, Seattle, WA).

Wolf, A. E., Garrison, M. M., Mills, B., Chan, T., & Rowhani-Rahbar, A. (2019). Evaluation of injury severity and resource utilization in pediatric firearm and sharp force injuries. *JAMA Network Open, 2*(10), e1912850.

Wolfe, A. (2017). *The Western genre and gun violence in United States culture: Using theatre as a laboratory for social critique* (Doctoral dissertation, University of Oregon, Eugene, OR).

Wojtowicz, A., French, M., Alper, J., & National Academies of Sciences, Engineering, and Medicine. (2019, February). Developing a culture of health care providers as interveners. In *Health systems interventions to prevent firearm injuries and death: Proceedings of a workshop.* Washington, DC: National Academies Press.

Wong, S. L., & Raphael, J. L. (2020). Are you listening to our children: Empowering youth advocates. *Pediatric Research, 87*(3), 432–433.

Wood, G., & Papachristos, A. V. (2019). Reducing gunshot victimization in high-risk social networks through direct and spillover effects. *Nature Human Behaviour, 3*(11), 1164–1170.

Woodside, A. G., Caldwell, M., & Calhoun, J. R. (2020). Service breakdown prevention. *International Journal of Contemporary Hospitality Management.*

Wortley, E., & Hagell, A. (2021). Young victims of youth violence: Using youth workers in the emergency department to facilitate "teachable moments" and to improve access to services. *Archives of Disease in Childhood-Education and Practice, 106*(1), 53–59.

Wray-Lake, L., & Abrams, L. S. (2020). Pathways to civic engagement among urban youth of color. *Monographs of the Society for Research in Child Development, 85*(2), 7–154.

Wright, A. W., Austin, M., Booth, C., & Kliewer, W. (2017). Systematic review: Exposure to community violence and physical health outcomes in youth. *Journal of Pediatric Psychology, 42*(4), 364–378.

Wright, R. G. (2019). *American violence: Survival, healing, and the failure of American policy.* Lanham, MD: Lexington Books.

Wright, V., & Washington, H. M. (2018). The blame game: News, blame, and young homicide victims. *Sociological Focus, 51*(4), 350–364.

Wu, C. (2020). How does gun violence affect Americans' trust in each other? *Social Science Research.*

Wu, J., & Pyrooz, D. C. (2016). Uncovering the pathways between gang membership and violent victimization. *Journal of Quantitative Criminology, 32*(4), 531–559.

Xie, M., & Baumer, E. P. (2019). Crime victims' decisions to call the police: Past research and new directions. *Annual Review of Criminology, 2*, 217–240.

Yang, Y. (2020). Youth violence prevention in Florida: A commentary. *Florida Public Health Review, 17*(1), 2–3.

Yang, Y., Liller, K. D., & Coulter, M. (2018). PW 2271 Photovoice and youth: A systematic review of violence and related topics. *BMJ: Injury Prevention, 24.*

Younge, G. (2016). *Another day in the death of America: A chronicle of ten short lives.* Lebanon, IN: Bold Type Books.

Yu, S.-s. V., Lee, D., & Pizarro, J. M. (2020). Illegal firearm availability and violence: neighborhood-level analysis. *Journal of Interpersonal Violence 35*(19–20), 3986–4012.

Zakrison, T. L., Puyana, J. C., & Britt, L. D. (2017). Gun violence is structural violence: Our role as trauma surgeons. *Journal of Trauma and Acute Care Surgery, 82*(1), 224.

Zakrison, T. L., Williams, B., & Crandall, M. (2021). Gun violence, structural violence, and social justice. In M. Crandall M, S. Bonne, J. Bronson, & W. Kessel (Eds.), *Why we are losing the war on gun violence in the United States* (pp. 11–14). New York, NY: Springer.

Zarilli, Z. (2019, August 19). Stop the bleed: Training citizens how to properly respond to a mass shooting event. *Surefire CPR.* https://www.surefirecpr.com/stop-the- bleed-training-citizens-how-to-properly-respond-to-a-mass-shooting-event/

Zebib, L., Stoler, J., & Zakrison, T. L. (2017). Geo-demographics of gunshot wound injuries in Miami-Dade County, 2002–2012. *BMC Public Health, 17*(1), 1–10.

Zeineddin, A., Williams, M., Nonez, H., Nizam, W., Olufajo, O. A., Ortega, G., … Cornwell, E. E. (2021). Gunshot injuries in American trauma centers: Analysis of the lethality of multiple gunshot wounds. *The American Surgeon, 87*(1), 39–44.

Zeoli, A. M., Goldstick, J., Mauri, A., Wallin, M., Goyal, M., Cunningham, R., & FACTS Consortium. (2019). The association of firearm laws with firearm outcomes among children and adolescents: A scoping review. *Journal of Behavioral Medicine*, *42*(4), 741–762.

Zhitny, V. P., Iftekhar, N., Moreno, S., & Stile, F. (2020). Abdominoplasty for treatment of abdominal gun-shot wound sequalae—A case report. *International Journal of Surgery Case Reports*, *72*, 365–368.

Zimring, F. E. (2020). Firearms and violence in American life—50 years later. *Criminology & Public Policy*, *19*(4), 1359–1369.

Zoller, H. M., & Casteel, D. (2021). # March for Our Lives: Health activism, diagnostic framing, gun control, and the gun industry. *Health Communication*, 1–11.

Zwislewski, A., Nanassy, A. D., Meyer, L. K., Scantling, D., Jankowski, M. A., Blinstrub, G., & Grewal, H. (2020). Erratum to "Practice makes perfect: The impact of Stop the Bleed training on hemorrhage control knowledge, wound packing, and tourniquet application in the workplace." *Injury*, *51*(3), 864–868.

INDEX

For the benefit of digital users, indexed terms that span two pages (e.g., 52–53) may, on occasion, appear on only one of those pages.